Human Factors Engineering

Human Factors Engineering

Jack A. Adams
University of Illinois at Urbana-Champaign

Macmillan Publishing Company
NEW YORK

TH
166
-A28
1989

Copyright © 1989 by Macmillan Publishing Company,
a division of Macmillan, Inc.

PRINTED IN THE UNITED STATES OF AMERICA

Macmillan Publishing Company
866 Third Avenue, New York, New York 10022

Collier Macmillan Canada, Inc.

LIBRARY OF CONGRESS CATALOGING-IN-PUBLICATION DATA

Adams, Jack A.
    Human factors engineering / Jack A. Adams.
        p.   cm.
    Includes index.
    ISBN 0-02-300370-7
    1. Human engineering.   2. Man-machine systems.   I. Title.
TA166.A28 1989   88-7033
620.8'2—dc19   CIP

Printing: 1 2 3 4 5 6 7   Year: 9 0 1 2 3 4 5

R1
cf 4.10
2-21-90

Preface

Belatedly perhaps, but in our time there is widespread appreciation that the behavior of personnel in human–machine systems is as fundamental to system effectiveness as are hardware components. A neglect of workplace design will produce human error and degrade system effectiveness just as surely as will neglect of the design of a machine.

Awareness of human factors in systems was heightened by the Three Mile Island incident which had human factors as its fundamental cause (Chapter 1), but the Three Mile Island incident was not where awareness of human factors began. Awareness of the importance of human factors stretches back decades, growing gradually with the appreciation that human–machine systems do not work well if personnel and machines are not seen as partners in the system enterprise.

This book is an undergraduate textbook. Historical background is included so that undergraduate students, in their first contact with human factors engineering, will know on whose shoulders they stand. Throughout there is reliance on laboratory and field research so that students will see the empirical origins of human factors principles. There is coverage of topics which are comparatively new to human factors engineering, such as human–computer interaction, automation, artificial intelligence, and reliability. Personnel selection and training, which are topics that have had small emphasis in previous human factors textbooks but which loom large in the world of human–machine systems, have chapters assigned to them. Traditional topics for human factors engineering, such as displays and controls, are, of course, included.

Students whose primary interests are elsewhere, such as mechanical or electrical engineering students, will find that this textbook will give general understanding of a discipline that is likely to impinge on their own. Students who develop an interest in human factors engineering and seek further training in it will find that this book will contribute to their preparation for advanced courses.

J.A.A.

Contents

Human Error in Human–Machine Systems

CHAPTER 1

Definition and Justification of Human Factors Engineering

There would be no justification for the field of human factors engineering if human beings did not make errors in their uses of machines. A light bulb can have its efficiency evaluated independently of human users, but not a system like an automobile or an airplane. The human operator is central to systems like these, and system efficiency depends as much on the capabilities of the operator as on the capabilities of the machine. The technicians who maintain these systems, as well as other support personnel, are also basic to system efficiency. We call such systems *human–machine systems*. Aeronautical, electronic, and mechanical engineers are among those who design the hardware of a human–machine system, computer scientists design the software, and *human factors engineers* focus on the humans who have roles in the system. Here is a formal definition of human factors engineering:

> The field of human factors engineering uses scientific knowledge about human behavior in specifying the design and use of a human–machine system. The aim is to improve system efficiency by minimizing human error.

There are a number of things to consider in minimizing human error, as the rest of this book will testify, but among them are understanding the sensory, motor, and cognitive capabilities of humans; design of controls and displays; layout of workplaces; human–computer interaction; personnel selection; and personnel training.

The design of a human–machine system is a team effort in which the requirements of the engineering specialties are balanced against each other. A mechanical engineer may have good reasons for the specifications chosen for a vehicle's control system, but it will not be a good control if the operator finds it difficult to use, no matter what its other virtues might be. Or, favoring an operator with a good control might result in a control system that would mate poorly with the rest of the system. The design of the system must conform to a prescribed list of system capabilities—all within time and cost constraints—and the cooperation of all the engineering specialists is required if the system is to be optimized. A balanced, cooperative engineering effort of this kind is called the *systems approach*. Cooperative efforts to attain a goal are not new, so it might seem that the systems approach is hardly worth mentioning. But the approach does deserve mention, because there was a time when it was neglected. Mechanical engineers would design a control that a human operator could not use very well, so the system did not work very well. A dial for a display might be selected solely for electromechanical and cost reasons, with no attention to its readability. Because a balance of engineering efforts in the design of a large system is not easy to maintain, special management methods have arisen to attain it. Keeping the development of a large system in balance and moving along on schedule is a management responsibility.

The highway transportation system in the United States is a good example of a poor system. The vehicles and the roads are built by two independent sets of engineers, and the traffic engineers and law enforcement personnel who regulate the use of the highway system are independent of both of them (DeGreene, 1970, p. 472). The United States is politically fragmented into states, and the rules of the road lack uniformity. The 50 states vary in their requirements for training and licensing. Nor is there any guarantee of the quality of vehicles on the road. It comes as no surprise to engineers of all specializations that a system like this produces thousands of deaths, several million injuries, and several billion dollars in property damage each year. The United States would declare a national emergency if a disease struck the country and killed as many as are killed on the highways each year. There have been wars that have done less damage.

Our air transportation system is a good system by comparison, and it is probably because of the regulation of the system at the national level. The Federal Aviation Administration controls the air traffic control system in the United States and so has uniform "rules of the road." Flight corridors for commercial aircraft are specified throughout the country. Checks on the proficiency of pilots are made by federally ap-

TABLE 1-1 Comparative Accident Data for Four Common Kinds of Transportation*

Year	Autos and Taxis	Buses	Trains	Scheduled Airlines	Airlines Safer Than Autos by a Factor of:
1973	1.70	.24	.07	.10	17
1974	1.50	.21	.07	.12	12.5
1975	1.40	.15	.08	.08	17.5
1976	1.34	.17	.05	.003	447
1977	1.33	.13	.04	.04	33
1978	1.30	.17	.13	.01	130
1979	1.31	.15	.05	.12	11
1980	1.32	.15	.04	.01	132

*Passenger fatalities per 100 million passenger-miles.

Source: FAA statistical handbook of aviation: Calendar year 1983 (Washington, D.C.: Federal Aviation Administration, December 31, 1984), Table 9.11.

proved examiners. The routines of aircraft maintenance are prescribed by the federal government. A few spectacular air crashes occur each year, but a case can be made that commercial aviation in the United States is efficient, swift, and safe. Table 1-1 compares the fatalities for different modes of passenger transportation during the period 1973–1980. All modes of public transportation are safer than the private automobile. The last column compares the air transportation system with the private automobile. Scheduled airliners are many times safer than private automobiles even in relatively bad years. The variability in the entries of the last column is because crashes of airliners are so rare that one or two more or less makes a big difference.

Human Factors Engineers and Their Background _____

Who are human factors engineers, and what are their backgrounds? Compared to electronics and mechanical engineers, human factors engineers can be varied in their professional training and scientific backgrounds. Human factors engineering had its origins in industrial engineering, and industrial engineering remains a primary contributor to systems. Psychology is also a primary contributor to human factors engineering because human factors turn fundamentally on scientific knowledge about human behavior. A physiologist can be a human factors engineer, contributing to such engineering requirements as life-support subsystems of manned space vehicles. Today, with software having a major role in system operations, a computer scientist might be concerned with the design of a computer program that will present information on a display in the most effective way, or will organize data so that they can be accessed readily in a useful format. System design is a practical undertaking, and system managers will use any specialist to optimize any human variable that will affect the system. Industrial engineering and psychology, however, are the specialties most closely identified with human factors engineering.

Historical Background

Origins of Human Factors in Engineering

The start of a scientific concern with human factors began about the turn of the twentieth century, and the pioneer was F. W. Taylor, who formulated his "principles of scientific management." Use of the principles became the origins of industrial engineering, a new discipline, although Taylor was a mechanical engineer by training. Taylor had four principles of work design (Taylor, 1911/1967, pp. 36–37):

1. Scientifically justify each element of a worker's job in terms of its contribution to efficiency.
2. Scientifically select the workers and train them.
3. Ensure that the job is done as prescribed.
4. Have active management participation in the job.

Taylor's principles seem straightforward today, but they were not self-evident to management at the time that Taylor formulated them. Work practices at the turn of the century were casual. Workers were hired with only superficial appreciation of their capabilities. They were allowed to work out their own methods of doing a job. Variability in efficiency was the result, and average proficiency was low. Taylor said that more attention should be given to worker capabilities and to the selection of workers for their jobs. Each element of a job should be analyzed for its contribution to productivity and only those elements that contribute to efficiency should be retained. A newly defined work method will seem unnatural to the workers, but it is scientifically rationalized, so the workers must be trained in it, and the supervisors must see that they do not return to their old ways. Management must do more than be concerned with the wide dimensions of production and marketing. Management must, in addition, concern itself with the detailed characteristics of jobs.

One of Taylor's first efforts was for the Bethlehem Steel Company in 1898, at the start of the Spanish-American War (Taylor, 1911/1967, pp. 40–48). Prices for iron had been so low that iron could not be sold for a profit, so Bethlehem Steel had 80,000 tons of raw iron ingots stored in an open field. At the start of the war the price of iron rose, so the sale of iron became profitable once again. A railroad switch line was laid to the storage field, and gangs of workers began to load ingots on the cars. Each ingot weighed 92 pounds, and each man on average loaded 12.5 tons per day. Taylor studied the workers and their work behavior and concluded that it was possible to load 47 tons per day. To prove it, Taylor chose a worker who was strong (the worker ran home every night after carrying iron ingots all day), who was not too intelligent (Taylor believed that bright workers found repetitive work monotonous and boring), and who would do exactly as he was told (so that the new steps of the restructured job would be followed). The man started to work, being told when to pick up an ingot, when to walk with it, and when to sit down and rest. At the end of the day he had moved 47.5 tons, and he continued to do so thereafter. His increase in productivity earned him a pay increase of 61 percent (from $1.15 to $1.85 per day in 1898). Management could easily see that the increase in wages was more than offset by the increase in productivity, and they were impressed. Taylor selected other workers and trained them in the same way.

Taylor and his followers are best remembered for their *motion and time study,* as the analysis of job elements and job restructuring came to be called, but it is almost forgotten that Taylor was also concerned with tool design. Shoveling was another of Taylor's projects at Bethlehem Steel, and he found that a worker could do his biggest day's work if a shovel load was about 21 pounds. A worker had always been allowed to select his own shovel, and Taylor changed that. He had the company provide 8 to 10 different types of shovels, and a worker was assigned one that handled a 21-pound load and was best suited for the particular kind of material that was being handled. Allowing a worker to select his own shovel seemed self-evident in 1898, but not to Taylor. For Taylor, there was a science of shoveling.

F. B. Gilbreth, a follower of Taylor's work, was also a mechanical engineer interested in work methods and efficiency. Gilbreth is best remembered for his studies of bricklaying (Gilbreth, 1909). Bricklaying is one of the world's oldest trades, with little change over thousands of years until Gilbreth came along. Gilbreth analyzed and specified the position of the bricklayer with respect to the pile of bricks, the mortar box, and the wall, making it unnecessary for the bricklayer to take a step or two to the pile of bricks and back again each time he laid a brick. Instead of the bricks and the mortar box being on the ground, requiring the worker to stoop about a thousand times a day, Gilbreth designed a scaffold for the bricks and the mortar so that the man, the bricks, the mortar, and the wall were all in convenient relationship to one another, with no stooping required. The number of motions in laying a brick were reduced from 18 to 5. The number of bricks laid per hour almost tripled. Management had no problem living with the increase in pay that the new productivity justified.

Motion and time study continues to be used in industrial engineering, and it has become sophisticated. More will be said about modern approaches to the topic in Chapter 8.

Origins of Human Factors in Psychology

An interest in the efficiency of human–machine systems came later to the field of psychology and was independent of industrial engineering. One could visualize a line of development where a psychologist would take his knowledge of such topics as training and fatigue and turn them toward a refinement of motion and time study, but it did not happen that way. A scattering of interest in work design is found in psychology before World War II, but it took the pressures of war to create a strong beginning for human factors in psychology.

War is a union of humans and complex machines dedicated to violent ends, and shortcomings in humans and machines are easily revealed when war exercises them under extreme circumstances. All able-bodied men, and many able-bodied women, were called to the armed forces during World War II, and psychologists were among them. Two kinds of psychologists became interested in how well humans and their war machines worked together. One group, the psychometricians, were concerned with the development and use of psychological tests for personnel selection and classification. Out of all the applicants, which will make the best pilots? The best navigators? A second group, the experimental psychologists, were concerned with determinants of human performance, such as the senses and learning. They found dials in aircraft cockpits that could not be read very well, controls that could not be manipulated very well, and personnel who were not trained very well. They put their knowledge to work, insofar as the demands of war allowed, and they conducted research to generate new knowledge. After the war some of these experimental psychologists returned to colleges and universities, carrying their new interests with them. New academic programs in human factors were begun. Other experimental psychologists started government laboratories for the investigation of human factors in military systems. Today, human factors is a visible dimension of psychology in universities, government laboratories, and industry.

Human Factors and Common Sense _____

So far, you might think that human factors is little more than glorified common sense. Common sense is knowledge that comes from everyday experience, independent of formal training. Because we all have had experience in the world, and because we all have worked and observed working, it would seem easy to prescribe the conditions of efficient work. There is some truth in this. From ordinary experience we all know something about human factors principles. We know that information on a display medium should be large enough and bright enough so that it can be read easily. We know that controls should be reachable. Desktops and chairs should be of heights that are comfortable for the user. These are useful human factors principles, and they will carry you some distance—but no further than common sense will carry you in physics. Thousands of years ago the people of early civilizations, using their common sense about objects and materials of the physical world, built good bridges, roads, and buildings. Yet nothing in that kind of conventional wisdom will ever lead to television sets or airliners. Systems like these depend on subtleties that systematic science brings. Research uncovers variables and the laws that relate them, and system subtleties become possible only when the body of scientific knowledge for a field becomes large and sophisticated. Common sense is left far behind when a scientific field matures.

Even seemingly simple human factors problems can defy common sense. You would think it would be easy for manufacturers of four-burner ranges for stove tops to specify which control knob should go with which burner so that wrong knobs would not sometimes be chosen. Chapanis and Lindenbaum (1959) conducted a study of preferences and found one design configuration that met the criteria. Shinar and Acton (1978) did a follow-up study, but this time the preferences were compared to the layout of 49 ranges on the market. Not a single range had the error-free layout. Perhaps worse, 27 percent of the ranges had a layout that was preferred by absolutely nobody.

Some years ago the Bell System had the engineering problem of specifying the layout of pushbuttons on a telephone, and a series of human factors experiments were conducted (Deininger, 1960). A total of 16 different arrangements of the 10 buttons were investigated. Aside from random arrangements, about any reasonable arrangement that you might think of was included—square arrays, horizontal arrays, vertical arrays, circular arrays, and so on. In addition, variables such as button–top size, letter size, the amount of force required to displace a button, and feedback in terms of an audio signal or "feel" to inform that the button had been depressed far enough were included. There was very little difference among variables in terms of errors or the time to dial a number. The major determinant of performance turned out to be how the user relied on memory while dialing. Some of the subjects dialed the entire number from memory, while others referred back to the number about halfway through the dialing. Those who relied on memory dialed faster but produced more errors than those who used referral. The pushbutton telephone is another example where common sense fails. Most of us probably believe that the arrangement of the pushbuttons makes a big difference—indeed, the investigator had a sufficiently strong belief in the importance of pushbutton arrangement to do experiments about it. Yet pushbutton arrangement is not an important variable for performance, contrary to common sense.

Does it matter much whether stove-top ranges and telephone pushbuttons are properly designed? What difference does an occasional error make for ordinary devices such as these? Turning the wrong control knob on the range may be no more than a minor frustration, and we can live with minor frustrations. On the other hand, the dinner could be lost if the clam sauce begins to boil over and you reach quickly to turn the heat down and grab the wrong knob. As calamitous as errors like these would be for a cook, the Bell System would not consider them as serious as errors in dialing or slow dialing. Errors and slowness by the millions of users of the telephone system at any moment tie up the system unnecessarily, and the amount of information that the system can process is reduced. Nevertheless, it would be hard to convince anyone outside the Bell System that dialing delays and errors matter much. There are, however, errors that have occurred in major systems which everyone would agree are serious, as the following examples will testify. They are errors that could have been prevented had proper human factors measures been taken.

Human Factors Issues in Major Systems

Aviation Accidents

Table 1-1 shows that accidents on scheduled airlines are rare but, when they do occur, pilot errors contribute to more than half of them. From 1970 through 1979, the pilot was a contributor to 61 percent of fatal accidents among certificated route air carriers, which are mostly passenger aircraft, although some are cargo carriers. Other personnel, including maintenance and traffic control, contributed to 48 percent of fatal accidents. These two human sources, together with weather, which contributed 46 percent, were the main factors in fatal accidents; all other causes were minor compared to them (National Transportation Safety Board, 1981). These percentages may seem a bit confusing to you because they do not add up to 100 percent, but remember that an accident can have several causes, such as a pilot's poor judgment in bad weather. An accident of this type would be tallied in both the pilot and weather categories.

The Accident at Three Mile Island

Sources of energy for mass consumption are a problem in our time. Some countries have an abundant supply of coal, but the abundance is a mixed blessing because coal is difficult to mine, bulky to transport, and a pollutant when used. Oil is easier to extract from the earth, but oil has pollution problems and its supply in the world is shrinking; oil will be in short supply in the twenty-first century, and expensive. Some countries have neither coal nor oil and depend on others, always living in fear of high energy prices that will hurt their economies or a war that will cut off their supplies.

The successful development of the atomic bomb in World War II hinged on splitting the atom and releasing the massive energy stored there. The release of energy is uncontrolled in the bomb, but the controlled release of energy by nuclear fission in a nuclear reactor raises the exciting possibility of a new and peaceful energy source. Heat is generated by nuclear fission, which is the splitting of atoms, usually uranium atoms. The nucleus of the atom is bombarded by neutrons, and when the nucleus is struck it splits and releases energy, much of it as heat. Some elements absorb neutrons, so they can be used to control the rate of nuclear fission, or to stop it entirely, by engaging the neutrons that would otherwise split atoms.

Nuclear fission gave us a new source of electrical power. Electricity, based on a limitless supply of fuel, competitively priced and clean, was a dream come true. We were awakened from the dream with an accident at one of two nuclear power plants on an island, called Three Mile Island, in a river near Harrisburg, Pennsylvania. The human factors dimensions of the accident can be appreciated if we know something of how a nuclear power plant works.

How a Nuclear Reactor Works. Figure 1-1 is useful in following this account of how a nuclear reactor works; it is a schematic of the nuclear power plant at Three Mile Island. The main function of a commercial power plant is to heat water. The hot water is turned into steam, which drives a turbine, which operates a generator that produces electricity. The conventional way of heating the water is with coal or oil, but the new way is with a nuclear reactor. The core of a nuclear reactor will have about 100 tons of uranium fuel in the form of

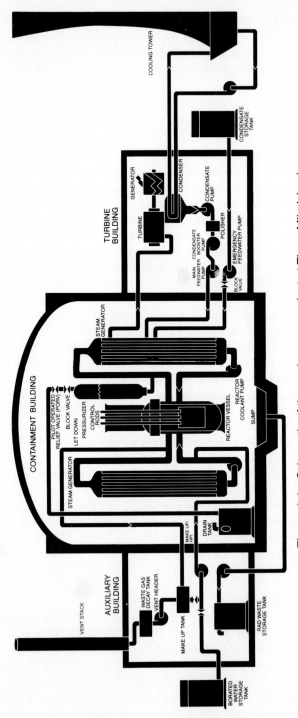

Figure 1-1 Schematic of the nuclear power plant at Three Mile Island. [Adapted from Kemeny (1979).]

pellets stacked in fuel rods. The core also contains control rods for absorbing neutrons and controlling the rate of fission. Withdrawing the control rods allows neutrons to bombard uranium atoms and split them, creating heat. The amount of withdrawal of control rods determines the amount of power generated.

Radioactive coolant water flows around the core, and loss of it, called a loss-of-coolant accident, could be very dangerous because of the high temperatures that can develop. If the temperature rises above 2000°F the fuel rods can have their casings damaged. At about 5200°F the uranium fuel itself will melt, destroying the reactor vessel and releasing large amounts of radioactivity. One frightening possibility is penetration of the earth's surface by melting uranium, contaminating the water supply and making a large region uninhabitable.

The water in the reactor coolant subsystem is moved by the reactor coolant pump to the steam generators. The steam generator has a series of pipes with water in them from a pump, called the main feedwater pump. The water in the pipes of the steam generator is surrounded by the very hot water of the reactor coolant system, which turns it to steam. The steam operates the steam turbine, which in turn runs the electrical generator. The water from the reactor coolant system, now having lost some of its heat in the heat exchange, is returned to the reactor vessel for more heat and to repeat the cycle. Both the steam generator and the reactor coolant system have emergency standby pumps.

The pressurizer is another principal subsystem. The system is classified as a pressurized water reactor. The coolant water around the core is pressurized to keep it from boiling and turning to steam. There is a large tank, called the pressurizer, which heats or cools coolant water to keep its pressure high enough to prevent boiling. Normally, the pressurizer is about half water and half steam. On top of the pressurizer is a valve, called the pilot-operated relief valve, which opens if the pressure becomes too great. Radioactive steam and water may then flow through a drain pipe to a tank on the floor of the building, easing the pressure. There is a standby valve in case the pilot-operated relief valve fails. These valves played a major role in the accident.

The Accident. At 4 A.M. on March 28, 1979 the main feedwater pump that supplies water to the steam generators stopped. Four operators were on duty. Several things happened, and the hardware dimensions of them were anticipated by designers of the system:

1. There was no water to the steam generators.
2. Without water for steam, the turbine and the electrical generator automatically shut down.

3. The water in the steam generators cools reactor coolant water in the heat exchange. Without water in the steam generators the temperature of the reactor coolant water increased.
4. The rapidly heating reactor coolant water expanded, forcing water up into the pressurizer. The steam compressed and the pressure increased, causing the pilot-operated relief valve at the top of the pressurizer to open and ease the pressure.
5. With the temperature of the reactor coolant water rising, the control rods automatically dropped down to halt nuclear fission. Nuclear fission stopped, but there was still plenty of heat in the system.
6. The emergency feedwater pump automatically started and began pumping water into the steam generator.

All of this would have been a minor mishap that no one would have heard of, but it did not work out that way. The automatic engineering solution to the emergency failed because of human error. Consider these human-centered problems:

The emergency feedwater pump to the steam generators came on as programmed when the main feedwater pump failed, and an operator in the control room noticed that they were operating. What he failed to notice were two lights which signaled that the valves on the two water-lines to the emergency pumps were closed; no water was getting to the steam generators. One of the lights was covered by a maintenance tag; nobody knows why the operator failed to see the other light. Why the valves were closed is a mystery. There was a routine test of the emergency pumps two days before, and it was thought that the valves might have been closed and never reopened.

The pilot-operated relief valve opened to ease the pressure and it should have closed in 13 seconds, but it did not because it stuck in the open position. This should not have been much of a problem because the operator could close the standby valve. An operator noticed the light on the control panel for the pilot-operated relief valve and interpreted it to mean that the valve was closed. *The logic of the circuitry was that the light was wired to indicate whether the electrical power was on or off, not whether the valve was open or shut. The light said that the power was off, which it probably was, but the operator interpreted it to mean that the valve was closed, which it was not.*

With the pilot-operated relief valve open, radioactive steam and water poured from the top of the pressurizer through a pipe into a drain tank. The operators might have made a timely diagnosis of the open valve if they had noticed that the drain tank was filling up. The key piece of information was the temperature of the drain pipe to the tank because of the hot water flowing through it, but no special significance was attributed to the value that was read and there was no inference that the pilot-operated relief valve was open. All this time radioactive water was pouring out of the open valve into the drain tank and the building. A loss-of-coolant accident was in progress.

Despite these problems you would think that the large control panel would have presented unmistakable and informative alarms. There were alarms, almost without end. In the first few minutes 100 alarms went off. One operator said he would have liked to have thrown the alarm panel away. In the designers' well-meaning attempts to inform about everything, they informed about nothing.

There was a computer in the control room that was printing out information about the system. The computer and its printer were slow. One item was printed every 4 seconds. The computer was sometimes running 2 hours behind. Up-to-date information was gotten by dumping computer memory, but that resulted in loss of data about events and trends that would have helped to diagnose the difficulty.

The steam generators boiled dry in a couple of minutes, so the reactor coolant water heated, expanded, and rose in the pressurizer. At about this time two standby pumps automatically began pouring water into the core at the rate of about 1000 gallons per minute. The level of water in the pressurizer began to rise, and an operator interpreted it to mean that the core had plenty of coolant water. He shut one pump down and reduced the flow of the other to less than 100 gallons per minute. Too much water in the pressurizer is not a good thing because it makes pressure control difficult, so if the water level was perceived as satisfactory, the proper action was to turn off the pumps. Following this line of reasoning led to an even more threatening situation as the reactor coolant water began turning to steam, which displaced more water into the pressurizer and continued to feed the misperception that the core had plenty of coolant water.

The system was on the threshold of the most dangerous of all situations for a nuclear reactor—uncovering of the core and excessive temperatures that could damage the fuel rods and even melt the uranium itself. Harrisburg, the capital of Pennsylvania, was only a few miles away. As much as two-thirds of the core was uncovered before the incident was closed. Temperatures in the core rose as high as 3500 to 4000°F. About 90 percent of the fuel rods had damaged casings. Some of the uranium fuel itself may have melted.

The end of the crisis did not bring the end of the accident because the cleanup will go on for years. More than 1 million gallons of radioactive water filled the containment building or auxiliary building tanks, and the badly damaged reactor core is radioactive, of course. A plan for the cleanup took two years to develop. The cleanup is scheduled for completion in 1988, although no one will be surprised if it takes longer. The cost of the cleanup is to be about $1 billion (Blake, 1983), and it will not be without risks.

Comprehensive studies of the accident were made. Two weeks after the accident President Carter appointed a commission to analyze the accident. The commission was chaired by J. G. Kemeny, president of Dartmouth College and a mathematician. Kemeny's commission concluded that "the fundamental problems are people-related problems and not equipment problems" (Kemeny, 1979, p. 8). They went on to say: "The equipment was sufficiently good that, except for human failures, the major accident at Three Mile Island would have been a minor incident" (Kemeny, 1979, p. 8). In other words, the equipment worked well except for the pilot–operated relief valve. The equipment failed safe; no one was hurt. The accident would not have become a disaster if it had not been for human intervention, or the lack of it. What were some of the human factors problems? They may be found throughout the report of the President's Commission, they appear in analyses by the U.S. Nuclear Regulatory Commission (1979), and they are documented by a study team of human factors engineers (Hopkins et al., 1982). Here is a summary of prominent human factors problems:

The design of the control room was poor for the management of an emergency from the standpoint of human factors. The President's Commission could not have been clearer when it said that "the information was presented in a manner to confuse operators" (Kemeny, 1979, p. 29). That over 100 alarms went off in the first minutes hampered diagnosis of the accident, but difficulties with the control room design ran deeper than that. Before the accident, the Electric Power Research Institute, which is supported by power companies, commissioned the human factors engineers of Lockheed Missiles & Space Company to conduct a human factors analysis of the design of control rooms in nuclear power plants. If their report (Seminara et al., 1977) had been taken seriously, the accident might conceivably have been averted. These analysts found dials that were difficult to read and controls that were difficult to manipulate. Banks of controls and displays with different functions were identical. In one control room the operators had put a beer can over the handle of a control so that they could discriminate it from others that looked the same. There were parallax problems (reading error as a function of viewing angle) in reading some of the instruments. There were mirror-image displays, where an instrument would have its mate on a companion display on another wall in the opposite position. Human factors design deficiencies of this sort filled a thick report.

An appreciable number of control room operators in U.S. nuclear power plants, although not all, were trained in nuclear programs of the U.S. armed forces. Even though experienced, these operators had to demonstrate their proficiency by passing a licensing examination of the U.S. Nuclear Regulatory Commission. The contents of the examination were judged to be poorly related to job requirements (Hopkins et al., 1982), implying that passing the examination did not necessarily mean that the applicant was qualified for the job. Moreover, a compensatory system of grading the examination was used, where the applicant could fail one part of the examination and make it up on another, just as long as an overall passing score was obtained (Kemeny, 1979, p. 49). It is as if a brain surgeon could use psychiatry to compensate for his deficiencies in neuroanatomy. The licensing examination for control room operators did not guarantee that they would know the procedures for resolving an accident like that which occurred at Three Mile Island.

The training was judged to be shallow (Kemeny, 1979, pp. 23, 49). There was a simulator, which is a major training device that imitates the operation of a nuclear power plant for training purposes (Chapter 18), and it differed from the way the plant actually operated; simulator training lacked pertinence in important respects (Kemeny, 1979, p. 23). Parts of the training program were subcontracted, and the contractors carried out their contractual responsibilities, but there was no overall responsibility for training (Kemeny, 1979, p. 22). Training was piecemeal.

There is no better illustration of the role of human error in human–machine systems than the Three Mile Island incident. It seems as if all the human factors mistakes that could be made were made. The more recent accident at the Chernobyl nuclear power plant in the Soviet Union may prove to be as instructive for human factors engineering as the Three Mile Island incident, but official details have yet to be released.

Reading of X Rays

An industrial or a military system comes to mind most readily when we think of a human–machine system, but there are good examples outside of industry and the military. The field of medicine provides good illustrations also.

Diagnosis, as the identification of ailments from signs and symptoms, is the cornerstone of medicine. Diagnosis requires accurate observation, knowledge, and the effective use of machines. The most important diagnostic procedure of this century has been the X-ray examination. As we all know, X rays passing through the body leave a kind of shadowgram on film. Sometimes the shadow is unmistakable, as with a broken bone. Other times the shadow can be vague or absent, as in the case of tissue. Today we have improved devices for diagnosis of this kind, such as nuclear magnetic resonance equipment, which does not rely on X rays, and the CAT (computerized axial tomography) scanner which is a computerized combining of X rays of body sections from various perspectives that reduces the reading difficulties arising from compacted, overlapping organs. These modern devices are more elaborate than the X-ray machine and much more expensive to buy and use, so the X-ray machine is assured a future in diagnostic medicine for a long time to come.

The problem with the X ray is that a physician's reading of the film is subject to error, and it is another example of how imperfect human interaction with a machine has implications for us all. When a patient has a disease and is told that he does not, it means trouble. When a patient does not have a disease and is told that she does, it means trouble. When a patient has a communicable disease and is told that he does not, it means trouble for us all. Even those who have a permissive attitude toward human error will find the evidence disturbing.

Chest X Rays. In 1944, the Veterans Administration convened a Board of Roentgenology (W. K. Roentgen, a German physicist, discovered X rays in 1895) to relate accuracy of reading X-ray films to four different kinds of equipment with four different film sizes that had been routinely used in the chest X raying of military personnel. The board conducted an experiment, collecting X-ray films from the four different kinds of equipment. Five readers read the films for signs of tuberculosis. They found that the differences between readers were so great that it was difficult to determine differences in equipment. The films were read several times, and a reader was found not only to differ from his colleagues on occasion (interreader variation) but also to differ from himself from reading to reading (intrareader variation). Moreover, readers missed positive instances of the ailment, and they sometimes declared false positives, where lesions were seen when there were none. Interpretation of chest X rays is not easy. Except for the ribs, the chest is a region of tissue, heavily compacted with organs. The discrimination of subtle shadows is difficult. Unreliability may be understandable, but it is not acceptable.

Yerushalmy et al. (1950) conducted one of the major studies of this topic. The data were chest X rays collected from entering students at the University of California at Berkeley in 1949, before a cure of tuberculosis was known. Tuberculosis is a contagious lung disease, and it was common for colleges and universities to X ray entering students.

The study used 1790 X rays. Seven radiologists and chest specialists examined the films separately and in conference, and there was unanimous agreement that 30 students had tuberculosis. Six other radiologists and chest specialists were then brought in as subjects of the study and asked to classify the films as positive or negative. Only 20 of the 30 positives were correctly identified, on average. Of the 1760 negatives, 30 were false positives. Put another way, 10 of the diseased students would have been declared healthy and admitted, potentially to infect other students. Thirty of the healthy students would have been denied admission and probably hospitalized.

In another study, Yerushalmy et al. (1951) had six physicians examine 150 pairs of chest X rays taken 3 months apart. The films were of known tubercular patients, and the issue was whether a patient had improved, remained the same, or gotten worse. Physicians knew which was the first and which was the second film. Any two readers disagreed about one-third of the time. A reader would contradict himself about 20 percent of the time from the first film to the second. Diagnosing whether a patient is getting better or worse is as difficult as diagnosing correctly in the first place.

You might think to yourself that these studies of Yerushalmy's are old. With belief in the inevitability of progress, a change for the better must have occurred. A recent study of Lesgold et al. (1981) shows that the reading of chest X rays is still unreliable. The ailment to be detected and diagnosed in his study was a collapsed middle lobe of the lung. A residency in medicine is postgraduate study after a medical degree has been awarded; 11 of the subjects were resident radiologists in their first and second years, and seven were residents in their third and fourth years. In addition, there were five expert radiologists with an average of 18 years of experience since their residencies. The experts had read thousands more films than the resident physicians; it was their full-time job. The procedure was first to look at a film briefly for 2 seconds and report any abnormalities. The film was then examined without time limit and a formal report was dictated. Then followed an examination of clinical data and dictation of a second report. Eighty-five percent of the experts correctly identified the lungs as the afflicted organ, but only about one-third of the residents were able to do it. The less experienced residents associated the abnormality with as many as seven different organs. Only two of the physicians diagnosed the ailment correctly. One was a resident physician, one an expert.

Other studies of recent years are no more encouraging. Pneumoconiosis is a hardening of the lungs that miners get from breathing coal dust, and Morgan et al. (1973) investigated the identification of it from chest X rays. Seven readers were 83 percent correct in the interpretation of 200 films, which is reasonably good, although the range among the seven was from 57 to 94 percent correct. Herman and Hessel (1975) had eight radiologists read 100 chest X rays that contained a variety of ailments. Fifty-seven percent of the interpretations had errors that would either affect the care of the patient or had a potential for affecting it. Rhea et al. (1979) found that 10 percent of the interpretations of 3300 X rays by resident physicians had to be changed when they were checked by staff radiologists. Parasuraman (1985) had eight radiologists read two sets of chest X rays, also with a variety of ailments. One set had a low proportion of abnormalities, and the other a high proportion. Correct disease identification was 57 percent in the set with low prevalence of disease and 56 percent in the set with high prevalence.

Dental X Rays. Research with particular methods or materials runs the risk that it may discover things that are unique to the methods or materials and not generalize. Is the unreliability of reading X rays peculiar to the difficult chest X ray? No, it would seem, because Goldstein and Mobley (1971) found the same thing for dental X rays. Their subjects were senior dental students who had completed course work in radiology and oral diagnosis and had had clinical experience. The correct detection of abnormalities was 73 percent. The percentage of false positives was 41 percent. Intrareader agreement was 90 percent. Interreader agreement was 70 percent.

Research Methods in Human Factors Engineering ———

The definition of human factors engineering at the beginning of this chapter stated that the design and use of a human–machine system relies on the scientific knowledge of human behavior. This section is a review of representative ways that scientific knowledge is acquired in human factors engineering. The vastly different characteristics of human–machine systems means that virtually all dimensions of human behavior are candidates for the deliberations, so the entire science of human behavior can be pertinent for human factors engineering. Notwithstanding, there is a body of scientific knowledge, generated by research-oriented human factors engineers, that is particularly pertinent because it involves variables that bear incisively on human behavior in human–machine systems. General experimental psychologists, often pursuing theoretical interests, may or may not collect data that are appropriate for the practical purposes of human factors engineering. Research methods overlap, however, whether or not the interests and motivations of investigators do.

Nonexperimental Methods

Laboratory experiments allow for careful control and manipulation of variables, and so is an approach honored by scientists, but it is not the only approach to reliable data. There are nonexperimental techniques that lead to data which are useful for the refinement of human factors in human–machine systems. Consider these various approaches, all of them nonexperimental, that can lead to useful data:

1. A worker's productivity is low and a human factors engineer hypothesizes that the worker's task is inefficiently organized. The engineer conducts a task analysis, systematically observing the movements that the worker makes and the time that he takes to make them. The task analysis confirms the hypothesis, the task is restructured for greater efficiency, and the worker's productivity is increased (Chapter 8).
2. An analysis of accident data for a system suggests that the readability of a particular instrument on the display panel is at fault. Interviews with system operators further suggest readability problems. The instrument is redesigned and replaced, and the accident rate decreases.
3. The operators of a system are surveyed with a questionnaire asking them about all phases of system operation and problems that they have had. Some of the questions turn up problems for a significant number of the operators, such as controls that are difficult to reach or controls that are easy to activate by mistake. Using the questionnaire data as point of departure, the operators are interviewed and the problems are explored in greater detail. Recommendations for system redesign are made.
4. Observation gives reasons for believing that system operators are not performing very well. A job knowledge test is designed, covering all phases of system operation, and it is administered to the operators. Their knowledge of basic system functions is found lacking. A retraining program is designed for the operators. The training program for new, incoming operators is revamped.

The methods used in these examples all represent approaches to reliable knowledge, yet they are all nonexperimental. Sometimes knowledge gained in ways like these may be preexperimental because it produces ideas for variables that deserve experimentation.

The Correlational Approach

The correlational approach is most often used in personnel selection and classification, although uses for it are found throughout the behavioral sciences. You hypothesize that success as an engineer in your company is related to mathematical ability, so you administer a mathematics test to all engineers and correlate the test scores with their salaries (assuming salary to be a good measure of success). You hypothesize that some workers do poorly in a factory because they lack enough physical strength, so you administer a test of strength and correlate the test scores with amount of work that each worker accom-

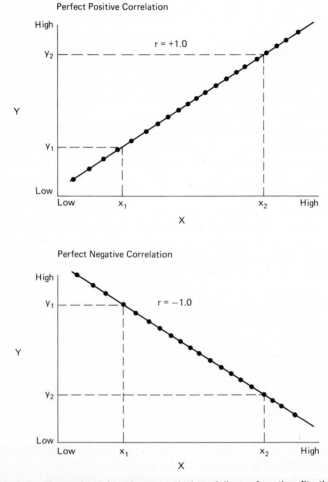

Figure 1-2 Illustration of perfect correlation. A linear function fits the data points exactly. A perfect correlation has a value of +1.0 or −1.0. A value of *y* can be predicted from a value of *x* exactly.

plishes in a month. The correlation coefficient is used in research of this kind to show the relationship between the two sets of scores, X and Y. A *correlation coefficient* is an index of fit, indicating how well a mathematical function fits empirical data points. The linear function is most commonly used:

$$Y = a + bX$$

The computation of the correlation coefficient, symbolized as r, fits the linear function to X and Y and indicates the degree of fit. The value

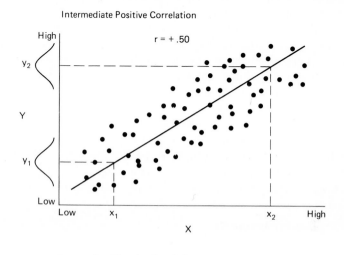

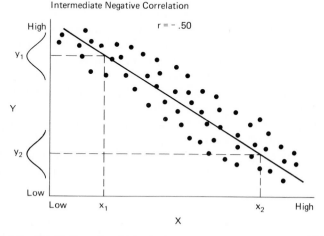

Figure 1-3 Illustration of an intermediate value of a correlation coefficient. The linear function is the best fit to the data points. The y values can be predicted from x values, but there is some error in the prediction as represented by the scatter around y.

of the correlation coefficient ranges from a perfect positive relationship of $+1.0$ to a perfect negative relationship of -1.0. A perfect correlation is when a straight line fits the X,Y pairs of data points exactly. Figure 1-2 illustrates perfect positive and negative correlations. When the relationship is perfect and positive, the score x_1, a low value of X, corresponds exactly to y_1, a low value of Y. The score x_2 is a high value of X and it predicts y_2, a high value of Y, exactly. The same exactness of prediction holds for a perfect negative correlation except that the low-valued x_1 predicts a high value of Y and the high-valued x_2 predicts a low value of Y. Whether positive or negative, the prediction is perfect.

Few relationships in the empirical world are perfect, and a useful characteristic of the correlation coefficient is that it represents imperfect relationships as well (in the examples above, the correlation of engineers' salaries and scores on a mathematics test, and the correlation of workers' production and scores on a strength test, will be imperfect). Figure 1-3 shows positive and negative correlations of intermediate value. A linear function is fitted to the somewhat scattered data points, and it is used to best estimate Y from X. Here a value of X does not predict an exact value of Y but rather a range of Y values; prediction is inexact. When $r = 0$, and there is no linear trend to the data whatsoever, no improvement in prediction for Y is possible. The best prediction that can be made of Y from any value of X is the mean, or average, value of the Y scores; discriminative prediction is lacking.

Experimental Methods

Nonexperimental research methods are descriptive, taking a situation as it is and learning more about it. As useful as nonexperimental methods are, the experimental approach is superior because variables of interest are systematically manipulated, unwanted variables controlled insofar as possible, and the effects of the variables on human behavior measured. The variables of interest that are manipulated are called *independent variables,* and the behavioral measures that are used to assess their effects are called *dependent variables.* A good experiment is *reliable* because its results can be repeated. It is desirable, also, that the results of an experiment be *generalizable* and apply to new situations and subjects. The results of some experiments have limited generality. An experiment might be designed to answer a narrow but important question for a particular system, and the results go no further. The best experiments, however, yield results that are far-reaching, going beyond the particular situation that generated the data. Without generalization each new problem requires a new experiment to answer it.

The amount of information that can be extracted from an experiment depends on the design of the experiment. A few illustrative designs will show the experimental approaches that might be used to generate human factors knowledge. Although not included in these illustrations, statistics are used to analyze the data of experiments like these, and human factors engineers are trained to use them.

Assume that a plant has a training program of 20 hours before workers are assigned to the production line. You have noticed that the transition from the training program to the line is not smooth. The quality of work tends to be low and there is waste—too many products have to be scrapped because of worker error. You hypothesize that doubling the training time would solve the problem, so you design the simplest of all experiments. One group of incoming workers is assigned to the standard training program of 20 hours, and acts as a control group. A comparable group is assigned to a new training program of 40 hours of training, acting as an experimental group. Your criterion of performance on which training is evaluated is the number of acceptable products made by each worker during the first week on the line. The average performance was found to be 100 acceptable products for the 20-hour training group and 150 for the 40-hour group.

You have learned something that will improve the efficiency of your plant, but you have not learned enough to optimize worker performance from the standpoint of training. Maybe 60 hours of training would have been better, or 80 hours. Having found that amount of training makes a difference, you decide to do a more comprehensive experiment with several amounts of training: 20, 40, 60, 80, and 100 hours. Five comparable groups of new workers are each given these different amounts of training. The 20-, 40-, 60-, 80-, and 100-hour groups average, respectively, 101, 149, 175, 173, and 174 acceptable products the first week on the job. Performance increases up to 60 hours, but not beyond, so a 60-hour training program is your recommendation to management. Knowing the optimum point is cost-effective. Too little training reduces the number of acceptable products, and too much training is costly in training time. Training is expensive in instructors and training equipment. Nothing is produced while a worker is in training.

Convinced of the importance of training, you might conduct a more elegant factorial experiment. A *factorial experiment* has two or more levels of two or more variables. In the case of this training example, there might be five levels of amount of training—20, 40, 60, 80, and 100 hours, and two training methods, A and B. Figure 1-4 shows how this factorial experiment is schematized. A group of subjects is assigned to each cell of the design, and all of them are evaluated on the criterion performance measure after training. A possible outcome might be like the data shown in Table 1-2. Essentially the same values as before are found for amount of training, but there is essentially no difference between the two methods of training. Amount of training is

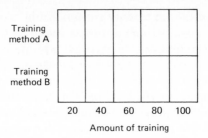

Figure 1-4 Hypothetical training experiment with a factorial design.

TABLE 1-2 Data of a Hypothetical Factorial Experiment on Training*

	Hours of Training					
Training Method	**20**	**40**	**60**	**80**	**100**	**Average**
A	98	150	174	176	177	155
B	96	148	176	174	175	154
Average	97	147	175	175	176	

*The entry is the average number of acceptable products produced during the first week on the job.

clearly the fundamental variable, so 60 hours of training, using the least expensive of the two training methods, is the best training program.

Which of these three hypothetical experiments should be run in the context of a practical training situation? The factorial experiment is the most costly in time and money, but it generates the most information. One might surmise that something less than a factorial experiment would be realistic in the practical context of most plants and their training programs.

Experimentation and Realism

Human factors engineering is a discipline of the real world, so it would seem desirable to do experiments in the real world, as close to the situation of interest as possible. The experiments of the foregoing examples of training research were conducted in the context of actual training operations and so would be classified as *field experiments*. Human factors engineering has done plenty of field experiments. Field experiments are to be contrasted to *laboratory experiments*, and human factors engineering have done plenty of them also. Laboratory experiments are more abstract than field experiments and less realistic. A laboratory experiment on amount of training might use a simple learning task on the assumption that relevant variables are easier to identify and control than with real-life tasks.

It would seem that field experiments would be the best because closeness of the experimental situation to the operational situation makes generalization easy. The comparative remoteness of the laboratory experiment from the operational situation makes generalization more tenuous. Things are not this simple, however. Laboratory experiments can be carefully controlled and the data correspondingly precise. Field experiments, done in the hurly-burly of practical situations, can have a good deal of error in their measures, so the findings are less reliable. Field experiments are also more expensive to run. It is trade-offs like these that a human factors engineer must consider when she turns to research data for the solution of a practical problem.

Final Comments

Human-centered difficulties in human–machine systems are the justifications for human factors engineering. A full range of remedies, perhaps none of them easy, is required for nuclear power plants like the one at Three Mile Island. The causes and correction of aircraft accidents and errors in the reading of X rays are less evident but not without possibilities. Ways of minimizing human error in human–machine systems, and research on them, constitute the subject matter of this book.

Summary

The justification for the discipline of human factors engineering is that the personnel of human–machine systems make errors, and these errors degrade system performance. Human factors engineers use their scientific knowledge about human behavior to minimize system error. Prominent among the bodies of knowledge that they use are design of the workplace, design of controls and displays, personnel selection, and personnel training.

The behavioral issues that can arise for a system are varied, and human factors engineers are correspondingly varied in their backgrounds. Industrial engineering is the historical root of human factors engineering and is still a major contributor. Psychology, concerned as it is with the science of behavior, is another major contributor.

Examples of human factors issues in major systems were given to illustrate how human behavior, when it is not considered scientifically and properly weighted as a variable of a system, can degrade system performance. Aviation accidents, the nuclear accident at Three Mile

Island, and the reading of X rays by physicians were reviewed. Representative ways of conducting research on human factors variables were summarized.

References _____

Blake, M. E. (1983). Three Mile Island: GPU, B&W settle suits; cleanup stretched out. *Nuclear News,* March, 43–44.

Chapanis, A., & Lindenbaum, L. E. (1959). A reaction time study of four control-display linkages. *Human Factors. 1,* 1–17.

DeGreene, K. B. (1970). *Systems psychology.* New York: McGraw-Hill.

Deininger, R. L. (1960). Human factors engineering studies of the design and use of pushbutton telephone sets. *Bell System Technical Journal, 39,* 995–1012.

Gilbreth, F. B. (1909). *Bricklaying system.* New York: Mryon C. Clark.

Goldstein, I. L., & Mobley, W. H. (1971). Error and variability in the visual processing of dental radiographs. *Journal of Applied Psychology, 55,* 549–553.

Herman, P. G., & Hessel, S. J. (1975). Accuracy and its relationship to experience in the interpretation of chest radiographs. *Investigative Radiology, 10,* 62–67.

Hopkins, C. O., Snyder, H. L., Price, H. E., Hornick, R. J., Mackie, R. R., Smillie, R. J., & Sugarman, R. C. (1982). Critical human factors issues in nuclear power regulation and a recommended comprehensive human factors long-range plan. Washington, DC: U.S. Nuclear Regulatory Commission, Technical Report NUREG/CR–2833 (Volumes 1–3), August.

Kemeny, J. G. (1979). *Report of the President's Commission on the accident at Three Mile Island.* Washington, DC: U.S. Government Printing Office, October.

Lesgold, A. M., Feltovich, P. J., Glaser, R., & Wang, Y. (1981). The acquisition of perceptual diagnostic skill in radiology. Pittsburgh, PA: Learning Research and Development Center, University of Pittsburgh, Technical Report PDS-1, September 1.

Morgan, R. H., Donner, M. W., Gayler, B. W., Margulies, S. I., Rao, P. S., & Wheeler, P. S. (1973). Decision processes and observer error in the diagnosis of pneumoconiosis by chest roentgenography. *American Journal of Roentgenology, 117,* 757–764.

National Transportation Safety Board. (1981). Annual review of aircraft accident data: U.S. air carrier operations 1979. Washington, DC: National Transportation Safety Board, Report NTSB-ARC-81-1, November 16.

Parasuraman, R. (1985). Detection and identification of abnormalities in chest X rays: Effects of reader skill, disease prevalence, and reporting standards. In R. E. Eberts & C. G. Eberts (Eds.), *Trends in ergonomics/human factors II.* Amsterdam: Elsevier.

Rhea, J. T., Potsaid, M. S., & DeLuca, S. A. (1979). Errors of interpretation as elicited by a quality audit of an emergency radiology facility. *Radiology, 132,* 277–280.

Seminara, J. L., Gonzalez, W. R., & Parsons, S. O. (1977). Human factors review of nuclear power plant control room design. Palo Alto, CA: Electric Power Research Institute, Technical Report EPRI NP-309, March.

Shinar, D., & Acton, M. B. (1978). Control–display relationships on the four–burner range: Population stereotypes versus standards. *Human Factors, 20,* 13–17.

Taylor, F. W. (1911/1967). *The principles of scientific management.* New York: W. W. Norton.

U.S. Nuclear Regulatory Commission. (1979). TMI-2 lessons learned: Task force final report. Washington, DC: U.S. Regulatory Commission, Technical Report NUREG-0585, October.

Yerushalmy, J., Harkness, J. T., Cope, J. H., & Kennedy, B. R. (1950). The role of dual reading in mass radiography. *American Review of Tuberculosis, 61,* 443–464.

Yerushalmy, J., Garland, L. H., Harkness, J. T., Hinshaw, H. C., Miller, E. R., Shipman, S. J., & Zerling, H. B. (1951). An evaluation of the role of serial chest roentgenograms in estimating the progress of disease in patients with pulmonary tuberculosis. *American Review of Tuberculosis, 64,* 225–248.

CHAPTER 2

Reliability and Human Error

The errors that personnel make in a human–machine system contribute as much to system unreliability as do hardware failures and software errors. Chapter 1, in its review of human error in aircraft accidents, the incident at Three Mile Island, and the reading of X rays, said that unreliable human participants in systems can have dangerous or catastrophic consequences. A less dramatic but nevertheless valuable illustration of how human error can affect system reliability is a study by Robinson et al. (1970) of 213 maintenance requirements in air defense equipment. Personnel were responsible for 25 percent of the malfunctions. Shapero et al. (1960) analyzed 3829 malfunction reports from seven missile systems and found that personnel were responsible for 29 percent of them. A report of the Comptroller General to the U.S. Congress (1981) on the reliability of weapons systems included a discussion of human reliability. The report said that human-induced failures in systems may be more common and widespread than ordinarily believed because human-initiated failures are sometimes hard to identify and it is easy to tally human error as equipment error. The concern with reliability in human factors engineering reflects the hope that scientific knowledge about human behavior can be used to minimize human error and raise the reliability of human–machine systems.

Reliability engineering began with hardware problems. There has always been concern with devices that worked reliably as a practical engineering matter, but it was not until the 1950s that reliability engineering emerged as a professional discipline (O'Connor, 1985, Chap. 1). The low reliability of some military equipment gave impetus to this new field. The complex equipment that the armed forces was buying was not working well; in addition, availability was low and maintenance costs high. The military is still struggling with the reliability problem. In 1983, Paul Thayer, Deputy Secretary of Defense, said that ". . . the cost of correcting defects in our weapons and equipment runs in the range of 10 to 30 percent. This represents enormous waste—billions of dollars when the overall budget is considered" (Thayer, 1983).

The formal structure that reliability engineering has given hardware reliability is a useful place to begin.

Hardware Reliability

Reliability is based on probability mathematics, and it is defined as probability of success for a system component under defined conditions of use:

$$\text{reliability} = \frac{\text{number of successful operations}}{\text{total number of operations}}$$

Being probability, reliability varies from 0 to 1. Probability of failure is 1 minus reliability. If the starter of your automobile fails once in 10,000 starts, its reliability is .9999. Theoretically, a four will appear one-sixth of the time in an infinite number of rolls of a die, so the starters will work .9999 of the time in an infinite number of tests. In practice, a reliability engineer will test samples of starters and use the findings to estimate reliability for the population of starters. Keep in mind that a component does not have one reliability value. Conditions of use are among the determinants of reliability, so a component has many reliabilities. Reliability of hardware decreases as temperature or vibration increase, for example.

Figure 2-1, which has been called the bathtub curve, shows the failure history of a piece of equipment. A disproportionate number of failures occur early, usually because of material and production deficiencies. The term *burn-in* derives from the electronics industry, where all parts of a production line may be operated for a specified number of hours and the weak components identified. The period of operational use for a component is the center part of the curve where failure is from random causes and has a constant rate. The *wear-out* part of the curve is where use has caused the component to perform unsatisfac-

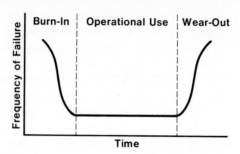

Figure 2-1 Characteristic stages in the reliability of a component. Adapted from Adams (1982), Figure 1. Copyright 1982, by the Human Factors Society, Inc., and reproduced by permission.

torily or to fail completely. In Figure 2-1, the burn-in and the wear-out segments are portrayed equivalently, but they need not be.

A basic equation for reliability theory relates reliability to time of use. The equation is derived from the period of operational use where the failure rate is constant (Figure 2-1):

$$\text{reliability} = e^{-ft} \qquad (2\text{-}1)$$

where f is failure rate and t is time. Equation 2-1 for two failure rates is shown in Figure 2-2. Getting from the flat function for constant failure rate in the middle of Figure 2-1 to the negative exponential function in Figure 2-2 might not be intuitively obvious, but Table 2-1 has an illustration that should make it clear. The constant failure rate means that a constant proportion of the number of components remaining at the end of each time period will fail. Table 2-1 assumes a failure rate of 10 percent per hour, so 10 percent of the components remaining will fail each hour. The number remaining decreases rather sharply at first, and then more gradually, which translates to the negative exponential function of Equation 2-1.

A commonly used reliability statistic is *mean time to failure,* which is the reciprocal of the failure rate. Using the example in Table 2-1 with a failure rate of .10, the average time of operation between failures of the component is 10 hours.

A system is a configuration of components, and the payoff comes in combining the reliabilities of components to yield the reliability of the system as a whole. How to proceed depends on the configuration of the system. Figure 2-3 shows four simple system configurations, and they indicate how system design and reliability can be related. One of the configurations is two components in a series—if either fails, the system fails. The failure of component A is assumed to be independent of the

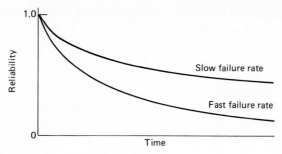

Figure 2-2 Reliability as a function of time for two failure rates of a component.

TABLE 2-1 Numerical Illustration of Failure Rate as a Function of Time*

Time (hours)	Number Remaining	Number of Failures per hour
0	100.0	——
1	90.0	9.0
2	81.0	8.1
3	72.9	7.3
4	65.6	6.6
5	59.0	5.9
6	53.1	5.3
7	47.8	4.8
8	43.0	4.3
9	38.7	3.9
10	34.8	3.5

*The failure rate is assumed to be 10 percent per hour.

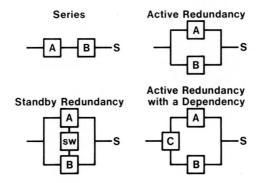

Figure 2-3 Four different arrangements of simple systems to illustrate the implications of design for reliability. See text for explanation. Adapted from Adams (1982), Figure 2. Copyright 1982, by the Human Factors Society, and reproduced by permission.

failure of component B, so the product rule for independent probabilities applies to give the reliability for the system, R_S:

$$R_S = (R_A)(R_B) \tag{2-2}$$

There are two consequences of the product rule for system reliability. One is that the higher the reliability of individual components, the higher the reliability of the system. The other is that the greater number of components in the system, the lower the system reliability. The multiplication of many component reliabilities, all with values less than 1, will give a relatively low system reliability. These two consequences are illustrated in Figure 2-4.

How can complex systems with thousands of components conceivably be made to work? As Figure 2-4 shows, even with component reliability high, increases in system complexity soon drives system reliability down when component failures are independent. The answer lies partly in Equation 2-2 and partly elsewhere. An implication of Equation 2-2 is that if component reliabilities are *very* high, system reliability can be high. The probability of the axle of your car failing is extremely small, for example, so if the reliabilities of the car's other components were all correspondingly high, the car's overall reliability would be high.

As instructive as it is, Equation 2-2 can be misleading in suggesting that it is difficult to make complex systems with many components work well. This is not entirely so. Obviously, there are reliable complex systems. Something else is required to make them work reliably, however, and the answer is redundancy. *Redundancy* is a spare that takes over when a component fails. Redundancy is used when a function is particularly critical for system success and warrants the extra equip-

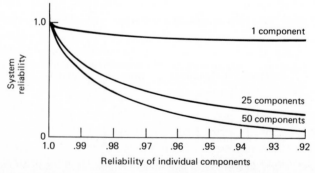

Figure 2-4   System reliability as a function of the reliability of the individual component and the number of components of the system.

ment. The spare tire in the trunk of your car is redundant, and it has all the advantages and disadvantages of a redundant component. The tire is added weight and cost, and it takes up space in the trunk. Notwithstanding, the difficulties of having a flat tire without a spare are so self-evident that none of us would be without one.

Figure 2-3 illustrates three kinds of system redundancies. All of them increase the reliability of a system. An *active redundancy system* has two components, both of them operating but only one needed to operate the system. The system fails only when both of them fail. The components are assumed to be independent, so if the probability of one failing is $1 - R$, the probability of both failing is

$$(1 - R_A)(1 - R_B) = 1 - R_A - R_B + R_A R_B$$

R_S is 1 minus the probability of the joint failure:

$$R_S = 1 - (1 - R_A - R_B + R_A R_B) \qquad (2\text{-}3)$$
$$= R_A + R_B - R_A R_B$$

A *standby redundant system* has only one component operating and its mate is switched on when it fails, so the reliability of the switch must be taken into account. The R_S for this system is

$$R_S = R_A + (1 - R_A)(R_B)(R_{sw}) \qquad (2\text{-}4)$$

where the reliability of the switch is R_{sw}. The addition of probabilities is the either–or occurrence of independent events, so R_S is the probability of component A succeeding, *or* the joint probability of component A failing, component B succeeding, and the switch succeeding.

An *active redundant system with a dependency* has both components operating, but they are dependent on the success of another. An example is two electric motors operating from the same generator. When one motor fails, the other takes over, but both motors fail when the generator fails. For this system

$$R_S = R_C R_A + R_C (1 - R_A)(R_B) \qquad (2\text{-}5)$$

R_S depends on the joint success of components C and A *or* the joint occurrence of the success of component C, the failure of component A, and the success of component B. A numerical example illustrates that Equations 2-3, 2-4, and 2-5, based on redundancy, result in higher system reliabilities than Equation 2-2, where redundancy is absent. Assuming that all component reliabilities are .90, R_S is .81 for components in a series (no redundancy), .99 for components with active redundancy, .98 for standby redundancy, and .89 for active redundancy with a dependency.

Not illustrated in Figure 2-3, but a common kind of redundancy, especially in electronic circuitry, is a triplex system with majority voting. Three identical channels for a circuit will be used, with their outputs continually compared in voting circuitry. When all outputs agree, the common signal is passed. If there is a malfunction in one channel, the other two outvote it and pass their common signal. If two circuits fail, all outputs differ, there is no majority for a decision, and the circuit fails. A complex circuit can have voting after each of many segments and have a comparatively high tolerance of failure in its elements. A complex circuit with three-channel redundancy in each of 20 segments is 17 times more reliable than the same circuit without redundancy, with a reliability of .98. Redundancy in 200 segments is 50 times more reliable (Klass, 1962).

Old aircraft had direct mechanical linkages between the pilot's input to the control and movement of the aircraft's control surfaces. Advanced jet aircraft of today now "fly-by-wire," where digital computers intervene between control input and hydraulic actuation of the control surfaces. Total failure of the flight control subsystem is catastrophic, so extremely high reliability is mandatory. An answer has been quadruplex redundancy with majority voting. The output of four computers is compared, and if one differs, it is shut down and the system then becomes a triplex system with majority voting. If one of the remaining three fail, the other two continue, and only when one of the two remaining fails does the flight control subsystem fail. The expectation is that loss of such aircraft through complete failure of the flight control subsystem will be 1 in 10 million (Warwick, 1985). Redundancy does not stop with the computers. The computers require electrical power, so the electrical subsystem is redundant also. Aircraft have long had standby electrical subsystems, but now the redundancy of electrical subsystems is more elaborate.

Software Reliability

These days it is a modest system that does not have a computer interfacing with other hardware, so the reliability of software becomes as fundamental to system reliability as the reliability of hardware. Software reliability lacks the theory and mathematical structure of hardware reliability, but that does not diminish the importance of it.

Software reliability differs from hardware reliability in a number of ways. Here are some things that are true of hardware reliability that are *not* true of software reliability (O'Connor, 1985): Hardware reliability is probabilistic, ranging in value from 0 to 1 (software reliability is normally 0 or 1—the program either works or it does not); hardware failures can be caused by deficiencies in design, production, use, and maintenance (when software does not perform it is program error, not failure in the hardware sense, because the program is the same before as after the error); hardware failures can be due to wear (software does not wear out); hardware failure can be related to burn-in and wear-out (burn-in and wear-out have no meaning for software); hardware failure is related to passage of time (software is time independent); hardware failure is related to environmental factors (the environment does not affect the reliability of software); and the reliability of a hardware system can be predicted from theory and empirical data (there is no theory and little empirical basis for predicting software failure). Some analysts believe that theory for hardware reliability can be adapted to software, while others think that an entirely new approach is needed (Amster and Shooman, 1975, p. 657). Whatever direction the topic of software reliability might take, it seems fair to say that at present there are few systematic empirical data on which theory can be based.

How to Have a Reliable Human–Machine System

The details are enough to support the professions of reliability engineering and human factors engineering, but the general principles of building a reliable human–machine system are only five. Three of the principles apply to hardware and two of them to personnel. There are no corresponding principles for software reliability.

It would be convenient if hardware engineers could concentrate on their principles and human factors engineers on theirs, but it is not so easy. The hardware conception of the system has trade-offs for personnel, which are often neglected (Chief of Naval Operations, 1977). Conversely, human factors has implications for system design.

Principles of Hardware Reliability in Systems

Keep it simple. Equation 2-2 says that the fewer the components in series, the more reliable the system. Table 2-2 illustrates some of the penalties of system complexity. The table compares two fighter bombers in the inventory of the U.S. Air Force. The A-10 aircraft is comparatively low in complexity, and the F-111D comparatively high. The A-

10 is an effective air weapon, but the F-111D has greater capabilities. The F-111D, however, because of its complexity, is often undergoing repair and is unavailable for its mission two-thirds of the time. The two columns on the right show that more of costly maintenance is required for the F-111D. Furthermore, the price of an F-111D is high because it is complex, so not many are procured. Thus, two-thirds of a relatively small number of F-111Ds are unavailable for their mission. Fallows (1981, p. 42) writes of the *real fleet* and the *phantom fleet*. The number of aircraft in the inventory is the phantom fleet, but all that matters is the real fleet that can be operating in time of war. The difficult design question is this, and reliability is central to it: Should we have fewer complex aircraft with extensive combat capabilities but which are unreliable with limited availability, or should we have more of simpler aircraft with less combat capability but with reliability and availability? A good argument can be made for the doctrine of make it simple, make it work, and make more of it (Spinney, 1980), although it runs counter to influential thinking in high U.S. military circles. A persuasive argument for some is that the United States cannot always match its potential adversaries person for person or tank for tank, so it is forced to rely on superior technology (e.g., Thayer, 1983). "Superior technology" usually translates to complex weapons systems with a potential for reliability problems.

Make each component as reliable as possible. This principle needs to be stated but not belabored. The merits of high component reliability for hardware are presented graphically in Figure 2-4 and implied in Equations 2-2 through 2-5.

Engineers are always on the lookout for developments that will improve reliability. The replacement of the unreliable vacuum tube with the solid-state transistor is probably the best example of modern times. *Glass cockpits,* where cathode ray tubes present flight data processed by a digital computer, are rapidly replacing electromechanical instruments in commercial and military aircraft. The reliability of glass cock-

TABLE 2-2   Reliability Data on Two Aircraft of the U.S. Air Force*

Aircraft (Fighter Bombers)	Comparative Complexity	Percent Planes Not Mission Capable	Average Flight Hours between Failures	Number of Repairs Required per Flight	Repair Man Hours Required per Flight
A-10	Low	32.6	1.2	1.6	18.4
F-111D	High	65.6	0.2	10.2	98.4

*The two aircraft differ in their relative complexity.

Source: Spinney, F. C. (1980), Page 41. Defense facts of life. Staff paper, Department of Defense. December 5. Reprinted by permission of F. C. Spinney.

pits is estimated to be three to four times that of electromechanical instruments (Nordwall, 1986).

Build it properly in the first place. Quality control engineering is monitoring production for defects, pinpointing the causes of defects, and doing something about them. In recent years Japanese manufacturers have been admired for the high quality of the electronics parts and components, photographic equipment, and automobiles which they have exported to the United States. Some U.S. manufacturers have been placed in uncomfortable competitive positions. In an informative study of this topic, Garvin (1983) compared the quality of Japanese and American home air-conditioners and related it to quality control practices. Ten U.S. companies and seven Japanese companies participated. They all made a standardized product with much the same manufacturing equipment, and they all used an assembly line, so the findings are not explained by differences in technology. Garvin visited all the companies and observed manufacturing methods, he interviewed company personnel, and he had each answer a questionnaire on manufacturing practices.

Using assembly-line defects as a measure, U.S. manufacturers had 67 times more defects than did the Japanese manufacturers. Another measure was failures in the first year that were covered by the warranty and required service calls. The U.S. manufacturers had 17 times more of these than did the Japanese. On both measures the *poorest* Japanese company had a failure rate less than half that of the best U.S. manufacturer.

A first reaction to these findings can be that the Japanese have a slow, methodical assembly line, and that they give compulsive attention to detail and earn quality at the cost of production rate. Or perhaps they are putting more money into production. Not so. The highest-quality producers had the highest output per labor hour. Also, quality control actually cut production costs. Fewer defects on the assembly line reduced expenditures on rework and scrap. Furthermore, the reduction of expensive service calls under the warranty lowered costs even more. Quality control pays.

What are the Japanese manufacturers doing? They work no magic. Their secret, if it can be called that, is a dedicated commitment to quality control. The Japanese in Garvin's study of air-conditioners use quality control procedures that U.S. manufacturers use indifferently, occasionally, or not at all:

1. The quality control manager of the plant was an important executive, with direct access to top management. He did not report to middle management, who may or may not report findings on defects to top management.
2. Defects were reviewed daily.

3. Product quality was more important than production schedules, producing at low cost, or increasing worker productivity.
4. Annual goals to reduce failures were set.
5. Service call statistics were quickly transmitted back to the plant for the information that they contained about defective components.
6. Reliability engineering was an established part of the manufacturing routine.
7. Training of workers was extensive. The workers were trained for all jobs on the assembly line, not just one job, so they were good at localizing defects that occurred at other workstations.
8. There were internal consumer review boards made up of employees whose role was to act as typical consumers in evaluation of the product. A board had final authority in product release. The consumer, not the design staff or management, had to be satisfied with the product's quality.

It is not surprising that Japanese products have a high reputation for quality. Noro (1984) writes that quality control in Japan is not a scientific technique that is practiced by a few specialists but a movement that pervades Japanese companies. Throughout all levels of a Japanese company there will be small groups of five or six workers, called *quality control circles,* who concern themselves with the quality of the product. Japanese industries have over 1.5 million quality control circles.

Principles of Human Reliability for Systems

1. *Manipulate human performance variables to minimize human error in the system.* Reliability engineers improve system reliability with their focus on hardware failures. Human factors engineers, with a corresponding emphasis on human errors, can also improve system reliability.

The various specialties within the discipline of human factors engineering conduct research on human behavior, and this scientific knowledge is applied to the minimization of human error in systems. What is known about the organization of workplaces, human–computer interaction, design and organization of displays, and design and organization of controls is applied in the design stage of a system so that users will work efficiently. The workplace has only some of the variables that affect human error. Psychological tests for the selection of personnel with favorable configurations of physical and mental abilities will contribute to error reduction. Attention to variables in the training of human skills is another major factor in error reduction.

The solution to human error will usually be found in the design, training, and personnel selection, but not always. Psychologists recently have been giving attention to errors attributable to "absentmindedness" (Reason, 1975, 1984; Norman, 1981; Reason and Mycielska, 1982). *Absentmindedness* is when we are busy thinking about one thing and automatically executing another and inappropriate act. William James (1890) tells of the absentminded person who goes into the bedroom to dress for dinner and then proceeds to undress fully and get into bed. Or the preoccupied individual who takes out his housekey on arriving at the door of a friend's house. These errors appear to involve highly learned acts that do not require conscious attention for their execution and which automatically run off when consciousness is preoccupied with other thoughts. Poorly understood though it is, absentmindedness is conjectured to be a source of errors and accidents. After analyzing British aircraft accident investigation reports, Reason (1975) found it reasonable to assume that pilot errors in some of the accidents had a close resemblance to absentminded mistakes. Evidence for Reason's assertion will be slow in forthcoming because identification of the cause of errors is not easy.

2. *Use the human being as a redundant subsystem to compensate for hardware and software failures.* Hardware and software failures will occur in the most reliable of systems and provision must be made for them if the loss of system operation is to be avoided. Automatic redundancies for critical hardware elements is one approach, as we have seen, and there are possibilities for software redundancies also. A complementary approach is *human redundancy,* where the system is designed so that an operator can substitute human responding for the hardware or software functions that have failed. A manned space vehicle may have an automatic control subsystem, but if it fails, an astronaut can take over if a manual control subsystem has been included and the astronaut has been trained to use it.

The concept of human redundancy in systems became clear in the early 1960s within the contexts of Project Mercury and the X-15 program, the pioneering efforts of the United States in manned space flight. *Mercury* was an orbital vehicle with one crew member, boosted to orbit by a rocket. The X-15 was an exotic aerospace plane with one crew member that rode piggyback on a B-52 bomber until it was released to climb to altitudes as high as 250,000 feet under the power of a rocket engine. The X-15 then returned to earth for a conventional landing. The reliabilities of both of these systems incorporated human redundancies.

The *Mercury* vehicle was designed to operate as an automatic system, and one possible version of it could have had the astronaut as an intelligent observer of space phenomena but a nonparticipant in system operation. Another way to look at the astronaut—and this is the design direction that was taken—was as a redundant subsystem that could take over when certain automatic subsystems failed, in addition to the role of an intelligent observer. Jones and Grober (1961, 1962) of the McDonnell Aircraft Corporation, developed a technique called *failure task analysis,* in which the consequences of hardware failures were documented for the astronaut as a redundant subsystem. The approach that Jones and Grober used is a version of *failure modes and effects analysis,* which is used in forecasting a system failure (Amstader, 1971, Chap. 10). A failure modes and effects analysis for hardware will consider the failure of components and estimate their impact on system failure. Failure modes and effects analysis is contrasted to *fault tree analysis* (Barlow & Lambert, 1975; Young, 1975). In fault tree analysis, system failure is the starting point, not the goal, and all potential sequences that could cause it are traced back. Fault tree analysis is the more exhaustive of the two approaches. The same system failure could arise from a number of causes, and an analyst is less likely to find them when, as with the failure modes and effects analysis, the causes and their importance are assumed at the start. Fault tree analysis makes no assumption about the causes of system failure, and it is more likely to uncover them, but the success can be costly in complexity, time, and money. Fault tree analysis has been criticized for being without reasonable bound and for engaging inconsequential faults. Search for causes of the tragic loss of the space shuttle *Challenger* and its crew on January 28, 1986 led to the criticism that a failure modes and effects analysis rather than a fault tree analysis was mostly used in reliability analyses (Diamond, 1986).

Jones and Grober considered the astronaut to have three functions in the event of hardware failure: (1) recognize that a problem exists from interpretation of the sensory cues that accompany a failure, (2) diagnose the failure, and (3) take corrective action. The failure task analysis was carried out for many kinds of hardware failures, and the results were incorporated in the system's design. The cockpit of the capsule was given the displays for the presentation of critical sensory cues required for diagnosis of emergencies, and controls were included so that the astronaut could resolve them. An approach that capitalizes on human capabilities for redundancy has an advantage over automatic hardware redundancy because the resolution of an emergency can depend on the situation. Automatic redundancy is used for conditions that have been anticipated and built into the system, but a trained, intelligent human operator is versatile and can adapt his responding to unanticipated situations.

Williams (1964) reported that this design conception paid off for the *Mercury* vehicle. Of the four *Mercury* flights that had been made up until that time, only one could have completed the mission in a fully automated mode. The X-15 had a similar design conception. The X-15 had some redundant subsystems in case of failure of primary subsystems, and it was the pilot's role in the redundancy sequence to sense and interpret the cues associated with the emergency and to select the redundant subsystem that would resolve it. Williams (1964) reported that this design conception resulted in 37 successful flights out of the first 44. No aircraft were lost. Examination of the emergencies that occurred allowed analysts to determine how alternative design conceptions might have worked. A wholly automatic system, with hardware redundancies and no human being in the loop, would have resulted in 18 lost aircraft in the 44 flights. The X-15 program never had 18 aircraft, so the entire program would have failed.

Summary

Reliability engineering is about hardware and software errors, and human factors engineering is about human errors, so reliability is a topic of general importance for human–machine systems.

The principles of hardware and software reliability were reviewed as background for an account of human reliability. Reliability for a hardware component is defined as probability of success. The reliability of a system with many components in series decreases as component reliability decreases and number of components increase. The trend of unreliability with system complexity can be offset by increasing the reliability of components, by keeping the system simple insofar as it is possible, and by using redundancy where spare components take over when there is failure.

There are two main principles of human reliability for minimizing human error in systems. One is to bring scientific knowledge to bear and manipulate human performance variables so that human error is minimized for a system. The other principle has the same rationale as redundancy in hardware reliability, and it says that the human being should be conceptualized as a redundant subsystem that can back up hardware and software when they fail.

References

Adams, J. A. (1982). Issues in human reliability. *Human Factors, 24,* 1–10.
Amstader, B. L. (1971). *Reliability mathematics.* New York: McGraw-Hill.
Amster, S. J., & Shooman, M. L. (1975). Software reliability and overview. In R. E. Barlow, J. B. Fussell, & N. D. Singpurwalla (Eds.), *Reliability and fault*

tree analysis. Philadelphia: Society for Industrial and Applied Mathematics. Pp. 655–685.

Barlow, R. E., & Lambert, H. E. (1975). Introduction to fault tree analysis. In R. E. Barlow, J. B. Fussell, & N. D. Singpurwalla (Eds.), *Reliability and fault tree analysis.* Philadelphia: Society for Industrial and Applied Mathematics. Pp. 7–35.

Chief of Naval Operations. (1977). Military manpower versus hardware procurement study (HARDMAN): Final report. Washington, DC: Office of the Chief of Naval Operations, Department of the Navy, October 26.

Comptroller General of the United States (1981). Effectiveness of U.S. forces can be increased through improved weapon system design. Gaithersburg, MD: U.S. General Accounting Office, Technical Report PSAD-81-17, January 29.

Diamond, S. (1986). NASA's risk assessment isn't most rigorous method. *New York Times,* February 5.

Fallows, J. (1981). *National defense.* New York: Random House.

Garvin, D. A. (1983). Quality on the line. *Harvard Business Review,* September–October.

James, W. (1890). *The principles of psychology.* New York: Holt.

Jones, E. R., & Grober, D. T. (1961). Studies of man's integration into the Mercury vehicle. St. Louis, MO: McDonnell Aircraft Corporation, Report 8276, June 15.

Jones, E. R., & Grober, D. T. (1962). The failure task analysis. St. Louis, MO: McDonnell Aircraft Corporation, June 21.

Klass, P. J. (1962). Redundancy utilized to boost reliability. *Aviation Week & Space Technology,* February 5.

Nordwall, B. D. (1986). Electronic cockpits are making conventional systems obsolete. *Aviation Week & Space Technology,* November 3.

Norman, D. A. (1981). Categorization of action slips. *Psychological Review, 88,* 1–15.

Noro, K. (1984). Ergonomics for quality control—present situation in Japan. *Ergonomics, 27,* 727–731.

O'Connor, P. D. T. (1985). *Practical reliability engineering.* Second Edition. New York: Wiley.

Reason, J. (1975). How did I come to do that? *New Behaviour,* April 24.

Reason, J. (1984). Absent-mindedness and cognitive control. In J. E. Harris & P. E. Morris (Eds.), *Everyday memory, actions and absent-mindedness.* New York: Academic Press. Pp. 113–132.

Reason, J., & Mycielska, K. (1982). *Absent-minded? The psychology of mental lapses and everyday errors.* Englewood Cliffs, NJ: Prentice-Hall.

Robinson, J. E., Jr., Deutsch, W. E., & Rogers, J. G. (1970). The field maintenance interface between human engineering and maintainability engineering. *Human Factors, 12,* 253–259.

Shapero, A., Cooper, J. I., Rappaport, M., Schaeffler, K. H., & Bates, C., Jr. (1960). Human engineering testing and malfunction data collection in weapon system test programs. Wright-Patterson Air Force Base, OH: Wright Air Development Division, Air Research and Development Command, United States Air Force, WADD Technical Report 36-60, February.

Spinney, F. C. (1980). Defense facts of life. Staff paper, Department of Defense, December 5.

Thayer, P. (1983). Address to Bottom Line Conference. Washington, DC: Fort McNair, June 1.

Warwick, G. (1985). Military avionics. *Flight International,* October 5.

Williams, W. C. (1964). The role of the pilot in the Mercury and X-15 flights. *Proceedings of the 14th AGARD General Assembly, NATO Advisory Group for Aeronautical Research and Development, September 16–17. Pp. 65–81.*

Young, J. (1975). Using the fault tree analysis technique. In R. E. Barlow & J. B. Fussell (Eds.), *Reliability and fault tree analysis*. Philadelphia: Society for Industrial and Applied Mathematics. Pp. 827–847.

PART II

Human Capabilities

CHAPTER 3

Audition

Part I gave you an awareness of the field of human factors engineering. Part II is to provide you with a background and an appreciation of basic human capabilities. The senses, primarily audition and vision, the motor system, and memory for information storage and processing are the foundation of human contributions in a human–machine system. Our scientific understanding of these capabilities comes primarily from laboratory research in psychology and physiology. Being pure science, these data were not collected with practical problems in mind; nevertheless, they can sometimes be turned to the solution of practical problems. In reading Parts III through VII of this book, which deal with practical matters, you will be reminded of the usefulness of the basic knowledge that is presented here in Part II.

The auditory system is the topic of this chapter. We begin with a description of the physical stimulus that produces the experience of hearing.

The Auditory Stimulus

Sound is transmitted through any medium that can propagate the vibrations of a solid object. For hearing the concern is with modulations of air pressure that are discernible to the human ear. The variations in atmospheric pressure which are sound waves travel away from the source at 344 m/sec (1130 ft/sec) at 20°C (68°F). The electrical signals from a radio cause elements of the loudspeaker to vibrate, which in turn disturbs the pressure of the surrounding air and impinges on the ear. Variation in atmospheric pressure caused by vibrations of the vocal cords is heard as speech.

Simple Sound Waves Described

A periodic kind of fluctuation is a *sine wave,* called that because it is described by the sine function of trigonometry. Sound waves are described by one or more sine waves. Two sine waves are illustrated in Figure 3-1, and as sound waves they represent air pressures that are slightly more or slightly less than the ambient atmospheric pressure. The sine wave at the top of Figure 3-1 is going through one cycle. The sine wave in the center of Figure 3-1 has three cycles. The *frequency* of the wave is the number of cycles per second, and it is called *hertz* (after Heinrich Hertz, a German physicist who studied electromagnetic waves). One cycle per second is equal to 1 hertz. Waves of 1 and 3 hertz are shown in Figure 3-1. Frequency determines our perception of pitch (see below). The wave's *amplitude* is determined by the intensity of sound pressure. Amplitude is defined as the height of the wave from baseline to peak. Amplitude determines the perception of loudness (see below). A sine wave is most easily produced by a tuning fork or a laboratory instrument called an oscillator.

Complex Sound Waves

A simple sine wave will produce a *pure tone,* which is an auditory experience of a whistle-like sound. Most auditory stimuli, however, are complex and produce complex sounds. When more than one wave is present, the resultant is a *complex wave.* A complex wave covers a band of frequencies. The range of the band is called the *bandwidth.* The two sine waves in Figure 3-1 are combined to produce the example of a complex wave shown in the lower part of Figure 3-1. Complex waves need not only the frequencies and amplitudes of their constituent waves described but also their *phases.* When the peaks and valleys of two waves coincide exactly they are *in phase;* otherwise, they are *out of phase.* The phase relationship of waves determines how they interact. The sound pressures of waves in phase will augment each other. Two waves that are fully out of phase, where the peaks of one coincide with the valleys of the other, will cancel each other.

Figure 3-2 shows a very complex waveform, and most sounds we hear derive from complex waves. The plucking of a violin string or a note on a saxophone are examples of complex waves. Speech involves many frequencies over the spectrum 100 to 10,000 hertz, the greatest involvement with frequencies below 1000 hertz. *Fourier analysis* (after

J. B. J. Fourier, a French scientist) is a mathematical method by which any complex wave can be analyzed into its constituent sine waves. *Ohm's law* in acoustics (after G. S. Ohm, a German physicist, who was also responsible for Ohm's law of electricity) says that the ear, to a limited extent, is capable of similar analyses. Two notes played simultaneously on the piano can be identified, for example.

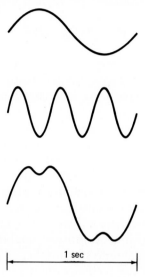

1 sec

Figure 3-1 Illustration of sine waves. The wave at the top is a 1-hertz wave, and the wave in the center is 3 hertz. The wave at the bottom is a complex wave that combines the 1-hertz and the 3-hertz waves.

Figure 3-2 Very complex waveform, made up of a number of sine waves.

The Decibel

The intensity of sound is proportional to the energizing force that actuates the sound source. The force itself is difficult to measure, so the emphasis is on the level of *sound pressure* that is produced by the force. Pressure is force applied to an area, and the unit of force is the dyne. The unit of pressure that is used is dynes per square centimeter (dyn/cm^2). The range of sound pressure, from the faintest sound that can be heard to a pressure so intense that it causes pain, can be 10 million to 1 (Goldstein, 1978). To avoid a large range of this kind, as well as units such as dyn/cm^2, a unit of sound pressure has been devised called the decibel. The decibel has a practical range of 0 to 140 for human auditory experience. The *decibel* is defined as

$$dB = 20 \log_{10} (P/P_0)$$

The decibel scale is a relative scale referenced with respect to P_0, which is a pressure of .0002 dyn/cm^2 above atmospheric pressure. P_0 is ordinarily the faintest 1000 hertz tone that a normal young adult can hear.

To illustrate decibel computation, a sound with pressure P that is 10 times P_0 corresponds to 20 decibels: $20 \log_{10} 10 = 20$. A sound P that is 1000 times P_0 corresponds to 60 decibels: $20 \log_{10} 1000 = 60$. Being a logarithmic scale means that decibel units do not correspond to atmospheric pressure differences in a direct linear way. The pressure difference between 30 and 40 decibels is not the same as the difference between 110 and 120 decibels, nor can the logarithmic units be added or subtracted like kilograms or ounces—doubling the intensity of the sound pressure does not double the number of decibels.

To give an idea of how the decibel scale relates to our everyday world, conversational speech at a distance of 3 feet is about 60 decibels. A soft whisper is about 20 decibels. Being near a jet engine at takeoff would be 130 decibels or more. Discomfort and tickling sensations occur around 120 decibels. There can be pain and damage to the ear above 140 decibels. The amount of damage to the ear depends on the amount of exposure. Loud sound is common for many jobs, and the human factors engineer must be prepared to find it and guard personnel against it.

Noise

Sounds that lack coherence are called *noise*. Noise could mean general auditory stimuli such as that of background conversations at a cocktail party, but the definition of noise for research is precise. A laboratory scientist might energize the bandwidth 100 to 20,000 hertz with equal intensities across all frequencies. Noise defined in this way is called *white noise*. White noise is undifferentiated and sounds like a waterfall, or water running from a faucet.

The Physiology of the Human Ear _____

Figure 3-3 shows a cross section of the human ear, divided into the outer ear, middle ear, and inner ear. The part of the ear that can be seen is called the *pinna.* The pinna captures the sound and channels it down the *auditory canal* to the eardrum, or *tympanic membrane.* The tympanic membrane, which divides the outer ear from the middle ear, responds to air pressure variations over a wide frequency range.

The middle ear contains the mechanisms for transmitting the vibrations of the tympanic membrane to the inner ear. The air in the middle ear is kept at the same pressure as the surrounding atmosphere by means of the *eustachian tube,* which opens to the back of the throat. Unequal pressure on the two sides of the tympanic membrane would distort the membrane and affect its responsiveness to sound. Three small bones, separately called the *hammer, anvil,* and *stirrup,* and collectively called the *ossicles,* mechanically transmit the variations of

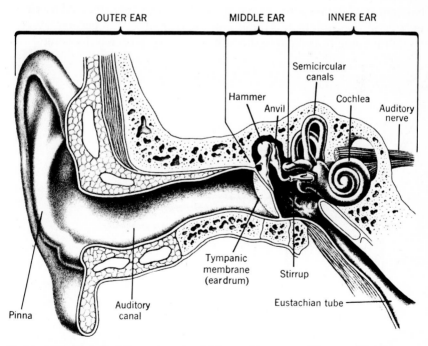

Figure 3-3 Sketch of the external, middle, and inner ear. [*Source:* White, *Our acoustic environment* (1975), Fig. 1, p. 106. Copyright 1975. Reprinted by permission of John Wiley & Sons, Inc.]

the tympanic membrane to the *oval window* covering the entrance to the inner ear.

The two principal mechanisms of the inner ear are the *cochlea* and the *semicircular canals* (which determine our sense of balance and do not concern us here). The cochlea is a system of spiral canals, coiled in about 2¾ turns. The cochlea has an upper gallery and a lower gallery separated by the *basilar membrane*. The galleries contain fluid which, like a kind of hydraulic system, is put in motion by the action of the tympanic membrane and the ossicles. Because the fluid in the cochlea is relatively incompressible, the *round window* at the base of the cochlea provides a release point for the pressure created in the fluid by the action of the ossicles at the oval window. Figure 3-4 is a schematic diagram showing the relationship between these various mechanisms.

The *Organ of Corti* rests on the basilar membrane and has hairlike sensory receptors which extend into the fluid of the cochlea. Movement of a traveling wave, with different characteristics for each frequency, is generated in the fluid by the ossicles at the oval window. The movement of the wave activates the hairlike receptors, which then send electrical signals along the auditory nerve to the brain. The mechanical energy of varying air pressure, mechanically transformed by mechanisms of the middle and inner ears, is transduced to become an electrical message that creates the subjective experience of hearing.

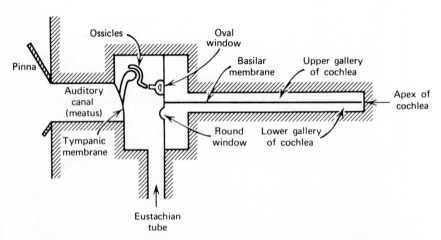

Figure 3-4 Schematic diagram of the relationship between the ossicles and the inner ear. [Adapted from White (1975), Fig. 3. Copyright 1975 by John Wiley & Sons, Inc. Used by permission.]

The Sensitivity of the Human Ear _____

A long-standing concept in sensory psychology is the *sensory threshold,* and it is a way of expressing sensitivity of the ear. A stimulus is detectable above the threshold and not detectable below it. A hearing test will measure the threshold at steps along the auditory spectrum to obtain an *audiogram,* which will show where and how much hearing loss there might be. A recommended technique (Schubert, 1980, p. 124) for assessing hearing loss is to present tones 1 second or so in length and have the listener raise a finger or press a button whenever a tone is heard. For a given frequency, the tester starts well above threshold and makes the tone softer and softer until a point is determined where the tone is heard about half the time in repeated tests. That point is the sensory threshold for the tone. The procedure is repeated for representative frequencies of the auditory spectrum, and a plot of the threshold points is the audiogram. There are jobs that require sensitive hearing, and the human factors engineer must see that personnel are measured and selected for it.

Psychophysics is the relationship between the experience of hearing, which is the realm of psychology, and the physical stimulus, which is the realm of physics. The *method of limits* for the measurement of threshold is a psychophysical technique that is a more elaborate version of the one described for the hearing test. The experimenter will have trials that start with the tone well above threshold. The decibel level of the tone will be gradually decreased by steps until the subject signals that the tone is no longer heard. On other trials the experimenter will start with the tone well below threshold and increase the decibel level by steps until the subject signals when the tone is first heard. There will be variation in threshold level as the tone is repeatedly tested, so an average of the decibel values at threshold will be used as the threshold measure. The procedure will be repeated for a number of tones across the auditory spectrum.

There are other psychophysical procedures for the determination of threshold, but the method of limits is representative and it is background for a classic study of sensitivity of the ear by Sivian and White (1933). Sivian and White used the method of limits to determine the threshold of tones ranging from 100 to 15,000 hertz. They measured thresholds of young adults with excellent hearing. Two kinds of measures were made: minimum audible field and minimum audible pressure. *Minimum audible field* is sound emitted from a loudspeaker in a soundproof room. *Minimum audible pressure* has the sound source fitted tightly to the ear and close to the eardrum. Minimum audible field measures were binaural, and minimum audible pressure measures were monaural. The results are shown in Figure 3-5. The minimum audible field procedure for the curve in Figure 3-5 had the loudspeaker directly in front of the subject. The notable finding is the relationship between threshold and frequency. The greatest sensitivity of the ear is in the general range 1000 to 4000 hertz (this range would be, approximately, the notes on the upper third of the piano keyboard). Low tones and high tones require a greater decibel level to be heard. The more natural binaural hearing by the method of minimum audible field shows more sensitivity than the minimum audible pressure measurements taken at the eardrum. The limits of hearing are 20 to 20,000 hertz in young adults with excellent hearing. Hearing sensitivity declines with age, particularly the higher frequencies.

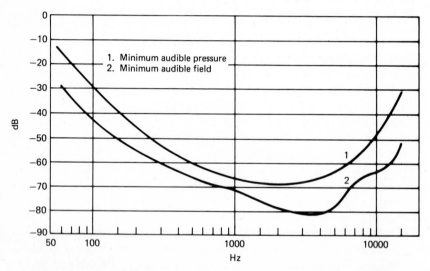

Figure 3-5 Auditory threshold curves as a function of frequency of the stimulus. The curves were determined in two ways: minimum auditory pressure (monaural) and minimum auditory field (binaural). Minimum auditory pressure had the sound source in the ear, close to the eardrum. Minimum auditory field had the sound presented in a anechoic chamber by a loudspeaker in front of the subject. [Adapted from Sivian, L. J. and White, S. D., *Journal of the Acoustical Society of America* 4, 288–321 (1933), Fig. 10. Used by permission of the American Institute of Physics.]

Pitch

Pitch and Frequency

Pitch is the subjective experience of hearing a sound stimulus. The experience of pitch does not have a linear relationship to the frequency of the stimulus that causes it. Using a psychophysical method called the *method of fractionation,* where a variable tone was adjusted to half the pitch of a standard tone, Stevens et al. (1937) conceived the *mel* as the numerical unit of pitch. The scale of the mel has the number 1000 anchored to the frequency of 1000 hertz. The 1000-mel tone that is heard when the 1000-hertz stimulus is presented would sound twice as high as a 500-mel tone, but the 500-mel tone would not necessarily be at 500 hertz. In further research, Stevens and Volkmann (1940) established that the function shown in Figure 3-6 is the relationship between pitch and frequency. The nonlinearity is evident.

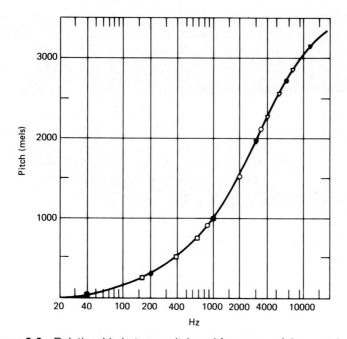

Figure 3-6 Relationship between pitch and frequency of the sound. [Adapted from Stevens and Volkmann (1940), Fig. 2. Used by permission of University of Illinois Press.]

Pitch and Intensity

Ask a singer to reproduce the pitch of a tuning fork with a frequency of 150 hertz. Move the tuning fork nearer the ear. The intensity of the tone will be increased and the pitch of the singer's voice will be lowered slightly. The singer hears the tone as lower. Stevens (1935) experimentally examined the relationship between pitch and intensity by presenting to a subject two tones that differed slightly in frequency. The subject was required to adjust the intensity of one of the tones until the pitch of the two tones was the same. Pitch for low tones of 150 and 300 hertz decreased with intensity increase. Pitch for high tones such as 8000 and 12,000 hertz increased with intensity increase. Tones in the middle range, 1000 to 2000 hertz, changed little with intensity.

Pitch and Stimulus Duration

When a tone is very brief, for example .01 second, it will be heard as a click. As the duration of the tone is increased the tone takes on pitch, but not the pitch of the tone when it is sounded for a longer time. Only with longer presentation time does a tone take on its characteristic pitch. The presentation time for a tone to have its characteristic pitch is not very long. Research (Stevens & Davis, 1938, p. 101) has found it to be only 40 milliseconds for a 1000-hertz tone.

Pitch and Complex Sounds

A complex auditory stimulus with a number of frequencies, like a note sounded on a musical instrument, will have its *fundamental* as the lowest frequency, and integral multiples of it as *harmonics*, which are higher frequencies. A note on a musical instrument with a fundamental of 300 hertz will have harmonics of 600, 900, 1200 hertz and so on. It is the harmonics that give the special quality to a complex sound.

The Missing Fundamental. Suppose that you wore a headset and an electronic filter was placed between you and a musical instrument that was sounding a note with a 300-hertz fundamental. The filter completely blocked the fundamental. Only the harmonics of 600, 900, and 1200 hertz, and so on, came through to your ear. What would you hear? Surprisingly, you would hear the same sound as without the filter. The human auditory system contributes the *missing fundamental,* as it is called. This is why pocket radios, with distressingly small speakers a scant 5 centimeters in diameter, sound so good. The speakers are incapable of reproducing low notes, but the auditory system fills them in by relying on the harmonics.

Beats. Another phenomenon of complex auditory stimuli is the *beat,* which occurs when two pure tones of slightly different frequencies and similar intensities are presented simultaneously. The beat is a third sound in the complex that is kind of a throbbing, or waxing and waning. The beat occurs at a frequency which is the difference between the two frequencies that gave rise to it. Two tones, one of 200 hertz and one of 203 hertz, will beat at 3 hertz. The beat is not contributed by the auditory system, like the missing fundamental. Rather, the beat is caused by the augmentation of energy when the two waves are in phase and their maxima coincide, and the cancellation of energy that occurs when the two waves are out of phase and the maximum of one wave coincides with the minimum of the other.

Loudness

Loudness and Intensity

Like pitch, loudness is perception, representing the psychological processing of auditory stimuli. Loudness does not have a one-to-one relationship with the intensity of the auditory stimulus, just as pitch does not have a one-to-one relationship with frequency. Stevens (1956) used a standard tone of 1000 hertz and 80 decibels and comparison tones of the same frequency but different intensities. A subject controlled the presentation of the standard tone and the comparison tone with two switches, one for each tone. On a trial he would turn on the standard tone for 1 to 2 seconds and then the comparison tone. Using a psychophysical technique called *magnitude estimation,* a subject would assign numbers to the comparison tone on the basis of perceived loudness. A value of 10 was assigned the standard tone. A tone that was half as loud was assigned a value of 5, and a tone twice as loud a value of 20. The relationship between perceived loudness and stimulus intensity is shown in Figure 3-7. Differences in decibel level at the low end of the intensity scale are only slightly perceived as loudness differences, but differences between tones at the high end of the scale are easily perceived as differences in loudness.

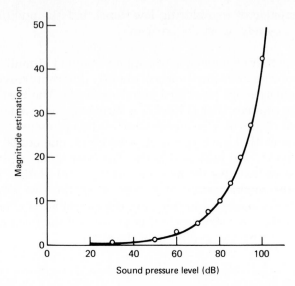

Figure 3-7 Loudness as a function of intensity of the sound. [Adapted from Stevens (1956), Fig. 8. Used by permission of the University of Illinois Press.]

Loudness and Frequency

Stimulus intensity is a strong variable for loudness, but loudness also varies with frequency. The loudness–frequency relationship has been determined using the psychophysical method called the *method of constant stimuli*. The procedure is to present a standard tone of given frequency and intensity before (or after) with a comparison tone of different frequency. Have the subject adjust the comparison tone's intensity until its loudness matches that of the standard tone. Do this for a number of comparison tones and an *equal loudness contour* can be specified, where the loudness of all comparison tones is the same as the standard tone. Change the intensity of the standard tone and repeat the procedure to get another equal loudness contour; Figure 3-8 has a family of them. These data were collected by Robinson and Dadson (1956). A *free field*, which means no reverberation of the sound, was used. The tones were presented over a loudspeaker in an anechoic (without echo) chamber.

An equal loudness contour in Figure 3-8 shows the decibel level for a tone to sound as loud as the 1000 hertz tone that was used as the standard. A 500 hertz–20 decibels tone and a 20 hertz–80 decibels tone are both on the 20 decibels contour and so are equally loud. Both tones

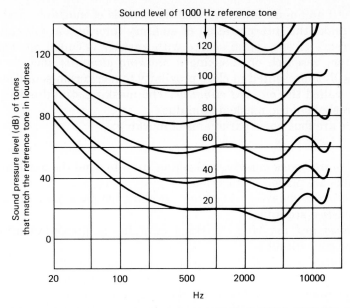

Sound level of 1000 Hz reference tone

Figure 3-8 Equal loudness-level contours. [Adapted from Robinson and Dadson, (1956), Fig. 8. Used by permission of IOP Publishing Ltd.]

are equal in loudness to the 1000-hertz reference tone at 20 decibels and so are equal in loudness to each other.

Loudness and Stimulus Duration

Very brief tones of 150 milliseconds or less do not sound as loud as tones of longer duration. A review of our understanding from research has concluded that it takes a tone 150 to 300 milliseconds to attain its full value of loudness (Scharf, 1978, p. 209).

Tones that are presented for a very long time show a decrease in loudness—a phenomenon called *auditory adaptation*. Perhaps that is why the clicking of a clock is scarcely noticeable.

Masking

Masking Defined

You have been to the noisy party where general talk and laughter make it difficult to hear the person with whom you are talking. In the world of human–machine systems the sounds of machines and other voices make the interpretation of auditory information difficult. The interference of one sound source with another is called *masking*. In laboratory research on masking a *masking stimulus* such as white noise will be presented in close proximity to a criterion sound such as a tone, called the *masked stimulus*. The masking stimulus will raise the threshold of the masked stimulus.

There are four basic paradigms of masking:

1. *Simultaneous masking.* Masking stimulus and the masked stimulus are presented together.
2. *Forward masking.* Masking stimulus is presented before the masked stimulus.
3. *Backward masking.* Masking stimulus is presented after the masked stimulus.
4. *Central masking.* Masked stimulus is presented in one ear and the masking stimulus in the other. (The other three types of masking can be binaural also, but binaural central masking raises provocative theoretical issues—see the section below on central masking and theory.)

A Representative Experiment on Masking

Egan and Hake (1950) performed a well-known study of masking. The stimuli were presented monaurally over a headset. The threshold for pure tones was first determined as a function of frequency. These tones then became the masked stimuli and had their threshold determined a second time in the presence of the masking stimulus, which made it a study of simultaneous masking. The masking stimulus in one of their experiments was a narrow band of white noise 90 hertz wide, with a center frequency of 410 hertz. *Threshold shift* was the measure of masking, which is the difference between a tone's threshold in quiet and its threshold in noise. Three intensities of the masking stimulus were used: 40, 60, and 80 decibels.

The results for one of the subjects are shown in Figure 3-9. The amount of masking increased as a function of intensity of the masking stimulus. The maximum effect was at the center frequency of the noise band, with spread in both directions. The masking curve was symmetrical for the 40- and 60-decibel levels of the masking stimulus, but became asymmetrical at the highest decibel level.

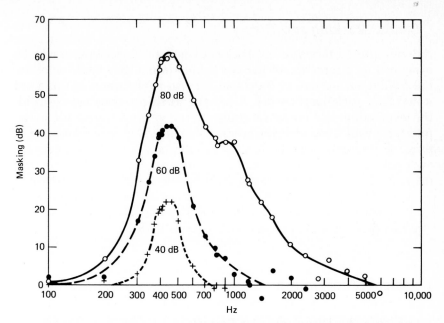

Figure 3-9 Masking capability of a narrow band of noise centered at 410 hertz. Three intensities of noise were used, which is the curve parameter. The measure is threshold shift, which is the difference between the threshold in silence and the threshold in noise. [Adapted from Egan, J. P. and Hake, H. W., *Journal of the Acoustical Society of America* 22, 622–630 (1950), Fig. 2. Used by permission of the American Institute of Physics.]

Forward and Backward Masking

Forward and backward masking are a function of the time interval between the masking stimulus and the masked stimulus. A typical experimental procedure for forward masking would be to follow a burst of noise (the masking stimulus) with a silent time interval and a tone (the masked stimulus). The threshold of the tone is measured. The experimental procedure for backward masking is the same except that the noise follows the tone. Forward masking disappears for time intervals longer than 300 milliseconds (Zwislocki, 1978, p. 309). Backward masking is essentially absent for intervals longer than 25 milliseconds (Elliott, 1962; Pickett, 1959).

Central Masking and Theory

Simultaneous forward and backward monaural masking might be explained by interaction of the two sounds when they both stimulate the basilar membrane of the same ear. An explanation of this kind would be a peripheral theory. A peripheral theory does not cover all the facts, however. Central masking effects, where a sound at one ear interferes with a sound at the other, is evidence that an explanation in terms of interaction at the basilar membrane is insufficient because each sound has a basilar membrane of its own. Without denying that interaction of stimuli at the level of the basilar membrane occurs when both stimuli impinge on the same ear, there are central effects also. Considerations like these complicate theorizing. There is yet no general theory of masking processes (Zwislocki, 1978).

Localization of Sound _____

Locating the Direction of a Sound

We have two ears on opposite sides of the head, which means that a sound can be heard at slightly different times in the two ears and have slightly different intensities. These time and intensity differences will influence the judgment of a sound's direction.

Figure 3-10 illustrates the reason for the time difference. When the sound source is at 0 or 180 degrees, the sound is equal in the two ears, but when the sound source is to the side of the listener, as it is in Figure 3-10, there will be a time difference because the sound to the far ear travels a longer path than to the near ear.

An intensity difference can result from the far ear being in the shadow of the head. At low frequencies the head is not a barrier because the sound waves are diffracted and bend around the head, so intensities at the two ears are about the same. At higher frequencies, however, the sound pressure will be reduced at the far ear because the waves are deflected by the head and the far ear falls in the shadow of the head. Sound travels at the speed of 344 m/sec (1130 ft/sec). A 100-hertz wave would be 3.44 meters long (11.3 feet), which would easily bend around the head. A 10,000-hertz wave, on the other hand, would be .0343 meter long (.113 foot, or 1.4 inches), and is easily deflected by the head.

There can also be a phase difference of the sound at the two ears, and it could conceivably be a cue for direction also. Analysts do not

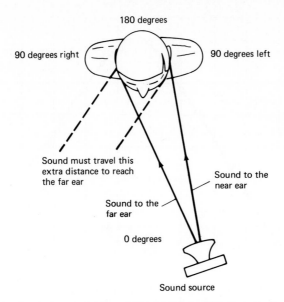

Figure 3-10 A sound displaced from the 0 to 180-degree plane takes a longer time in traveling to one ear than the other.

consider phase information to be a useful directional cue for sound because the discrimination of phase for higher frequencies is not reliable (Gulick, 1971, pp. 190–193).

The Experiment of Feddersen et al. Feddersen et al. (1957) investigated both time and intensity differences of sound at the two ears. Time differences were measured by two small microphones in the subject's ear canals. The sound was a click, and its arrival time at each ear was measured for various positions around the subject's head. The results are shown in Figure 3-11. There is no difference in arrival time when the sound is directly in front of the head (0 degrees) or directly behind it (180 degrees) because the distance to the two ears is the same, but angular departures from the 0- to 180-degree plane produce systematic differences. The time difference is maximum—about 65 milliseconds—when the sound is directly opposite an ear (90 degrees) and the distance to the other ear is the longest.

Feddersen et al. measured intensity from the output of a small microphone in the ear canal as a tone was moved around the subject's head. The frequency of the tone was an experimental variable. The reduction in decibel intensity for the far ear is plotted in Figure 3-12 as a function of direction of the sound source and frequency. There is

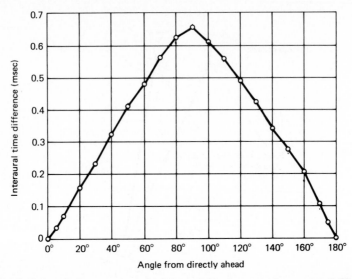

Figure 3-11 Difference in arrival time at the two ears for a sound. The maximum difference is when the sound source is directly opposite an ear. [Adapted from Feddersen, W. E., Sandel, T. T., Teas, D. C., and Jeffress, L. A., *Journal of the Acoustical Society of America* 29, 988–991 (1957). Used by permission of the American Institute of Physics.]

no reduction in intensity when the sound source is directly in front of the subject or directly behind, which is expected. Nor is there intensity reduction when the frequency of the tone is low and the wave can bend around the head, which is expected also. But there can be as much as a 20-decibel reduction in intensity when the sound source is at 90 degrees and the frequency is high and cannot bend around the head.

The Precedence Effect

A human being can do a respectable job of localizing a sound in the laboratory, but how is a sound localized in the real world, where it reflects from surfaces and comes at the listener from several directions and at different times? You would think that the listener would be awash in a sea of sound. Moreover, we hear only one sound in the presence of reverberation, not a succession of sounds as the reflecting sound keeps coming to the listener from various surfaces. The answer is the precedence effect (Wallach et al., 1949).

The *precedence effect* asserts that two sounds separated by a brief time interval will be heard coming from the location of the sound that arrives first. Wallach et al. arranged two loudspeakers equidistant from a listener. The tone from one loudspeaker was presented 7 milli-

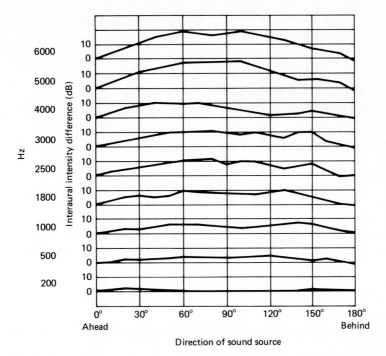

Figure 3-12 Difference in sound intensity at the two ears as a function of frequency and position of the sound source with respect to the head. There is no difference when the sound is directly ahead of the subject (0 degrees), directly behind the subject (180 degrees), or when the frequency is low. The maximum difference is at the high frequencies and when the sound source is directly opposite the subject's ear (90 degrees). [Adapted from Feddersen, W. E., Sandel, T. T., Teas, D. C., and Jeffress, L. A., *Journal of the Acoustical Society of America* 29, 988–991 (1957). Used by permission of the American Institute of Physics.]

seconds before the same tone was presented from another loudspeaker. The sound appeared to come from the loudspeaker whose tone was presented first.

Wallach et al. also arranged two needles to play in the same groove of a phonograph record. The two needles were placed very close together so that the output of one preceded the other by only 35 milliseconds. Each needle had its own loudspeaker for output. The sound was perceived as coming from the loudspeaker whose needle was ahead in time. The effect was found for both human speech and musical instruments. Tests with this experimental arrangement and an interval of 70 milliseconds produced an echo, so there is a limit to the precedence effect. Wallach et al. concluded that the limit of the precedence effect for clicks is about 5 milliseconds, but that it could be as much as 40 milliseconds for complex sounds.

The Sensory Threshold and Signal Detection Theory

The Threshold Revisited

Psychophysics, as the relationship between the experience of sensation and the physical stimulus, had the sensory threshold as its foundation for about 100 years. The use of psychophysical techniques for threshold was a fundamental approach of experimental psychologists, and valuable research knowledge has been based on it, as we have seen in some of the experiments that have been reviewed in this chapter. Like a neuron which fires or not when it is stimulated, so it was believed that a stimulus was detected or not when it was presented to a subject. The transition point between detection and no detection was the sensory threshold.

Two things were realized about the threshold almost from the beginning:

1. The threshold was variable and basically statistical in nature. A subject would not consistently say "Yes, I detect it" or "No, I do not detect it" when a stimulus was presented. In an earlier section of this chapter on the sensitivity of the ear, where the psychophysical method of limits was discussed, an average of several threshold values for a stimulus was taken as a way of getting a better estimate of the threshold. The explanation of the unstable threshold was that the experimental stimulus was not the only stimulus impinging on the subject. Internal and external stimuli of all kinds bombard the subject, and they can conspire to influence the judgment about the experimental stimulus.

2. The threshold was affected by the attitude of the subject. A subject could adopt a strategy of saying "yes" on all trials, guaranteeing that all stimuli would be detected. The subject would generate a lot of false alarm responses because there would be a response of "yes" even when no stimulus was present on a trial. A decision of this sort has more to do with the subject's decision-making behavior than with sensory behavior. Classical psychophysics would occasionally inject *catch trials*, where no stimulus was presented, to see if the subject would say "yes." The subject would be warned of laxness if "yes" occurred on too many catch

trials in an effort to change the decision-making behavior. Notwithstanding, decision behavior and sensory sensitivity were confounded in the methods of classical psychophysics.

Signal Detection Theory

The answer to these difficulties with the threshold was signal detection theory. *Signal detection theory* asserts that the subject in a psychophysical situation is a decision maker, and that the sensitivity to stimuli and the criterion of judgment should be separated (Swets, 1964). The criterion for decision might be very lax, as in the example above where the subject said "yes" on all trials. Or the criterion could be very conservative, with "no" on all trials. Both criteria are unlikely to be correct and mistakes will be made. A quality control inspector in a factory who must discriminate good from bad products and who says "yes" for all products that come off the assembly line and accepts them all, or who says "no" for all products and rejects them all, has her criterion set wrong and will make mistakes because products are seldom all good or all bad. A physician who says "yes" or "no" to all X-ray films will make mistakes (see an application of signal detection theory to recognition memory and the reading of X rays in Chapter 6).

An assumption of signal detection theory is that signal and sensory noise are the two states of the situation where a detection is being made. Noise can be in the input, as static on the radio, or it might be neural and affect the signal as it is being internally processed. Whatever its source, noise is the matrix in which a signal is embedded and it affects the detection situation.

Table 3-1 shows the response that can occur in a signal and noise situation. The subject can correctly detect a signal and have a *hit,* fail to detect and *miss,* indicate a signal when none is present and have a *false alarm,* or have a *correct rejection* by correctly indicating that the signal is absent. Each response will be tallied in one of these cells, and the proportion of responses in each cell will be calculated.

Figure 3-13 shows a weak and a strong signal in noise. Points A and

TABLE 3-1 Fourfold Response Classification Used by Signal Detection Theory for the Detection of Signals in Noise

Response	Signal and Noise	Noise
Yes	Hit	False alarm
No	Miss	Correct rejection

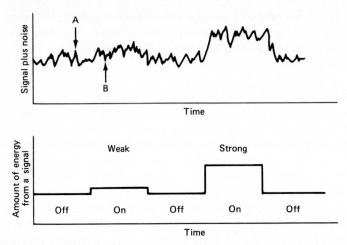

Figure 3-13 Effect of noise on weak and strong signals. Noise can virtually bury the weak signal, making its detection difficult and the mistaking of noise for signal easy. Point A is noise only, and point B is noise plus signal. The momentary surge of noise at point A could easily be taken for signal. The strong signal in noise is more discernible. [Source: from Figure 2.2 in *Engineering Psychology and Human Performance* by C. D. Wickens. Copyright © 1984 by Scott, Foresman and Company. Reprinted by permission.]

B on the top curve indicate how randomly fluctuating noise can influence the judgment of a weak signal. The strength of the input at point A when the signal is absent is stronger than at point B when the signal is present and could elicit a "Yes" response, which would be a false alarm. When a signal is strong relative to the noise level, there is much less chance of mistaking signal and noise.

The Criterion. Figure 3-14 shows how signal detection theory schematizes the detection situation. The quantity X on the horizontal axis, or abscissa, is the level of sensory activity, which varies randomly from moment to moment as a function of psychological and physiological factors. The probability of any sensation level occurring is represented by the bell-shaped probability function, called the *normal distribution function* in statistics. The vertical axis, or ordinate, is the probability of any particular sensation level occurring. The subject sets the *criterion,* called β (beta), as the basis for saying "yes" or "no" about the presence of the signal. Moving the criterion to the right, which would be a conservative criterion where the subject would say "yes" only when she was quite sure, would decrease the number of hits and false alarms and increase the number of misses and correct rejections. Conversely, moving the criterion to the left, which would be a relaxed criterion where the subject would freely say "yes" much of the time, would increase the number of hits and false alarms and decrease the number of misses and correct rejections.

Sensitivity. Sensitivity is continuous for signal detection theory, not discrete as it is for a threshold conception. The distributions in Figure 3-14(a) represent a highly sensitive subject who separates signal from noise very well. The subject illustrated in Figure 3-14(b) has low sensitivity. The measure of a subject's sensitivity is d' (d prime), and it is distinct from β. The computation of d' and β is easily done from tabled values (Hochhaus, 1972; Theodor, 1972). A major contribution of signal detection theory is that it allows separation of sensitivity and the criterion.

The weakness of the threshold conception from classical psychophysics is now apparent. The proportion of hits and misses can vary widely depending on how β is set. Different values for the threshold would be found for different values of β, which would make no sense because the subject, after all, has only one sensitivity. What is needed

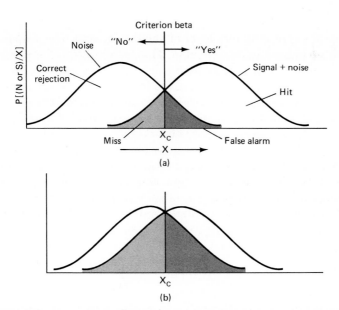

Figure 3-14 How signal detection theory schematizes the detection of a signal in noise. See text for explanation. [Source: From Figure 2.3 in *Engineering Psychology and Human Performance* by C. D. Wickens. Copyright © 1984 by Scott, Foresman and Company. Reprinted by permission.]

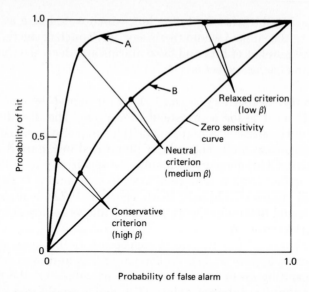

Figure 3-15 Receiver operator characteristic (ROC) curve. See text for explanation. [Source: From Fig. 2.4 in *Engineering Psychology and Human Performance* by C. D. Wickens. Copyright © 1984 by Scott, Foresman and Company. Reprinted by permission.]

is a way of expressing one level of sensitivity and how responding is affected by changes in the criterion. The presentation of one sensitivity over different values of β is with the *receiver operator characteristic curve,* or *ROC curve.* The ROC curves for high d' (curve A), and low d' (curve B) values, each plotted over three levels of β are shown in Figure 3-15. Larger values of d' mean greater sensitivity. Curves A and B, respectively, could be either two observers of high and low sensitivity or strong and weak signals for one observer; the bow in the curve is perceived signal strength. Larger values of β mean greater conservatism in judgment. The curves show that a conservative observer will have comparatively few hits and false alarms, and an observer with a relaxed criterion will have a greater number of hits and false alarms.

Summary

Sound is transmitted through any medium that can propagate the vibrations of a solid object. For hearing we are concerned with modulations of air pressure that are discernible to the human ear. A simple sound is one whose modulation of air pressure is described by a sine wave. A complex wave, like a note from a musical instrument or speech, is a composite of more than one sine wave. Frequency and intensity are the two basic characteristics of a sine wave. Frequency is the number of cycles per second. Intensity is amplitude of the wave and is measured by the decibel scale.

The ear is divided into the outer ear, the middle ear, and the inner ear. Sounds entering the outer ear actuate the ear drum, which divides the outer ear and the middle ear. The mechanisms of the middle ear transmit the sound vibrations to the inner ear, where they stimulate sensory receptors that translate the mechanical energy of the sound into electrical energy for the nervous system. The electrical messages to the brain produce the experience of hearing.

The frequency of a sound produces the subjective experience of pitch, and the intensity of a sound the subjective experience of loudness. Pitch and loudness have multiple determinants. Judgments of pitch do not have a one-to-one relationship with frequency, and judgments of loudness do not have a one-to-one relationship with intensity. Pitch also varies with intensity and stimulus duration. Loudness varies with frequency and stimulus duration.

The interference of one sound with another is called masking. Simultaneous sounds can interfere with one another, as can sounds that precede or follow one another by brief intervals. A masked sound must have its intensity increased to be heard.

We have two ears on opposite sides of the head, which means that a sound can be heard at slightly different times in the two ears and have slightly different intensities. When the sound source is not directly in front of the listener or directly behind, the time difference occurs because the sound takes slightly longer to reach one ear than the other. The intensity difference occurs because the far ear is in the shadow of the head and at some frequencies will not bend around the head. These time and intensity differences determine the judgment of a sound's direction. The localization of a sound in the presence of other sounds is not as confusing as it might seem because we respond to the sound that arrives first. This is called the precedence effect.

Psychophysics, as the relationship between the experience of sensation and the physical stimulus, had the sensory threshold as its foundation for about 100 years. The threshold is a discrete conception of sensation; the listener either hears a stimulus or does not in a detection situation. Threshold was doubted by scientists primarily because of the judging standards that a listener imposes. The listener can have a strict criterion of judgment and assert the occurrence of a signal only when she is very sure, or the listener can have a relaxed criterion and respond when she is not so sure. In recent times signal detection theory has replaced threshold in the thinking of scientists. Signal detection theory allows separation of the sensitivity of the listener from the listener's criterion of judgment.

References

Egan, J. P., & Hake, H. W. (1950). On the masking pattern of a simple auditory stimulus. *Journal of the Acoustical Society of America, 22,* 622–630.

Elliott, L. L. (1962). Backward masking: Monotic and dichotic conditions. *Journal of the Acoustical Society of America, 34,* 1108–1115.

Feddersen, W. E., Sandel, T. T., Teas, D. C., & Jeffress, L. A. (1957). Localization of high-frequency sounds. *Journal of the Acoustical Society of America, 29,* 988–991.

Goldstein, J. (1978). Fundamental concepts in sound measurement. In D. M. Lipscomb (Ed.), *Noise and audiology.* Baltimore: University Park Press. Pp. 3–58.

Gulick, W. L. (1971). *Hearing: Its physiology and psychophysics.* New York: Oxford University Press.

Hochhaus, L. (1972). A table for the calculation for d' and β. *Psychological Bulletin, 77,* 375–376.

Pickett, J. M. (1959). Backward masking. *Journal of the Acoustical Society of America, 31,* 1613–1615.

Robinson, D. W., & Dadson, R. S. (1956). A re-determination of the equal-loudness relations for pure tones. *British Journal of Applied Physics, 7,* 166–181.

Scharf, B. (1978). Loudness. In E. C. Carterette and M. P. Friedman (Eds.), *Handbook of perception.* Volume 4. *Hearing.* New York: Academic Press. Pp. 187–242.

Schubert, E. D. (1980). *Hearing: Its function and dysfunction.* New York: Springer-Verlag.

Sivian, L. J., & White, S. D. (1933). On minimum audible sound fields. *Journal of the Acoustical Society of America, 4,* 288–321.

Stevens, S. S. (1935). The relation of pitch to intensity. *Journal of the Acoustical Society of America, 6,* 150–159.

Stevens, S. S. (1956). The direct estimation of sensory magnitudes—loudness. *American Journal of Psychology, 69,* 1–25.

Stevens, S. S., & Davis, H. (1938). *Hearing: Its psychology and physiology.* New York: Wiley.

Stevens, S. S., & Volkmann, J. (1940). The relation of pitch to frequency: A revised scale. *American Journal of Psychology, 53,* 329–353.

Stevens, S. S., Volkmann, J., & Newman, E. B. (1937). A scale for the measurement of the psychological magnitude pitch. *Journal of the Acoustical Society of America, 8,* 185–190.

Swets, J. A. (Ed.). (1964). *Signal detection and recognition by human observers: Contemporary readings.* New York: Wiley.

Theodor, L. H. (1972). A neglected parameter: Some comments on "A table for the calculation of d' and β." *Psychological Bulletin, 78,* 260–261.

Wallach, H., Newman, E. B., & Rosenzweig, R. (1949). The precedence effect in sound localization. *American Journal of Psychology, 62,* 315–336.

White, F. A. (1975). *Our acoustic environment.* New York: Wiley.

Wickens, C. D. (1984). *Engineering psychology and human performance.* Columbus, OH: Charles E. Merrill.

Zwislocki, J. J. (1978). Masking: Experimental and theoretical aspects of simultaneous, forward, backward, and central masking. In E. C. Carterette & M. P. Friedman (Eds.), *Handbook of perception.* Volume 4. *Hearing.* New York: Academic Press. Pp. 283–336.

CHAPTER 4

Vision

Vision is the primary system through which information is transmitted to the personnel of human–machine systems. The sensitivity of the human visual system, and its capabilities for perceiving detail, color, texture, depth, size, pattern, and items that are spatially distributed, make vision eminent among the senses. Vision has more demands placed on it in a human–machine system than any of the other senses. A description of the physical stimulus, as the source of visual experience, is the place to begin.

The Physical Stimulus

The Nature of Light

Each of our sensory systems has a category of physical stimuli that activates it. We saw in Chapter 3 that variation in air pressure was the adequate physical stimulus for the ear, for example. The adequate stimulus for the eye is light.

Particles and waves are considered to be the properties of light. A particle of light is called a *photon* to distinguish it from other physical particles, such as protons or neutrons. Photons are emitted by hot sources such as the sun or an electric light bulb, as well as a variety of specialized sources: cold light from the tail of a firefly or a laser, for example. Photons travel in a straight line at the speed of 186,200 miles/sec. A photon may be absorbed when it strikes a surface, or it may bounce off.

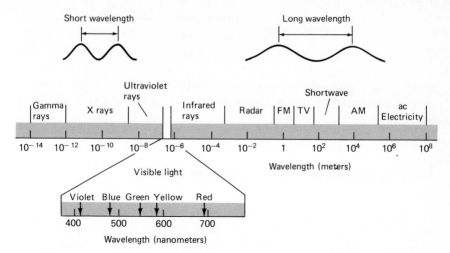

Figure 4-1 Electromagnetic energy spectrum, with the segment that produces visible light amplified. [Adapted from Coren et al. (1984), Fig. 3.1. Used by permission of Harcourt Brace Jovanovich, Inc.]

Light is also considered to act as an oscillating wave, and both the particle and the wave conceptions of light have been scientifically fruitful. The spectrum of electromagnetic energy has wavelengths that vary from a tiny fraction of a meter (10^{-14} meter) to many kilometers (10^8 meters). Figure 4-1 presents the electromagnetic energy spectrum. The electromagnetic energy that stimulates the eye is a very small segment of the total spectrum, extending from about 380 to about 760 nanometers (a nanometer is one billionth of a meter—.000000039 inch). Wavelength variation in the visible portion of the electromagnetic energy spectrum produces our sensations of color, as Figure 4-1 shows.

Units of Light

Electromagnetic waves are described by their frequency and amplitude, but measurement of the consequences of light is known as *photometry*. Light is either emitted from a source or reflected from a surface. All photometric units are based on an internationally standardized light source called the *candle*. *Radiance* is the amount of light coming from a source. Its unit is the *lumen*. *Illuminance* is the amount of light falling on a surface. *Luminance* is the amount of light reflected from a surface. *Reflectance* is the proportion of light that is reflected from a surface:

$$\text{reflectance} = \frac{\text{luminance}}{\text{illuminance}} \times 100$$

Brightness is our subjective reaction to the intensity of light. Brightness is measured by psychological techniques such as psychophysical methods.

The Physiology of the Human Eye

   The eye is heavily protected by being recessed in sockets in the skull and by a strong outer membrane called the *sclera* (Figure 4-2). The *cornea* is continuous with the sclerotic coat, but it is a transparent opening through which the light enters. The *choroid* is a second layer that is rich with blood vessels, and the inner layer is the *retina,* which contains the photoreceptors, which transduce light to neural electrical signals for transmittal to the brain.
   The eye and a camera have their similarities. Light is focused by the

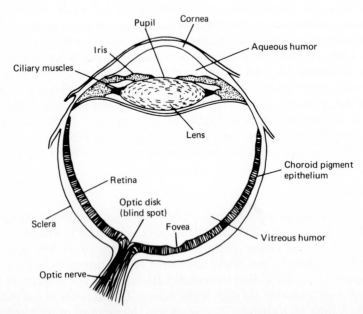

Figure 4-2   Cross section of the human eye. [Adapted from Coren et al. (1984), Fig. 3.2. Used by permission of Harcourt Brace Jovanovich, Inc.]

lens of the eye to produce an inverted image on the retina, just as a camera lens produces an inverted image on film. The *pupil* of the eye, located in the center of the colored membrane that is the *iris* (whose color is "the color of your eyes"), reflexively increases in diameter when the amount of light is low and decreases in diameter when the amount of light is high, just as the iris diaphragm of a camera is adjusted for amount of light. A camera focuses light rays on the film by moving the lens in and out. The *lens* of the eye focuses light rays on the retina by changing shape. Changing the shape of the lens is called *accommodation,* and it is accomplished by the *ciliary muscles.* The chambers in front and behind of the lens are filled with watery fluid, called *aqueous humor* and *vitreous humor,* respectively.

The outermost layer of the retina is a thin layer of neural tissue with two kinds of *photoreceptors* that are anatomically distinguished by their shapes: *cones,* which are somewhat tapered, and *rods,* which are thicker and shorter. The eye has about 125 million rods and about 6.5 million cones (Geldard, 1972). The most important part of the retina is the *fovea,* which is directly behind the pupil in the line of sight. The fovea is densely packed with cones; it has no rods. The number of cones decrease rapidly outside the fovea, and conversely, as the number of cones decrease, the number of rods increase. The cones and the rods have different visual functions, as we will see when photopic and scotopic vision are discussed in the next section.

The responses of the photoreceptors are transmitted to *bipolar cell* and *ganglion cell* layers in the retina. The ganglion cells are junctures for a number of photoreceptors and bipolar cells, and their axons (neural fibers) become the *optic nerve* that transmits impulses to the brain. The axons that form the optic nerve come together and exit from the eye at a hole in the retinal wall. The point of exit, called the *optic disk,* is about 15 degrees toward the nasal side, and it has no photoreceptors. Without photoreceptors there is no response to light, so it is the eye's *blind spot.* In addition, the retina has cells with extensive lateral connections, called *horizontal cells* and *amacrine cells.*

The Sensitivity of the Eye

Photopic and Scotopic Vision

Cones and rods differ in their physical characteristics and their distribution on the retina. Various lines of evidence have been building up for over a century that cones and rods also differ in their function. Cone vision, called *photopic vision,* is concentrated in the center of the retina for daylight color vision. Rod vision, called *scotopic vision,* is concentrated in the periphery of the retina and is used for sensitive seeing at night when levels of light are low. Scotopic vision is without color. Acuity for the resolution of detail is greater for photopic vision than for scotopic vision (see the section below on vision acuity). That there are different functions for cones and rods is called the *duplicity,* or *duplexity theory* of vision.

Dark Adaptation. Sometimes we go from darkness to the light and require light adaptation, but it takes only a minute or so and is no problem. *Dark adaptation,* in going from light to darkness, can take nearly a half hour and be a problem. We all have had the experience of a daytime movie and the difficulties in finding a seat when going from bright sunlight to the darkened theater. Dark adaptation has received extensive experimental study.

Typical results of a dark adaptation experiment (Hecht & Shlaer, 1938) are shown in Figure 4-3. The procedure was to light adapt the subject by having him fixate a bright field, and then put him in darkness and measure the threshold periodically. The test stimulus for measuring threshold was a small bluish light (460 nanometers). The threshold function is in two distinct parts. The initial part, where adaptation is rapid, is photopic vision, where the subject sees the color of the test stimulus. The second part, which proceeds more slowly, is scotopic vision, where the test stimulus is without color. Scotopic vision is clearly more sensitive when the level of light is low, and the sensitivity gradually increases with time in the dark, for about 30 minutes or so.

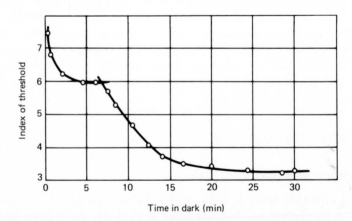

Figure 4-3 Course of human dark adaptation following adaptation to light. [Adapted from Hecht and Shlaer (1938), Fig. 1. Copyright 1938 the Optical Society of America. Used by permission.]

Anyone who must see at night, such as a security guard or a ship's lookout, will do better with dark-adapted eyes and off-center viewing that capitalizes on the rods of the scotopic system.

The Purkinje Shift. The wavelength of light affects not only our perception of color but our perception of brightness as well. Different colors of equal intensity will not necessarily have the same brightness. The amounts of energy to produce equal perceptions of brightness are plotted as a *luminosity curve*. The procedure is to use a split field with a standard color in one half of it and a comparison color in the other half. The viewer adjusts the brightness of the comparison color until it has the same brightness as the standard color. Carrying out this pro-

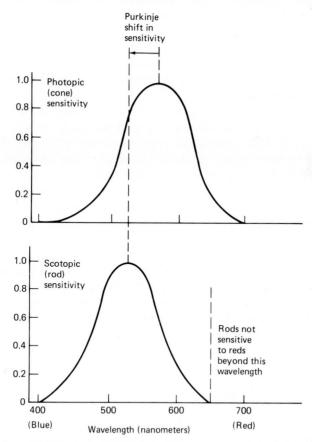

Figure 4-4 Differences in relative sensitivity under conditions of photopic and scotopic illumination. [Adapted from Coren et al. (1984), Fig. 6.3. Used by permission of Harcourt Brace Jovanovich, Inc.]

cedure for various combinations of colors gives the relative amounts of energy needed to produce perceptions of equal brightness for the different wavelengths of light. A plot of these energy values, with the wavelength requiring the least amount of energy given the value of 1.0 and the others smaller values, is the luminosity curve. Figure 4-4 displays the luminosity curves for photopic and scotopic vision, and they are not the same. The photopic curve has peak sensitivity at 555 nanometers, but the scotopic curve is shifted to the left, peaking at 505 nanometers. Measures for the photopic curve are taken at daylight levels of illumination where the colors are seen. Low levels of illumination are used for the scotopic curve, and the viewer does not see colors, although wavelength affects the perception of brightness nevertheless.

The luminosity curves in Figure 4-4 explain other visual phenomena as well. The Czechoslovakian scientist J. E. Purkinje noticed that colors changed in their brightness as the light dimmed at twilight. Reds, which had been brighter than blues, now appeared darker, and the blues became brighter. As darkness set in the reds became black and the blues became gray. These visual effects are known as the *Purkinje shift,* and are explained by the transition from photopic vision to scotopic vision (Figure 4-4).

Military pilots on alert duty wait in a ready room until a signal to run to their aircraft and fly a mission. When alert duty is at night, there can be a dark-adaptation problem as the pilot goes from the lighted ready room to the darkened cockpit. Dark adaptation takes minutes, and there is no time for it. The recommendation is that red goggles be worn in the waiting room, and Figure 4-4 shows the reason why. Photopic vision is sensitive to red, so pilots with goggles would be able to see well enough while in the lighted waiting room. Scotopic vision is insensitive to red, so dark adaptation for it will proceed unimpaired while the pilots are using their photopic vision for routine seeing at the same time. The pilots will have no problem with the darkened cockpits when the call comes to fly. Hecht and Hsia (1945) measured dark adaptation after subjects had been preadapted to white or red light. The speed of dark adaptation after exposure to red light was three times faster than after white light.

Spatial Factors in Vision

Visual Acuity

How small a stimulus that can be seen, or how small a difference between stimuli that can be detected, are questions of visual acuity.

Visual acuity is the ability of the eye to resolve details. Many of us are acquainted with procedures that are used by ophthalmologists and optometrists for the measurement of visual acuity. Figure 4-5 shows a *Snellen letter,* found in the familiar eye chart. The letters are arranged in rows, and visual acuity is calculated from the smallest letter that the viewer can read. Visual acuity measured in this way is commonly expressed as 20/20, or 20/60. Normal vision is expressed as 20/20, and it means that the examinee can read at 20 feet what any person with normal vision can read at 20 feet. The 20/60 rating is degraded acuity and means that the examinee can read at 20 feet what a person with normal vision can read at 60 feet.

Some of the Snellen letters can be confused, which can affect the calculation of acuity. *C* and *O* have similar shapes and can be mistaken for each other, for example. Another problem with the Snellen letters is that it requires literacy. Consequently, other tests of visual acuity have been developed, and representative ones are shown in Figure 4-5. The *Landolt ring* is a circle with a gap in it. The gap will be oriented up, down, right, or left, and the examinee is required to indicate the gap's orientation. The circles vary in size, and the measure of visual acuity is based on the smallest circle whose gap can be detected. Other kinds of tests are *Vernier acuity,* which requires the discrimination of a break in a line, and *resolution acuity* and *grating acuity,* which require the detection of separation between lines.

A general way of expressing visual acuity is the minimum *visual angle* of a detail that can be detected. Figure 4-6 illustrates visual angle for viewing a U.S. 25-cent coin, with size *(S)* of 2.4 centimeters in diameter, at a distance *(D)* of 70 centimeters. Tangent of the angle is $S/D = 2.4/70 = .034$, which is an angle of about 2 degrees.

Visual Acuity and Retinal Position. Figure 4-7 shows relative visual acuity for various positions across the retina. Acuity is best in the fovea and falls off sharply outside the fovea. Because cones are concentrated in the foveal region of the retina and rods in the periphery, it is evident that cones are the agents of high acuity. The reason that we frequently move our eyes is that we are striving to keep stimuli of interest in the center of the eye, where acuity is best. Chapters 9 and 10 will indicate that having high visual acuity in the fovea is pertinent for design. Time-consuming eye movements are used in display scanning as operators strive to keep visual information in the effective foveal region. Visual displays should be designed to minimize these eye movements.

Scotopic vision and photopic vision are responsible for a relationship between visual acuity and illumination level. If visual acuity is poor in the periphery of the retina where rods function, and excellent in the

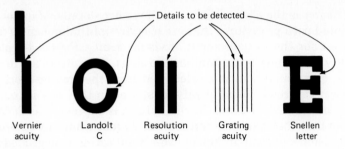

Figure 4-5 Tests of visual acuity. [Adapted from Coren et al. (1984), Fig. 6.9. Used by permission of Harcourt Brace Jovanovich, Inc.]

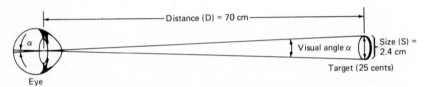

Figure 4-6 Computation of visual angle for a U.S. 25-cent coin. See text for explanation. [Adapted from Coren et al. (1984), Fig. 6.8. Used by permission of Harcourt Brace Jovanovich, Inc.]

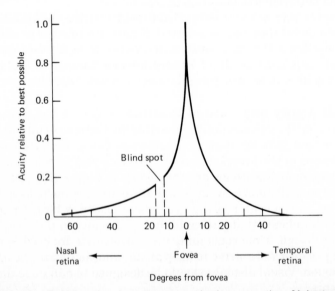

Figure 4-7 Relative visual acuity across the human retina. [Adapted from Coren et al. (1984), Fig. 6.10. Used by permission of Harcourt Brace Jovanovich, Inc.]

fovea where cones are concentrated (Figure 4-7), poor visual acuity should be expected when illumination is low and rods are active, and good visual acuity should be found when illumination is high and cones are operating. Just such a relationship holds.

Stimulus Area and Intensity

For pinpoints of light, where the stimulus area is 10′ (10 minutes) of visual angle or less, there is a trade-off between area and stimulus intensity for threshold detection:

$$A \times I = C$$

where A is stimulus area, I is stimulus intensity, and C is a constant. This relationship is known as *Ricco's law*. For somewhat larger areas, up to 24 degrees or so, the square root of the A term is taken and it is known as *Piper's law*. Detection depends only on stimulus intensity beyond 24 degrees.

Brightness Contrast

Vision becomes the wider topic of perception when the interaction of sensory stimuli occurs, and brightness is a good example of it. The brightness of a visual stimulus is a function of amount of light from a stimulus reaching the eye, but it is a function of background stimuli as well. This phenomenon is called *simultaneous brightness contrast*. View the same gray patch on a gray background and a black background. In both cases the same amount of light from the gray patch strikes the eye, but the gray patch on the black background will appear the brighter of the two.

Temporal Factors in Vision _____

Stimulus Duration and Intensity

A short exposure is required to expose a photographic film properly under strong illumination, and a long exposure is required when illumination is dim. The same amount of light is required in both cases, but it takes longer to collect it on the film in the case of dim illumination. This relationship for photochemical reactions is called the *Bunsen–Roscoe law*. The same relationship holds for photoneural reactions in vision, and it is known as *Bloch's law*. Bloch's law says that there is a trade-off between the intensity and the duration for threshold detection of light. Specifically,

$$T \times I = C$$

where T is stimulus duration, or time, I is stimulus intensity, and C is a constant. The detection of a stimulus can be determined by manipulating either its duration or its intensity. The law is limited to flashes of .10 second or less. Stimulus intensity is the primary determiner of threshold beyond .10 second.

Color Vision

Isaac Newton, the seventeenth-century British physicist and mathematician, deserves credit for demonstrating that different wavelengths of light produce different sensations of color. The bending of light when it passes through a prism is called *refraction,* and the amount of refraction depends on wavelength. Newton passed sunlight through a prism and found that it split into its constituent wavelengths and appeared as the color spectrum on a piece of white paper held behind the prism. The spectrum ranged from violet and blue on one end, through green and yellow, to red and purple on the other (Figure 4-1). A second prism placed behind the first, which refracted the wavelengths in the opposite direction, recombined the wavelengths into white light.

Trichromatic Theory of Color

The cones are the basis of color, but how? The human observer can distinguish about 200 different hues. Is there a class of cones for each hue? Or perhaps all the cones are alike but different colors cause them to react differently and send different neural codes to the brain. At the beginning of the nineteenth century Thomas Young (1773–1829), a British physician and physicist, theorized that it was unrealistic to assume a receptor for each hue that could be seen. Furthermore, a tiny spot of light of any color on a small area of the retina could be correctly perceived, and it was unreasonable to assume that the small area contained a receptor for each hue. Young hypothesized that the retina had three kinds of receptors, each responsive to a different color, and that all of our perception of color is based on them. The three kinds of receptors were assumed to be red, green, and blue. John Dalton (1766–

1844), a British chemist and physicist, made the first systematic observations of color blindness (his own). Dalton told how the spectrum was divided into yellow and blue for him, and Young explained it by the deficiency of receptors for red. In the middle of the nineteenth century, Hermann von Helmholtz (1821–1894), a German physiologist and physicist, and James Clerk Maxwell (1831–1879), a British physicist, conducted experiments on the mixture of monochromatic (one *chroma*, or color) light sources and concluded that any color could be obtained by a proper mixture of the three primary colors, red, green, and blue, giving additional credence to Young's theory. That any color can be obtained from three basic retinal processes is called the *trichromatic theory* of color vision.

Physiological Evidence for Trichromatic Theory. Physiological evidence for three kinds of retinal receptors as mediators of color vision did not come until the 1960s with the pioneering research of Marks et al. (1964) and Brown and Wald (1964). Consistent with trichromatic theory, they found evidence of three different kinds of cones in the human retina. A more recent study by Bowmaker and Dartnall (1980) is a good illustration of the experimental techniques and findings. Concerned that early work might have been biased because of uncontrolled exposure of the cones to light, pathology, or tests made too long after death, Bowmaker and Dartnall controlled all of these likely sources of bias by using the eye of a patient who had the eye removed because of cancer of the choroid pigment (Figure 4-2). The retina was healthy and normal. The eye was light protected during and after surgery, and tests on it began shortly after removal. A microspectrophotometer was used. Samples of cones were taken in the vicinity of the fovea and mounted on slides. Tiny beams of light of different wavelengths were passed through a cone, and the amount of light absorption at each wavelength was measured. The more a wavelength is absorbed, the more sensitive is that particular wavelength. The average results for a number of cones tested are summarized in Figure 4-8. In striking confirmation of trichromatic theory, there are three unmistakable categories of cones: "blue," "green," and "red" cones peaking at 420, 534, and 564 nanometers, respectively.

Color Blindness. Color blindness impairs only color perception; visual acuity is unaffected. Color blindness occurs in 8 percent of males and .5 percent of females (Coren et al, 1984). The principal kinds of color blindness that occur are consistent with trichromatic theory and are another line of evidence for it. One, two, or all three kinds of cones can be deficient.

A person with no cones or one kind of cone functioning is called a

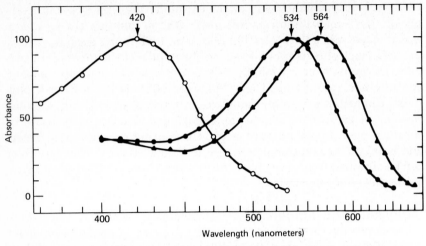

Figure 4-8 Average absorbance spectra for the blue, green, and red cones of the human eye. The function labeled 420 nanometers is an average curve for blue cones, 534 nanometers is for green cones, and 564 nanometers is for red cones. [Adapted from Bowmaker and Dartnall (1980), Fig. 2. Used by permission of Cambridge University Press.]

monochromat. Monochromats with no cones functioning have the severest deficiency. They see only gradations of intensity, without hue discrimination. Everything is seen in shades of gray like a black-and-white photographic film because they are responding only with the rod system, which is insensitive to color. Completely lacking a photopic system, they find daylight uncomfortable. Slightly less deficient are monochromats who have one cone system functioning in addition to the rods. They lack hue discrimination also, distinguishing intensities of only one color, like the monochromat without cones, who sees only shades of gray.

The *dichromat* has two of the three cone categories functioning. John Dalton, mentioned above, had dichromia with a red deficiency. The red deficiency is called *protanopia.* (The prefix *pro* is derived from the Greek, meaning "first." Red, green, and blue are the first, second, and third primary colors respectively, so protanopia is a deficiency in the first primary.) What does a protanope see? A color-blind person might not be easy to detect because she can learn to name the correct colors of objects as a normal person can do. A red object would look a muddy yellow to her, but the object is called red by everyone else and the protanope can call it red also.

Deuteranopia is a deficiency in green, the second primary (the prefix *deuter* is from the Greek, meaning "second"). Graham et al. (1961) had the rare opportunity to study the capabilities of a deuteranope when they evaluated a woman who was deuteranopic in one eye and normal in the other. She was given a binocular matching test. A color was presented to the normal eye, and the requirement was to adjust a knob and match it with the color that was seen simultaneously in the deuteranopic eye. She matched all wavelengths from 516 nanometers and longer in the deuteranopic eye to near 565 nanometers. The long wavelengths, in the yellow–orange–red range of the spectrum, all appeared yellowish to her. Wavelengths 489 nanometers and shorter were matched by wavelengths near 470 nanometers. These short wavelengths, in the blue–violet range of the spectrum, all appeared blue to her. A neutral point, near 502 nanometers where green lies, was seen as gray.

The other form of dichromia is *tritanopia* (a deficiency in the third primary), which is rare. Tritanopia is failure of the blue cones. The longer wavelengths appear red and the shorter ones green or bluish green, with perhaps a gray band in the region of 570 nanometers (Geldard, 1972, p. 110).

Opponent Processes Theory. For a long time, particularly with the color theorizing of the German physiologist Ewald Hering (1834–1918), there was the suspicion that the trichromatic theory of color vision is not the whole story. Why is it that no one ever reports a reddish-green hue? We can see bluish green or reddish orange, but not reddish green. Nor do we perceive bluish yellow. An appropriate balance of the three kinds of cones should allow any mix of colors, it would seem, according to trichromatic theory. Hering speculated that colors might be arranged in opposing pairs, and that the presence of one inhibits the other. Thus, a cell associated with red and green cones would increase its activity in the presence of either a red light or a green light but would be inhibited if the green cone and the red cone occurred together. With no signal to the brain, no reddish-green hue would be perceived. On the surface this thinking might seem troublesome for trichromatic theory, but actually it is complementary to it. The resolution is trichromatic theory as the first stage of color vision, with opponent processes occurring at a later stage of processing.

Physiological research on opponent processes has strengthened Hering's hypothesis; the evidence originated in studies of Svaetichin and his associates (for a review, see Boynton, 1972, pp. 323–326). Using the retina of the goldfish, Svaetichin used microelectrodes to record the electrical activities of retinal cells near photoreceptors, such as horizontal cells which are connected to a number of photoreceptors. A red light would elicit a full response from the cell, as would a green light.

But when red and green lights stimulated the cell simultaneously, no signal was transmitted to the brain. Red and green opposed each other, as Hering had surmised from psychological data years earlier. Subsequent research has secured the pioneering findings of Svaetichin, and opponent processes theory has now joined trichromatic theory as an established part of color theory. Signals from the red, green, and blue cones are subject to interactions at other retinal layers.

Color Mixing

The eye does not have the power to perform a frequency analysis and estimate the wavelengths that comprise the mixture. The ear can differentiate two tones that are presented together, but the eye cannot separate two wavelengths that are presented together. Present a 500-hertz tone and a 2000-hertz tone together and the ear will perceive a low tone and a high tone. Mix a red and a green light and the viewer cannot infer that the resultant yellow is based on red and green.

There are two kinds of color mixture: additive and subtractive. *Additive color mixture* occurs when lights are mixed. *Subtractive color mixture* is the mixing of paints, and it will not produce the same results as the mixing of lights because pigments work differently. A surface has the color that it does because it reflects the wavelengths of the color that you see and absorbs the others.

Additive Color Mixture. The *primary colors* are defined as the three colors of which no two can be mixed to produce the third. Primary colors are basic reference stimuli, and they are red, green, and blue. *Complementary colors* are any two colors whose mixture is white (or grayish).

Colors have hue, saturation, and brightness. *Hue* corresponds to our normal meaning of color. A change in wavelength of the light will change the hue. *Saturation* is determined by the amount of white light that has been added to a color. A pure color becomes washed out in appearance as white light is added to it. *Brightness* is our subjective reaction to the intensity of light.

Color mixing would be simple if it was all as intuitively obvious as a red light and a blue light combining to make purple. It is not intuitively obvious, however, that green and red produce yellow when combined, or that complementary colors produce white. The simplest way of organizing many of the facts of additive color mixing is with the *color wheel,* or the *color circle.* The color wheel is shown in Figure 4-9.

The hues of the spectrum are placed on the color circle, with complementary colors opposite each other. Moving around the circle gives the

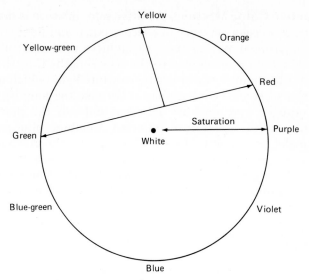

Figure 4-9 Color wheel, or circle.

various hues. White is at the center. Moving toward the center for any particular hue specifies the amount of saturation. The outcome of mixing any two colors is determined by drawing a straight line between them. Drawing a line through the center of the circle, between complementary colors, will produce white. Draw an off-center line between noncomplementary colors and the mix of two colors is indicated also. The line drawn between red and green in Figure 4-9 will produce yellow. The particular yellowish hue is determined by the amounts of red and green that are used—the location of the arrow at a right angle to the red–green line. Brightness is not represented on the color wheel. The two-dimensional circle would have to become a three-dimensional figure to include brightness.

A more exact approach to additive color mixing is with a *colorimeter* (Boynton, 1972, pp. 350–353). The colorimeter presents a test color in the upper half of the visual field. Three knobs are used to mix various amounts of the three primaries in the bottom half of the visual field until the observer has achieved a match with the upper half. If the test color was orange, the observer would have to turn the blue off completely and adjust only the red and green. The tests are continued through a large number of test colors. The outcome is an exact specification of the relative amounts of each of the primaries required to produce any color.

Subtractive Color Mixture. Additive color mixture is much easier than subtractive color mixture to conceptualize and predict. Subtractive color mixture is not as exact as additive color mixture because pigments of paints are complex in the wavelengths that they absorb and reflect. Blue paint looks blue when a white light (which has all the colors of the spectrum) is shined on it because the long wavelengths (yellows, oranges, and reds) are absorbed and only the shorter wavelengths (blues and violets) are reflected. Combine blue and yellow paints to make green paint. Shine a white light on the green paint. The blue paint absorbs the long wavelengths (yellows, oranges, and reds), and the yellow paint absorbs the short wavelengths (blues and violets). What remains to be reflected are intermediate wavelengths, which are green.

Color Contrast

Simultaneous brightness contrast was discussed earlier in this chapter, where the spatial area surrounding a stimulus influenced its brightness. A corresponding effect is true for color also. A simple stimulus such as a light shining in the eye will have its color commanded by wavelength, but a stimulus in a colored surround will have its color determined by the color of the surround as well. This phenomenon is called *color contrast.*

The tendency is for the color of a stimulus to be the complement of the surrounding color (Geldard, 1972, pp. 146–153). Put a gray patch on a blue background and it will appear slightly yellow. Put a gray patch on a yellow background and it will appear slightly blue.

Summary

Vision is the primary system through which information is transmitted to the personnel of human–machine systems. The human eye is a remarkable system and is eminent among the senses.

Light, which is the adequate stimulus for visual experience, is only a narrow segment of the electromagnetic spectrum, ranging from about 380 nanometers to about 760 nanometers. Wavelength variation within this segment produces our sensations of color.

Light passes through the lens of the eye to the retina at the back of the eye. The retina has photoreceptors which transduce light to electrical signals for the brain. Cones and rods are the two kinds of photoreceptors. Cones are concentrated in the central foveal region of the retina. Rods predominate in the retina's periphery. Cone vision, called photopic vision, is responsible for daylight and color vision. Rod vision, called scotopic vision, is used for sensitive seeing at night and is without color. Dark adaptation, in going from light to darkness, is primarily a function of the scotopic system.

Visual acuity is the ability of the eye to resolve small details. Various tests of visual acuity have been devised: Snellen letters are used in a test that calculates visual acuity on the basis of the smallest letters that the viewer can read. Other tests are the Landolt ring, which requires discrimination of a gap in a circle, vernier acuity, which requires discrimination of a break in a line, and resolution acuity and grating acuity, which require the detection of separation between lines. The highest visual acuity is at the fovea, where cones are concentrated.

For almost 200 years scientists have hypothesized that color vision is determined by three basic mechanisms of the eye. Called the trichromatic theory of color, there is now direct evidence for it. Three kinds of cones—blue, green, and red—have been identified with physiological research. The various kinds of color blindness also provide support for trichromatic theory. The signals from the red, green, and blue cones are complicated by interactions at other retinal layers, which is called opponent processes theory.

Color mixing has been of scientific and practical interest for a long time. There are two kinds of color mixing: additive color mixture, which is the mixture of lights, and subtractive color mixture, which is the mixture of paints. The mixing of paints will not produce the same results as the additive combining of lights because a painted surface reflects the wavelengths that you see and subtracts the others by absorption.

References

Bowmaker, J. K., & Dartnall, H. J. A. (1980). Visual pigments of rods and cones in a human retina. *Journal of Physiology, 298,* 501–511.

Boynton, R. M. (1972). Color vision. In J. W. Kling & L. A. Riggs (Eds.), *Woodworth & Schlosberg's experimental psychology.* Volume 1. *Sensation and perception.* New York: Holt, Rinehart and Winston.

Brown, P. K., & Wald, G. (1964). Visual pigments in single rods and cones of the human retina. *Science, 144,* 45–52.

Coren, S., Porac, C., & Ward, L. M. (1984). *Sensation and perception.* Second Edition. New York: Academic Press.

Geldard, F. A. (1972). *The human senses.* Second Edition. New York: Wiley.

Graham, C. H., Sperling, H. G., Hsia, Y., & Coulson, A. H. (1961). The determination of some visual functions of a unilaterally color-blind subject: Methods and results. *Journal of Psychology, 51,* 3–32.

Hecht, S., & Hsia, Y. (1945). Dark adaptation following light adaptation to red and white lights. *Journal of the Optical Society of America, 35,* 261–267.

Hecht, S., & Shlaer, S. (1938). An adaptometer for measuring human dark adaptation. *Journal of the Optical Society of America, 28,* 269–275.

Marks, W. B., Dobelle, W. H., & MacNichol, E. F. (1964). Visual pigments of single primate cones. *Science, 143,* 1181–1183.

CHAPTER 5

Motor Behavior

The output of an operator in a human–machine system can be a verbal response, but it can be a movement of one or more of the limbs also. The concern might be the speed of getting a movement started, the speed of a movement once it has started, or the skill of regulating graded movements.

Physiological Essentials of the Regulation of Movement

Closed-Loop Control Defined

The physiological regulation of movement is by closed-loop control. Closed-loop control is contrasted to open-loop control. In an *open-loop system* the controlling agent receives no input from the thing controlled. A woodburning stove is an open-loop system. There is no automatic adjustment of the fire if the temperature becomes too hot or too cold; the system lacks a compensatory capability. A *closed-loop system* is a compensatory system whose elements are feedback about the thing controlled, error detection, and error correction. There is a reference of correctness and the output of the system is fed back and compared to it for error. Error is corrected if it is found. The automatic home furnace is a closed-loop system. The thermostat setting is the reference of correctness and the heat output of the furnace is compared to it. If error is detected, the furnace cuts in or out until the error is reduced to zero.

How Muscles Are Controlled

The regulation of movements is closed-loop, and the physiological source of the feedback is *proprioception.* Proprioception is a term attributable to the British neurophysiologist C. S. Sherrington (1906). Sherrington distinguished three main kinds of sensory receptors on the basis of the stimuli that activate them: (1) *exteroceptors* on the outer surfaces of the body, for sensing stimuli of the outside world; (2) *interoceptors* on the body's inner surfaces, whose stimuli are primarily chemical; and (3) *proprioceptors,* which lie between the inner and outer surfaces of the body in the muscles, tendons, and joints, and whose stimuli are mechanical forces like pressure, tension, stretch, and torque. Sherrington also classified the labyrinthine mechanism of the inner ear as proprioceptors because of its role in equilibrium. *Efferent* signals go from the brain and spinal cord to the muscles of the limbs and activate them, but proprioceptors initiate *afferent* signals, which travel on sensory fibers to the spinal cord and the brain. These afferent signals from proprioceptors are feedback information about the position and movement of the limbs. A research approach for establishing the role of proprioceptors in movement regulation is to sever the sensory fibers that lead from the proprioceptors to the spinal cord and the brain, and then evaluate effects of the surgery on movement. This research, which is done on animals, is called *deafferentation* (Taub, 1977; Taub & Berman, 1968). Deafferentation produces movement disorganization, leaving no doubt that proprioceptors play a major role in movement control.

Figure 5-1 diagrams the essential elements in muscle control. The proprioceptor attached to a muscle is called a *muscle spindle.* The muscle is called the *extrafusal fiber* and is activated by the *alphamotoneuron.* The *intrafusal fiber* runs parallel to the extrafusal fiber and contains the muscle spindle, and so the muscle spindle is sensitive to the length of the extrafusal fiber. Consider the muscle–spindle relationship when you are holding a brick in your upturned hand and someone suddenly adds a second one. When the arm is stretched the muscle is stretched also and the intrafusal fiber in parallel is stretched along with it. Muscle spindles are activated and signals are sent along sensory nerves, called *1a afferents.* The signals from the 1a afferents connect with *alphamotoneurons* in the spinal cord which send signals along *alpha efferent nerves* to the muscles, causing them to contract more and compensate for the extra weight on the arm. Notice that the process is closed-loop, where the extra brick on the outstretched arm generates an error signal for correction, and the message is sent along the 1a afferent to the alphamotoneuron for compensatory action. Without compensation the arm would give way and the bricks would fall.

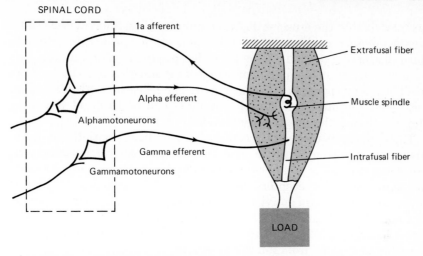

SPINAL CORD

1a afferent

Extrafusal fiber

Alpha efferent

Muscle spindle

Alphamotoneurons

Gamma efferent

Intrafusal fiber

Gammamotoneurons

LOAD

Figure 5-1 Simplified diagram of the physiology of muscle length control. [*Source:* Adapted from Kelso (1982), Fig. 2.4. Used by permission of Lawrence Erlbaum Associates, Inc.]

Figure 5-1 also shows *gammamotoneurons* as another efferent output. Gammamotoneurons are influenced by the brain. The *gamma efferent* connects to the muscle spindle, affecting its rate of discharge. The muscle spindle is not only affected by the length of the extrafusal fiber but by the gamma system as well.

There are, therefore, two neural pathways that affect muscle length: the alpha system and the gamma system. The fibers on the alpha system are larger and faster than those of the gamma system. Why two systems? The alpha system is reflexive, for fast response to sudden changes in the environment, like the second brick placed on the hand. The slower gamma system, receiving signals from the brain, contributes to voluntary movement.

How Position and Movement of Limbs Are Known

A view that prevailed for a long time is that joint receptors are the proprioceptors that inform about the position and movement of a limb (e.g., Goldscheider—see Sherrington, 1900, pp. 1015–1016; Mountcastle, 1966; Boyd & Roberts, 1953; Adams, 1977). Joint receptors fire differentially as a function of joint angle and movement velocity, so it was believed that they are the source of signals about limb position and movement (Adams, 1977).

That joint receptors communicate about limb position and movement has been controversial. The alternative position, which is an old one also, is that the muscles do the communicating. Sherrington (1900) was a foremost spokesman for this point of view. Muscles, after all, have distinctive configurations for different positions and movements as joints do, so it is equally plausible that muscles are the locus of feedback about position and movement. The research difficulty is that both joint receptors and muscle receptors can be firing at the same time, so their influences are confounded.

The hypothesis that joint receptors inform about position and movement ran into trouble in the mid-1970s. Physiological investigators, recording from afferent fibers of joint receptors in the cat and the monkey, found that the receptors mostly fired at the extremes of limb displacement, with little activity in the midrange of movement (Clark & Burgess, 1975; Clark, 1975; Grigg, 1975; Grigg & Greenspan, 1977; Rossi & Grigg, 1982). Receptor activity in the midrange of movement were found to come from muscle receptors.

The position that is prevailing today is that a joint does not have much of an effect on its receptors except at the extremes; joint receptors are limit detectors. Muscle receptors are the influential source of signals about position and movement in a joint's operating range. The two rival hypotheses of the past argued that signals about position and movement must come from either the joints or the muscles. The new hypothesis is a compromise, finding a role for both joint and muscle receptors, with a tilt toward a larger influence of muscle receptors. How well the new hypothesis will survive future tests remains to be seen.

Reaction Time _____

Reaction time is speed of initiating a response to the onset of a stimulus. The scientific interest in it goes back to the nineteenth century.

Historical Origins of Research on Reaction Time

Astronomy. The origin of a scientific interest in reaction time was astronomy (Woodworth, 1938, pp. 300–302). The concern was with the exact time that a star passed the meridian of an observatory. The observer would watch the star enter the field of his telescope and proceed across grid lines, one of which was the meridian. As the star neared the meridian the observer would read the time from the clock to the nearest second and then count additional seconds by listening to the ticks of the clock. The time of crossing the meridian could be calculated to 1/10 second by marking the position of the star at the tick just before and just after the meridian was crossed. The story goes that Nevil Maskelyne, who was head of the Greenwich observatory in Great Britain in the nineteenth century, fired an assistant because his transit times were a half second later than his own (which he assumed to be correct). Later the German astronomer Bessel tested a number of astronomers and found individual differences in transit times. Bessel assumed that differences between individuals were constant and could be used to correct the transit times, and he called them the "personal equation." The differences proved not to be constants but a function of viewing conditions, however.

Physiology. Helmholtz, the German physiologist and physicist, was interested in the time of nerve conduction. He used the muscle of a frog that had motor nerves intact but was severed from the body (Helmholtz, 1853). He measured the time between electrical stimulation of a nerve and the time that the current was broken by muscle contraction, and he concluded that the motor nerves of the frog propagated impulses at the rate of 26.4 m/sec (87 ft/sec).

Helmholtz asked the same question about nerve conduction in humans, and he used a similar research method, although his approach was behavioral. His approach was the first use of the reaction-time paradigm as we know it today. The stimulus was a slight electrical shock to the skin, and the subject was instructed to make a movement with the hands or the teeth as soon as the shock was sensed, which interrupted the electrical current. Time between onset of the current and its interruption was reaction time. Reaction times were recorded for stimulation of different parts of the body, some near the responding effector, such as the face, some distant, such as the toe. Helmholtz reasoned that the differences between reaction times for near and far points would give the speed of nerve transmission. The reaction time from stimulation of the big toe was 1/30 second longer than stimulation of the ear or face, for example. Helmholtz concluded that the velocity of human sensory nerve transmission was 60 m/sec (197 ft/sec).

Reaction time found other applications in the nineteenth century. During a reaction-time sequence the neural impulse must pass through the brain from the point of stimulation to the responding effector, and it can be asked how much central processes contribute to reaction time. The Dutch physiologist F. C. Donders used reaction-time methodology to deduce the contribution of central processes (Donders, 1868/1969; for a summary, see Woodworth, 1938, pp. 302–303). Speech syllables were used both as stimuli and responses in an experiment of Donders'. The syllables were *Ka, Ke, Ki, Ko,* and *Ku.* Three basic conditions were defined:

a reaction The stimulus was always *Ki* and the response was always *Ki*.

b reaction The stimulus was any one of the five syllables and the response was with the same syllable.

c reaction The stimulus was any one of the five syllables but the response was always *Ki*.

The *a reaction* was a simple reaction. The *b reaction* involved discrimination among the five stimuli and selection among the five responses. The *c reaction* required discrimination among stimuli but no selection among responses. Donders reasoned that he could deduce central processing times by subtraction:

$$c - a = \text{time required by sensory discrimination}$$
$$b - c = \text{time required by motor selection}$$

Average reaction time values were found to be

$$a = 197 \text{ milliseconds}$$
$$b = 285 \text{ milliseconds}$$
$$c = 243 \text{ milliseconds}$$

Thus,

$$c - a = 46 \text{ milliseconds}$$
$$b - c = 42 \text{ milliseconds}$$

The Donders approach has been instrumental in the revival of an interest that modern experimental psychologists have in the use of reaction time for calculating the time of mental processes.

Simple Reaction Time

The reaction-time paradigm is formalized in Figure 5-2. A *ready signal* alerts the subject, there is a waiting period called the *foreperiod,* the *stimulus* is presented, and the subject then makes a rapid *response.* The speed of response initiation to stimulus onset is *reaction time.*

The two principal types of reaction time are simple reaction time, which is featured in this section of the chapter, and choice reaction time. *Simple reaction time* has a single stimulus and a single response option. Pressing a key when a light comes on is an example of simple reaction time. *Choice reaction time* has two or more stimulus options, with a response for each. An example of choice reaction time is any one

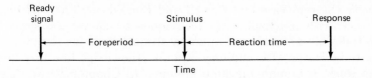

Figure 5-2 Reaction-time paradigm. A ready signal alerts the subject and starts a waiting period, called the foreperiod. The stimulus that occurs at the end of the foreperiod is the cue for the subject to respond speedily. The time between stimulus onset and the response is reaction time.

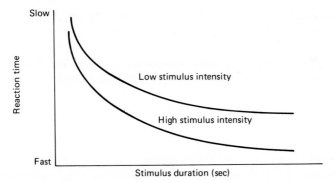

Figure 5-3 Simple reaction time as a function of stimulus intensity and stimulus duration.

of 10 lights that can occur and 10 keys for responses to them, one for each finger.

Prominent Stimulus Effects. Stimulus intensity and stimulus duration are strong variables for visual reaction time (Teichner & Krebs, 1972). When the stimulus is a flash of light striking the fovea, without temporal or spatial uncertainty, reaction time decreases as stimulus intensity increases. The same function holds for stimulus duration. Figure 5-3 shows the negatively accelerated decreasing functions that are the relationships between reaction time, stimulus intensity, and stimulus duration. The stimulus duration values are less than 1 second.

An early finding of research on reaction time was that reaction time was faster for auditory than visual stimuli (Woodworth, 1938, p. 324), and the finding is confirmed in more recent studies (e.g., Botwinick & Brinley, 1962; Niemi, 1979). A criticism that has been made of this line of research is that the auditory and visual stimuli are not equated for intensity. Stimulus modality is the apparent variable, with stimulus intensity the real variable. Howell and Donaldson (1962) addressed this issue with psychophysical techniques to equate the subjective magnitudes of the visual and auditory stimuli. Reaction time was still found to be faster with auditory stimuli. Knowledge like this can be put to work in human–machine systems. In Chapter 10 we discuss signals that warn of emergencies, and reaction time to them can be critical for system functioning or even system survival. Everything else equal, auditory warning signals, with their faster reaction time, could have a slight advantage.

Foreperiod. A *constant foreperiod* is when the interval between the ready signal and the stimulus is the same on each trial. A *variable foreperiod* is when the interval varies from trial to trial. Research has established that reaction time increases as constant foreperiod increases, and decreases as variable foreperiod increases (Niemi & Nää-tänen, 1981).

An experiment by Klemmer (1956) demonstrates the relationship between visual reaction time and constant foreperiod. Klemmer used a click as a warning signal, a neon light stimulus that came on for .02 second, and a telegraph key for response. The constant foreperiods that he used were .25, 4.25, or 8.25 seconds. Reaction time values for these foreperiods are plotted in Figure 5-4. The longer a constant foreperiod, the longer the reaction time.

Findings by Niemi (1979, Experiment 1) illustrate the effects of variable foreperiod on visual reaction time. Niemi had the presentation of events controlled by a computer and displayed to the subject on a cathode ray tube, as on your television set. A small arrow was the ready signal, and a small rectangular light bar was the stimulus. Response was pressing a key with the finger. Foreperiods of .50 to 3.0 seconds were randomized over the trials. Reaction times as a function of the foreperiods are shown in Figure 5-5, and they decrease as foreperiod increases. Range of foreperiods is also a variable. Klemmer (1956) found that reaction time is increased if a foreperiod is embedded in a wider range of variable foreperiod values.

The explanation of foreperiod influences on reaction time is in terms of temporal expectancy. *Expectancy,* in the context of reaction time, means a subject's estimate of time of stimulus occurrence and its effect on readiness to respond. The subject's timekeeping can develop with some accuracy when a constant foreperiod is being used because the same time interval is experienced each time. We can accurately learn to tap our finger at a 1-second rate to the click of a metronome, for example. The accuracy of timekeeping that develops over trials with constant foreperiod is not the same for all intervals. Figure 5-4 shows that reaction time is faster when a constant interval is short than

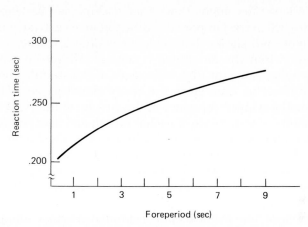

Figure 5-4   Simple reaction time as a function of duration of a constant foreperiod. A constant foreperiod remains unchanged from trial to trial. [*Source:* Data from Klemmer (1956), Table 1.]

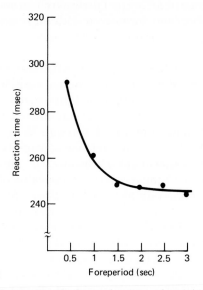

Figure 5-5   Simple reaction time as a function of variable foreperiod. A variable foreperiod changes from trial to trial. [Adapted from Niemi (1979), Fig. 2. Used by permission of *Acta Psychologica.*]

when it is long. You might think that nothing can be learned about timekeeping when the foreperiod is different on each trial, yet the function in Figure 5-5 shows that behavior is orderly. What is going on, apparently, is that the foreperiod is "aging" (Nickerson & Burnham, 1969). By *aging* is meant that occurrence of the stimulus becomes increasingly likely as time since the ready signal passes. The subject learns something about the range of the foreperiod intervals and about the probability of occurrence of each interval, and uses this knowledge to govern readiness to respond. The subjects in the Niemi (1979) experiment (Figure 5-5) know that brief intervals of .5 second occur but that the chances are greater for the interval to be longer, so it is adaptive to prepare for a long interval. Reaction time is less for longer intervals.

An outcome of faulty timekeeping is premature responding. The subject, being sure that the stimulus is about to occur, responds ahead of time, in error. Catch trials are used to avoid premature responding. A *catch trial* omits the stimulus and trains the subject away from premature readiness.

Choice Reaction Time

Choice reaction time is typified by choice among stimulus and response alternatives. Lacking stimulus–response alternatives, simple reaction time has been an arena for research on such variables as stimulus characteristics and foreperiod. Stimulus and response alternatives have been a focus of research on choice reaction time.

Number of Stimulus–Response Alternatives. That number of stimulus–response alternatives is a major variable for choice reaction time has been known for a long time. Using visual choice reaction time, Merkel in 1885 assigned the Arabic numerals 1 through 5 to the left hand and the Roman numerals I through V to the right hand. The fingers rested on keys, and key release as rapidly as possible was the response. Merkel's experimental variable was number of stimulus–response alternatives. Using a reaction-time value of 187 milliseconds as the no-choice baseline (simple reaction time), Merkel found that two alternatives increased reaction time to 316 milliseconds. The reaction time for 10 alternatives was 622 milliseconds. (For an account of Merkel's work, see Woodworth, 1938, pp. 332–334; Hick, 1952.) Merkel's pioneering work was revived in modern times by Hick (1952), whose data, which are similar to Merkel's, are shown in Figure 5-6.

The Merkel–Hick function is commonly found (e.g., Brainard et al., 1962; Hyman, 1953), but there are exceptions that raise doubts about it as an unqualified generalization for choice reaction time. Mowbray

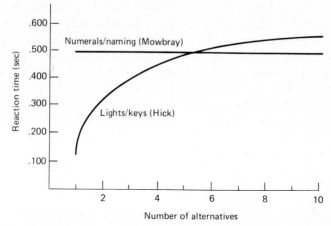

Figure 5-6 Choice reaction time as a function of number of alternatives for two kinds of tasks. The function depends on the kinds of stimuli and responses that are required. [*Source:* After Hick (1952), Fig. 1, and Mowbray (1960), Fig. 1. Used by permission of the Experimental Psychology Society, United Kingdom.]

(1960) was one of the first to demonstrate that the Merkel–Hick function is qualified by the kinds of stimuli and responses that are used.

Instead of a typical reaction time task with lights and keys, Mowbray had his subjects name a lighted numeral as fast as possible. A clock was started when a numeral appeared and was stopped when the subject spoke. The number of stimulus–response alternatives varied from 2 to 10. Mowbray's results are shown in Figure 5-6. The flat function contrasts sharply with Hick's findings. Other studies have since confirmed Mowbray's finding (for a summary of them, see Longstreth et al., 1985).

Teichner and Krebs (1974) interpret the differences between curves in Figure 5-6 in terms of *stimulus–response translation* (Welford, 1976, Chapter 4), or the compatibility between stimulus and response. An example of straightforward translation is when each stimulus light is directly above the finger that responds to it. A subject immediately knows which finger goes with which light, almost as if a line were drawn between them. A more difficult translation, which lengthens reaction time, is when the stimuli and the responding fingers lack compatible one-to-one correspondence, such as a stimulus on the right applying to a leftmost finger, and vice versa (Duncan, 1977a,b). There are other kinds of translations also. Hick (1952) had his stimulus lights arranged in an irregular circle and the responses were with 10 telegraph keys beneath the 10 fingers, so the spatial positions of the lights had to be translated into the spatial positions of the keys. Symbolic stimuli such as numerals create another kind of translation. Merkel used numerals which designated key positions, so the numerals had to be translated into spatial positions. The concept of translation helps to understand the findings shown in Figure 5-6. Reaction in Hick's task required translation, and the greater the number of stimulus–response alternatives, the more difficult the translation. Reaction time was a function of the number of stimulus-response alternatives. Mowbray's task, which required the subject to name the numeral, was highly compatible and did not require translation. Mowbray found no effect of stimulus–response alternatives on reaction time.

The stimulus–response translation process has also been called *stimulus–response coding* (Longstreth et al., 1985). If coding is the basic process, there must be other direct codes besides numeral naming. Longstreth et al. (1985) used a choice reaction time task where duration of a key press was defined by the stimulus. The numeral 2 would define a press of two time units, for example. The function relating reaction time to number of stimulus–response alternatives was flat, presumably because the timing code did not require translation.

Speed–Accuracy Trade-off. A subject in a choice reaction-time situation can emphasize speed at the expense of accuracy, accuracy at the expense of speed, or compromise between speed and accuracy. Maximum speed and maximum accuracy appear to be incompatible. In a task with, say, 10 stimuli and 10 responses, the subject might adopt a speed set and respond fast but occasionally press the wrong key because of indifference to accuracy. With an accuracy set the subject is cautious and responds more slowly to ensure that an error is seldom made, so speed suffers. That subjects can exchange speed for accuracy, and vice versa, is called the *speed–accuracy trade-off*.

The function that describes speed–accuracy trade-off is called the *speed–accuracy operating characteristic* (Pew, 1969), and it is illustrated in Figure 5-7. Although there can be special circumstances where the extremes of the curve—fast speed (with low accuracy) or slow speed (with high accuracy)—are acceptable, a reaction time situation ordinarily requires both speed and accuracy. A subject can be expected to elect intermediate speed and reasonably high accuracy in the absence of an urging to do otherwise. Notice in Figure 5-7 that the curve does not go to zero percent accuracy for very fast reaction time because by chance a subject will occasionally respond correctly. Similarly, it is unlikely that slow responders will always be 100 percent correct because even the most experienced subjects will make an occasional error (Pachella, 1974). Slow, deliberate responders will come close to 100 percent, however.

There are several ways of studying speed–accuracy trade-off exper-

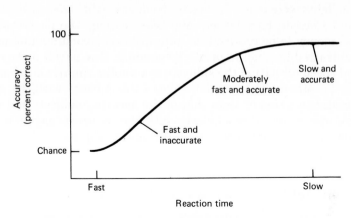

Figure 5-7 Speed–accuracy operating characteristic.

imentally (Wickelgren, 1977), and instructions to the subject is a prominent one. A typical reaction-time experiment will use compromise instructions that urge subjects to be as fast and as accurate as they can, but an experiment on speed–accuracy trade-off can use instructions that deliberately urge speed or accuracy. An experiment by Howell and Kreidler (1963) demonstrates the instructional approach. A choice reaction-time task with 10 stimulus lights on a panel and 10 response keys was used. Speed subjects were told: " . . . although you would like to make correct responses, speed is the important thing and accuracy is definitely a secondary consideration." Accuracy subjects were told: ". . . make as few errors as possible; speed is definitely a secondary consideration." Speed/accuracy subjects with compromise instructions were told to respond ". . . as fast and as accurately as you can; speed and accuracy are of equal importance in performing your task." On the final trials the speed subjects had an average reaction time of .543 second, with 87 percent correct responses. The accuracy subjects were slower but more accurate, with average reaction time of .581 second and 98 percent correct. The speed/accuracy subjects were between the other two groups, with average reaction time of .568 second and 96 percent correct.

Another prominent research approach to speed–accuracy trade-off is a payoff matrix where subjects receive rewards for appropriate speed or accuracy behavior, and perhaps have rewards taken away for inappropriate behavior. In accordance with Figure 5-7, accuracy should decrease as a subject's reaction time becomes fast when speed is rewarded, and accuracy should increase and reaction time slowed as accuracy is rewarded. Pachella and Pew (1968) used a reaction-time apparatus with four stimulus lights and four response keys. Below the stimulus lights were four informative feedback lights labeled Fast Correct, Slow Correct, Fast Wrong, and Slow Wrong. The payoff matrix specified money to be received for appropriate responding and money to be subtracted for inappropriate responding. One payoff matrix emphasized speed for part of the subjects and another matrix emphasized accuracy for others. The reaction time of the subjects with the speed payoff matrix was faster than that of the accuracy subjects and their accuracy was poorer. The subjects with the accuracy payoff matrix were more accurate but slower than the speed subjects.

Movement Time

Reaction time is speed of initiating a movement. *Movement time* is the speed of a movement once it has started, and it is not necessarily related to reaction time. Fitts' Law (Fitts, 1954; Fitts & Peterson, 1964) is the most substantial relationship that psychology has for movement time. The law says that movement time is a joint function of the extent and precision of the movement that is required. The frame of reference is a movement from a starting position to a target, where the movement can be any length and the target any size. Fitts defined an *Index of Difficulty* for tasks of this kind:

$$\text{index of difficulty} = \log_2 2A/W \tag{5-1}$$

where A is the amplitude (extent) of the movement measured from the starting point to the center of the target, and W is the width of the target at the end of the movement. The index of difficulty was then related to movement time to give *Fitts's law:*

$$\text{movement time} = a + b(\log_2 2A/W) \tag{5-2}$$

where a and b are constants.

Figure 5-8 shows movement time plotted as a function of index of difficulty. The task was reciprocal tapping, where the subject moved a stylus back and forth between two target plates as fast as possible. The amplitude of movements varied from 2 to 16 inches, and width of the target plates varied from .25 to 2.0 inches. Linear equation 5-2 fits the data points quite well. Fitts's law is a kind of speed–accuracy trade-off in reciprocal tapping. With target amplitude held constant, and target width decreased so that accuracy requirements increase, movement time will increase. If target width is increased and accuracy requirements loosened, movement time will decrease. Notice that amplitude

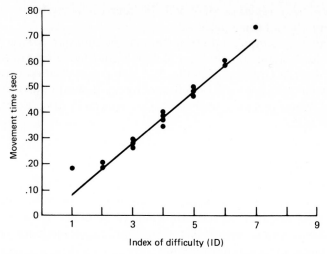

Figure 5-8 Fitts's law is a linear relationship between movement time and an index of difficulty whose variables are movement extent and target width. The data above, from a reciprocal tapping task, fit the linear function well. [*Source:* Data from Fitts (1954), Table 1.]

and target width are compensatory. If distance and target width are proportionately decreased or increased, movement time will remain the same.

A sound scientific principle has generality, and Fitts's law qualifies because it fits data from a variety of tasks. Tasks where Fitts's law applies are Fitts' own studies of reciprocal tapping (above), transferring washers from one pin to another and transferring pins from one set of holes to another (Fitts, 1954), microscopic movements performed under a stereoscopic microscope (Langolf et al., 1976), manipulating a control stick in one or two dimensions to move a cursor to a target on a display (Jagacinski et al., 1980; Jagacinski & Monk, 1985), head movements (Jagacinski & Monk, 1985), and a stylus-to-target movement by 4 to 5 year-old children (Wallace et al., 1978).

Some Principles of Learning Motor Skills

Reaction time is a ballistic, all-or-none response, and ordinarily we do not call it a skill. Skills are more complex, involving graded responses. Tennis is a skill. Operating an industrial machine like a lathe is a skill. Even comparatively simple responses, like moving the hand 10 inches, is a skill. Skills may or may not depend on fast reaction time.

Ordinarily, fast reaction time will be more important for a tennis player than for a lathe operator.

Skills may be simple, but often they can be complex, taking years to learn. Knowledge of results and observational learning are basic operations for the training of skills that have been studied experimentally. A review of them will be followed by an account of distribution of practice, or how learning trials should be spaced in time. First, knowledge of results.

Knowledge of Results

Knowledge of results is information that is presented to a learner about the accuracy of a response. Another term for knowledge of results is *informative feedback*. Informative feedback is distinguished from proprioception, which is response-produced feedback that is a sensory consequence of movement.

Knowledge of results, commonly shortened to KR, is usually in the form of error information and is familiar to us all. The teacher says "wrong" to the student after a response, and the student makes an effort to correct it on the next try. The golfer drives a golf ball and slices it to the left. The error is seen and an effort is made to improve on the next shot.

Knowledge of results is the most potent determiner of human learning that we know. A layperson will say that practice is the most potent determiner of learning, not KR. Practice is response repetition, and the layperson knows that if the response is made over and over, there will be improvement. Does not practice make perfect? "Yes" is the answer, but only if the responses are accompanied by KR. The golfer practices his drive and eventually corrects his slice, but only because he saw the error after each shot. Bartlett (1948, p. 86) said it best: "The common belief that 'practice makes perfect' is not true. It is practice *the results of which are known* that makes perfect."

There are a number of variables associated with KR (Adams, 1971, 1978, 1987; Newell, 1976; Salmoni, et al., 1984; Schmidt, 1982, Chapter 13), and a familiarization with the main ones gives an appreciation of KR.

Response Acquisition and KR. Edward L. Thorndike (1874–1949), a psychologist, was a giant of the psychology of learning. His ideas were influential in his lifetime and remained influential after his death. One of his main principles was the *law of effect*. The law of effect said it was the consequences of a response, such as reward and punishment, that were fundamental to learning. This emphasis on events that follow a response and are instrumental in producing its learning is called *instrumental learning*. Thorndike was led to this principle by his doctoral dissertation (Thorndike, 1898), in which he rewarded animals when they figured out how to trip a door latch and escape from a box. Thorndike began his career as an animal psychologist but soon turned to human learning and education, and there he found the law of effect useful also. Thorndike considered his law of effect a general principle that applied to both animal and human learning. In human learning the event that follows the response and produces its learning is KR.

Thorndike did much of his research on the instrumental learning of verbal responses, but he also did research on motor behavior. A classic experiment of Thorndike's on the simple motor skill of line drawing illustrates the importance of KR (Thorndike, 1927). The experimenter had the blindfolded subject draw lines of specified lengths, and the experimenter said "right" if the movement was within a tolerance band around the correct length, or "wrong" otherwise. "Right" and "wrong" were the KR. The percent "right" rose from 13 percent to 54.5 percent after 4200 lines had been drawn. Thorndike also asked in this experiment if practice repetitions alone, without KR, would produce learning. The percent "right" remained unchanged over the drawing of 5400 lines. Thus, there was learning with KR and no learning without it. It is from research roots like this that modern analysts continue to emphasize the role of KR in human learning and believe, with Bartlett, that it is not practice that makes perfect but practice the results of which are known that makes perfect.

Why KR works as it does to produce learning is theory. One point of view, which was held by Thorndike, is that KR produces an increment of *habit,* and when the habit strength is great enough, the response will occur reliably. Habit was considered a conceptual state that generated the response. Habit has lost favor to the informational view, which modern-day analysts prefer. Habit theory is noncognitive and does not rely on an aware, thinking learner as the informational view does. The *informational view* says that a subject receives KR after a response and then, in the interval between KR and the next occurrence of the response, weighs the error in her last response and plans the next response to eliminate it. The golfer who slices to the left will observe this KR and decide that his stance was wrong and should be changed the next time.

The Withdrawal of KR. The effects of taking KR away, after it has been applied and some learning has occurred, is demonstrated in an experiment by Newell (1974). The task was to move a slide along a linear track of 24.03 centimeters in 150 milliseconds. All subjects were given 77 trials. Different groups of subjects had 2, 7, 17, 32, and 52

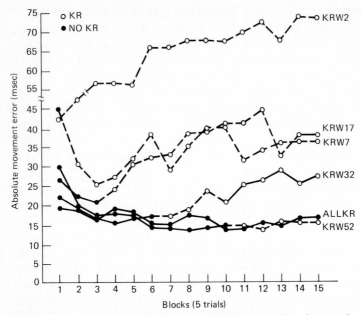

Figure 5-9 Effect of withdrawing knowledge of results after varying numbers of trials with knowledge of results. KRW2 is the condition that had knowledge of results withdrawn after two trials, KRW7 had knowledge of results withdrawn after seven trials, etc. [*Journal of Motor Behavior*, 6, 235–244. Adapted from Newell (1974), Fig. 2. Reprinted with permission of the Helen Dwight Reid Educational Foundation. Published by Heldref Publications, 4000 Albemarle St., N.W., Washington, D.C. 20016. Copyright © 1974.]

trials with KR, with KR withdrawn on the remainder of the trials. Newell's results appear in Figure 5-9. The measure of performance that was used is absolute error, which ignores the direction of an error. The comparison is with a control group that had KR throughout. The principal finding was that performance deteriorates substantially when KR is withdrawn after a small numbers of trials but not when KR is withdrawn after a relatively large number of trials. Many KR trials not only refine the accuracy of the response but give the response stability after KR withdrawal as well.

KR Precision. The KR can be partly informative or detailed. "Right" and "wrong" after responses are examples of *qualitative KR*, which is partly informative, and is contrasted to *quantitative KR*, where more detailed information about direction and amount of error is given.

Trowbridge and Cason (1932) conducted an experiment on KR precision. Their procedures were similar to the ones that Thorndike (1927) used. A subject was blindfolded and was required to draw 3-inch lines. The main comparisons were between a no-KR condition, qualitative KR of "right" and "wrong," and quantitative KR such as "plus 2" if the response was ⅜ of an inch too long, or "minus 7" if the response was ⅞ of an inch too short. There was learning with qualitative KR and no learning without KR, which confirmed Thorndike (1927), but quantitative KR produced the most rapid learning of all. In terms of the informational view of KR, quantitative KR gives the subject more information with which to plan the next response.

That quantitative KR is best does not mean the more precise the KR, the better. Rogers (1974) used an apparatus which had, from the subject's view, a knob mounted in a panel. The subject's task was to learn a specified rotation of the knob. The KR was either qualitative (too long or too short), or quantitative: one digit (e.g., +3), two digits (e.g., +3.2), or four digits (e.g., +3.214). Qualitative KR produced relatively poor performance, one-digit KR was better, and two-digit KR was better still. Four-digit KR was a reversal, however, performing at the level of qualitative KR. Some precision of KR is desirable, but it can be overdone.

KR Intervals. The time between responses is divided into the interval between the response and KR, called the *KR delay interval,* and the interval between KR and the next response, called the *post-KR interval.*

An established principle of animal learning is that the longer the delay of reward, the slower the learning of the response. The expectation was that delay of KR would be the same for humans, but it did not work out that way. Lorge and Thorndike (1935) were the first to show that KR delay made no difference for human learning, and the finding has since been replicated a number of times (e.g., Bilodeau & Bilodeau, 1958).

The *post-KR interval* is the period in which the subject processes the KR and plans the response, according to the informational view, so time must allow for these mental operations. Using a line-drawing task, Weinberg et al. (1964) found that learning was impaired if the post-KR interval was too brief. How long the post-KR interval should be is not clear, but it should be generous. Although without evidence, it would seem that a more complex response would need a longer post-KR interval because more time should be required to relate the KR to it.

Kinematic KR. The KR that has been discussed so far is delivered at the end of the sequence and is called *outcome KR.* The only concern was whether the endpoint was reached or not, not whether the movement was fast or slow, erratic or smooth. Yet there are many movements where the patterning and timing of the sequence is as important as reaching the correct endpoint. A diving coach wants the trainee to enter the water properly, but he also wants an ideal pattern of movements throughout the dive. KR for features of the movement sequence in its entirety is called *kinematic KR* (kinematics is the branch of mechanics that deals with pure motion without reference to the masses or forces involved in it). Outcome KR will influence the sequence to some extent as well as the endpoint, but kinematic KR, as additional information about dimensions of the movement sequence, improves learning of the sequence the most (Adams, 1984, 1985, 1987). Systematic research on kinematic KR is relatively new, although coaches and other instructors of skills have been using kinematic KR for a long time.

Rothstein and Arnold (1976) examined the research literature on videotaping as a way of training athletic skills. The best procedure is for the instructor to draw the trainee's attention to aspects of his performance and to comment about them (a form of kinematic KR) rather than have the subject merely watch the tape. Lintern (Lintern, 1980; Lintern & Roscoe, 1980) found that error information throughout the sequence was the best procedure for teaching the landing of the aircraft simulated by a flight simulator (which had a dynamic visual display of the runway scene as part of it). Error cues appeared on the display only when the trainee's performance exceeded acceptable limits. As the trainee improved, error cues became fewer and fewer.

Wallace and Hagler (1979) studied the skill of shooting a basketball through a hoop from a fixed position. One group saw the results of each shot, which is outcome KR, but a second group received information about stance and motion in addition, which is kinematic KR. The KR was then withdrawn and the two groups compared in a series of test trials. Kinematic KR produced the best performance.

Observational Learning

Observational learning is defined as learning to perform a skill by sensory experience. Observational learning has a model and an observer. A *model* demonstrates the way to perform a skill and the *observer* watches and learns. The model will often be live, but she could be on film (instructional films for teaching skills are common in training programs for system personnel). Observational learning can be auditory also, as with a musician who learns something about a selection when he hears it played. Observational learning is usually visual, however.

Observational learning is a case of perceptual learning. We do a great deal of learning by sensory experience. Recognition is a prominent kind of perceptual learning, where we judge whether we have experienced an object or a scene before (Chapter 6). Gibson (1969), a perceptual learning theorist, believes that perceptual learning is not imprinting a template, or copy, of what has been experienced but is a matter of extracting critical features that distinguish an object or scene from others. Observational learning has perceptual learning as its foundation but it goes beyond, with the requirement of translating the products of perceptual learning into motor activity.

Observational learning is an underdeveloped area in the psychology of learning (Adams, 1987). One reason is that the founding fathers of the psychology of learning in the United States, like Thorndike, did not have much luck with observational learning, so they cast their lots with learning operations that worked, such as instrumental learning. Developmental psychology has sustained the research interest in observational learning over the decades because of the reasonable belief that children learn many of their language and social skills by observation. The layperson also believes in the worth of observational learning and relies on it in everyday learning. Observational learning is regularly used in practical situations, as with a factory supervisor who demonstrates the steps in operating a machine to a worker, a training film that demonstrates the steps in operating a machine to a trainee, or a coach demonstrating the proper movements to an athlete.

The problem with observational learning is that it works for the learning skills but no one knows how well (Adams, 1987). One limitation is that there are dimensions of skills, such as pressures, muscular tensions, and dimensions of movement, that are out of view and which cannot be learned by watching. This point was made in an experiment by Carroll and Bandura (1982), who used a television system to make the unobservable sides of a complex movement observable, and learning benefited from it. The Carroll and Bandura findings are an advance, but it is still far from clear how much can be learned by observation. Newell et al. (1985) hypothesized that observational learning can teach the order of the movement sequence and the bounds of their operation, but thereafter KR must take over for refinement of response details. Maybe so. An important research direction for the future will be the limits of observational learning.

Distribution of Practice

We all know from common experience that the quality and quantity of responses decline when too many responses are made in too short a time. A fatigue-like state develops. Adams (1954) used the Rotary Pursuit Test for a systematic demonstration of the effects of massed practice on motor performance. The Rotary Pursuit Test has a disk like a phonograph turntable which turns at 60 revolutions per minute. The disk has a target embedded in it, about the size of a small coin, and the subject's task is to keep a hand-held stylus on the target as much as possible. The score on a trial is time on target. The degree of massed practice was defined by the amount of time between trials. Using 30-second trials, various groups had intertrial intervals of either 0, 3, 10, 20, or 30 seconds. The results are shown in Figure 5-10. The greater the concentration of trials the poorer the performance.

That performance is lowered with massed practice is useful to know, but it is not the issue for learning. The issue for learning is the extent to which massed practice affects the characteristics of the skill being learned. Will the quality of a skill acquired under massed practice be as high as when it is acquired under distributed practice? The *performance* of the skill will be impaired during massed practice, but will the *learning* be?

Adams and Reynolds (1954) put this hypothesis to test. Different groups received four different amounts of massed practice and then switched to distributed practice after a rest. Comparison was with a control group that had distributed practice throughout. The groups that had massed practice should perform less well after the switch than the control group if massed practice affects learning. The results

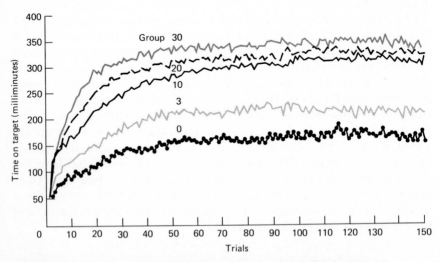

Figure 5-10 The less the spacing between trials, the poorer is motor performance. The designation for a group indicates the intertrial interval in seconds. Trial duration was 30 seconds. The Rotary Pursuit Test was used. (Adapted from Adams (1954), Fig. 1.)

are shown in Figure 5-11. After a brief period of transition, the groups that had massed practice performed as well as the control group. Massed practice affects performance, not learning. The degrading effects of massed practice are temporary.

Textbooks often assure students that learning is better with distributed practice than with massed practice. The generalization is wrong; it is a confusion of performance and learning. Performance, not learning, is affected by massed practice. Distributed practice is a good way to learn a skill, but all else equal, massed practice will yield as much learning in less time. Practical training methods are discussed in Chapter 18. A message for programs of training skills in practical situations is that distributed practice does not have to be considered seriously. Training programs in industrial and military settings are expensive, and it is useful to know that training time can be minimized with massed practice.

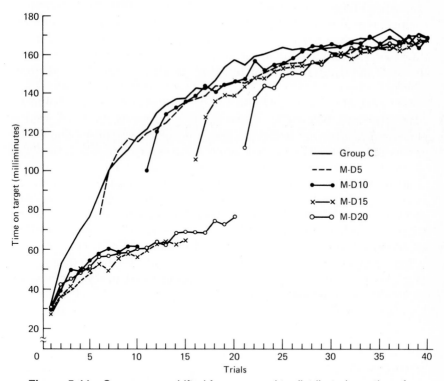

Figure 5-11   Groups were shifted from massed to distributed practice after 5, 10, 15, or 20 trials. Transition to the level of group C, a control group that had distributed practice throughout, occurred readily. These data indicate that massing of practice affects performance of a skill, not its learning. The Rotary Pursuit Test was used. [Adapted from Adams and Reynolds (1954), Fig. 1.]

Summary ———————————————————————

Proprioceptors, which are sensory receptors in the muscles and the joints, feed information to the spinal cord and the brain about positions and movements of the limbs. Whether the locus of the information is fundamentally receptors in the muscles or in the joints was a long-standing controversy in physiology. Physiological research is resolving the controversy in favor of the muscles, with joint receptors playing a lesser role.

Reaction time is the speed of initiating a response to the onset of a stimulus. Prominent variables that determine reaction time are stimulus intensity and duration, number of stimulus response alternatives, the kinds of stimuli and responses, stimulus modality, duration and type of foreperiod, and whether the subject emphasizes speed or accuracy when choice is involved.

Movement time is the speed of a response in attaining a goal, once the response has been initiated. Fitts' Law, which relates movement time to extent of movement and width of the target that is the goal, is the most substantial relationship that psychology has for describing movement time.

The most effective variable for the training of a motor skill is knowledge of results. Knowledge of results is informative feedback that is presented to a learner about the adequacy of the response. Knowledge of results is often given to the learner as an error score. The knowledge of results can be delivered as outcome information about the accuracy of attaining a goal, or it can be delivered throughout the sequence as kinematic information to shape a particular pattern of movement. Observational learning, which is learning by sensory experience, is another way of learning skills. We have less scientific knowledge of observational learning than knowledge of results at this time.

Too many motor responses in too short a time results in fatigue and degrades performance, but does the concentrated practice affect the quality of the skill being learned? Research has established that massing of practice degrades the performance of a skill, not its learning.

References ———————————————————————

Adams, J. A. (1954). Psychomotor performance as a function of intertrial rest interval. *Journal of Experimental Psychology, 48,* 131–133.

Adams, J. A. (1971). A closed-loop theory of motor learning. *Journal of Motor Behavior, 3,* 111–149.

Adams, J. A. (1977). Feedback theory of how joint receptors regulate the timing and positioning of a limb. *Psychological Review, 84,* 504–523.

Adams, J. A. (1978). Theoretical issues for knowledge of results. In G. E. Stelmach (Ed.), *Information processing in motor control and learning.* New York: Academic Press. Pp. 229–240.

Adams, J. A. (1984). Learning of movement sequences. *Psychological Bulletin, 96,* 3–28.

Adams, J. A. (1985). The use of a model of movement sequences for the study of knowledge of results and the training of experts. *Journal of Human Movement Studies, 11,* 223–236.

Adams, J. A. (1987). Historical review and appraisal of research on the learning, retention, and transfer of human motor skills. *Psychological Bulletin, 101,* 41–74.

Adams, J. A., & Reynolds, J. B. (1954). Effect of shift in distribution of practice conditions following interpolated rest. *Journal of Experimental Psychology, 47,* 32–36.

Bartlett, F. C. (1948). The measurement of human skill. *Occupational Psychology, 22,* 83–91.

Bilodeau, E. A., & Bilodeau, I. M. (1958). Variation of temporal intervals among critical events in five studies of knowledge of results. *Journal of Experimental Psychology, 55,* 379–383.

Botwinick, J., & Brinley, J. F. (1962). An analysis of set in relation to reaction time. *Journal of Experimental Psychology, 63,* 568–574.

Boyd, I. A., & Roberts, T. D. M. (1953). Proprioceptive discharge from stretch-receptors in the knee joint of the cat. *Journal of Physiology, 122,* 38–58.

Brainard, R. W., Irby, T. S., Fitts, P. M., & Alluissi, E. A. (1962). Some variables influencing the rate of gain of information. *Journal of Experimental Psychology, 63,* 105–110.

Carroll, W. R., & Bandura, A. (1982). The role of visual monitoring in observational learning of action patterns: Making the unobservable observable. *Journal of Motor Behavior, 14,* 153–167.

Clark, F. J. (1975). Information signaled by sensory fibers in medial articular nerve. *Journal of Neurophysiology, 38,* 1464–1472.

Clark, F. J., & Burgess, P. R. (1975). Slowly adapting receptors in cat knee joint: Can they signal joint angle? *Journal of Neurophysiology, 38,* 1448–1463.

Donders, F. C. (1868/1969). On the speed of mental processes. In W. G. Koster (Ed.), *Attention and performance II.* Amsterdam: North-Holland. Pp. 412–431. (English translation of the 1868 article.)

Duncan, J. (1977a). Response selection rules in spatial choice reaction tasks. In S. Dornic (Ed.), *Attention and performance VI.* Hillsdale, NJ: Lawrence Erlbaum. Pp. 49–61.

Duncan, J. (1977b). Response selection errors in spatial choice reaction tasks. *Quarterly Journal of Experimental Psychology, 29,* 415–423.

Fitts, P. M. (1954). The information capacity of the human motor system in controlling the amplitude of movement. *Journal of Experimental Psychology, 47,* 381–391.

Fitts, P. M., & Peterson, J. R. (1964). Information capacity of discrete motor responses. *Journal of Experimental Psychology, 67,* 103–112.

Gibson, E. J. (1969). *Principles of perceptual learning and development.* New York: Appleton-Century-Crofts.

Grigg, P. (1975). Mechanical factors influencing response of joint afferent neurons from cat knee. *Journal of Neurophysiology, 38,* 1473–1484.

Grigg, P., & Greenspan, B. J. (1977). Responses of primate joint afferent neurons to mechanical stimulation of knee joint. *Journal of Neurophysiology, 40,* 1–8.

Helmholtz, H. v. (1853). On the methods of measuring very small portions of time, and their application to physiological purposes. *Philosophical Magazine, 40,* 313–325.

Hick, W. E. (1952). On the rate of gain of information. *Quarterly Journal of Experimental Psychology, 4,* 11–26.

Howell, W. C., & Donaldson, J. E. (1962). Human choice reaction time within and among sense modalities. *Science, 135,* 429–430.

Hyman, R. (1953). Stimulus information as a determinant of reaction time. *Journal of Experimental Psychology, 45,* 188–196.

Howell, W. C., & Kreidler, D. L. (1963). Information processing under contradictory instructional sets. *Journal of Experimental Psychology, 65,* 39–46.

Jagacinski, R. J., & Monk, D. L. (1985). Fitts' law in two dimensions with hand and head movements. *Journal of Motor Behavior, 17,* 77–95.

Jagacinski, R. J., Repperger, D. W., Moran, M. S., Ward, S. L., & Glass, B. (1980). Fitts' law and the microstructure of rapid discrete movements. *Journal of Experimental Psychology: Human Perception and Performance, 6,* 309–320.

Kelso, J. A. S. (Ed.). (1982). *Human motor behavior: An introduction.* Hillsdale, NJ: Lawrence Erlbaum.

Klemmer, E. T. (1956). Time uncertainty in simple reaction time. *Journal of Experimental Psychology, 51,* 179–184.

Langolf, G. D., Chaffin, D. B., & Foulkes, J. A. (1976). An investigation of Fitts' law using a wide range of movement amplitudes. *Journal of Motor Behavior, 8,* 113–128.

Lintern, G. (1980). Transfer of landing skill after training with supplementary visual cues. *Human Factors, 22,* 81–88.

Lintern, G., & Roscoe, S. N. (1980). Visual cue augmentation in contact flight simulation. In S. N. Roscoe (Ed.), *Aviation psychology.* Ames, IA: Iowa State University Press, Pp. 227–238.

Longstreth, L. E., El-Zahhar, N., & Alcorn, M. B. (1985). Exceptions to Hick's law: Explorations with a response duration measure. *Journal of Experimental Psychology: General, 114,* 417–434.

Lorge, I., & Thorndike, E. L. (1935). The influence of delay in the after-effect of a connection. *Journal of Experimental Psychology, 18,* 186–194.

Mountcastle, V. B. (1966). The neural replication of sensory events in the somatic afferent system. In J. C. Eccles (Ed.), *Brain and conscious experience.* New York: Springer.

Mowbray, G. H. (1960). Choice reaction times for skilled responses. *Quarterly Journal of Experimental Psychology, 12,* 193–202.

Newell, K. M. (1974). Knowledge of results and motor learning. *Journal of Motor Behavior, 6,* 235–244.

Newell, K. M. (1976). Knowledge of results and motor learning. In J. Keogh & R. S. Hutton (Eds.), *Exercise and sport sciences reviews*. Santa Barbara, CA: Journal Publishing Affiliates. Pp. 195–228.

Newell, K. R., Morris, L. R., & Scully, D. M. (1985). Augmented information and the acquisition of skill in physical activity. In R. L. Terjung (Ed.), *Exercise and sport sciences reviews*. New York: Macmillan. Pp. 235–261.

Nickerson, R. S., & Burnham, D. W. (1969). Response times with nonaging foreperiods. *Journal of Experimental Psychology, 79,* 452–457.

Niemi, P. (1979). Stimulus intensity effects on auditory and visual reaction processes. *Acta Psychologica, 43,* 299–312.

Niemi, P., & Näätänen, R. (1981). Foreperiod and simple reaction time. *Psychological Bulletin, 89,* 133–162.

Pachella, R. G. (1974). The interpretation of reaction time in information-processing research. In B. H. Kantowitz (Ed.), *Human information processing: Tutorials in performance and cognition*. Hillsdale, NJ: Lawrence Erlbaum. Pp. 41–82.

Pachella, R. G., & Pew, R. W. (1968). Speed-accuracy tradeoff in reaction time: Effect of discrete criterion times. *Journal of Experimental Psychology, 76,* 19–24.

Pew, R. W. (1969). The speed–accuracy operating characteristic. *Acta Psychologica, 30,* 16–26.

Rogers, C. A., Jr. (1974). Feedback precision and postfeedback interval. *Journal of Experimental Psychology, 102,* 604–608.

Rossi, A., & Grigg, P. (1982). Characteristics of hip joint mechanoreceptors in the cat. *Journal of Neurophysiology, 47,* 1029–1042.

Rothstein, A. L., & Arnold, R. K. (1976). Bridging the gap: Application of research on videotape feedback and bowling. *Motor Skills: Theory into Practice, 1,* 355–386.

Salmoni, A. W., Schmidt, R. A., & Walter, C. B. (1984). Knowledge of results and motor learning: A review and critical appraisal. *Psychological Bulletin, 95,* 355–386.

Schmidt, R. A. (1982). *Motor control and learning*. Champaign, IL: Human Kinetics.

Sherrington, C. S. (1900). The muscular sense. In E. A. Schäfer (Ed.), *Textbook of physiology*. Volume 2. Edinburgh, Scotland: Pentland.

Sherrington, C. S. (1906). *The integrative action of the nervous system*. New York: Scribner.

Taub, E. (1977). Movement in nonhuman primates deprived of somatosensory feedback. In J. Keogh & R. S. Hutton (Eds.), *Exercise and sport sciences reviews*. Volume 4. Santa Barbara, CA: Journal Publishing Associates. Pp. 335–374.

Taub, E., & Berman, A. J. (1968). Movement and learning in the absence of sensory feedback. In S. J. Freedman (Ed.), *The neuropsychology of spatially oriented movement*. Homewood, IL: Dorsey Press. Pp. 173–192.

Teichner, W. H., & Krebs, M. J. (1972). Laws of the simple visual reaction time. *Psychological Review, 79,* 344–358.

Teichner, W. H., & Krebs, M. J. (1974). Laws of visual choice reaction time. *Psychological Review, 81,* 75–98.

Thorndike, E. L. (1898). Animal intelligence: An experimental study of the associative processes in animals. *Psychological Review Monograph Supplement, 2,* (4) (Whole No. 8).

Thorndike, E. L. (1927). The law of effect. *American Journal of Psychology, 39,* 212–222.

Trowbridge, M. H., & Cason, H. (1932). An experimental study of Thorndike's theory of learning. *Journal of General Psychology, 7,* 245–258.

Wallace, S. A., & Hagler, R. W. (1979). Knowledge of performance and the learning of a closed motor skill. *Research Quarterly, 50,* 265–271.

Wallace, S. A., Newell, K. M., & Wade, M. G. (1978). Decision and response times as a function of movement difficulty in preschool children. *Child Development, 49,* 509–512.

Weinberg, D. R., Guy, D. E., & Tupper, R. W. (1964). Variations of postfeedback interval in simple motor learning. *Journal of Experimental Psychology, 67,* 98–99.

Welford, A. T. (1976). *Skilled performance: Perceptual and motor skills.* Glenview, IL: Scott, Foresman.

Wickelgren, W. A. (1977). Speed-accuracy tradeoff and information processing dynamics. *Acta Psychologica, 41,* 67–85.

Woodworth, R. S. (1938). *Experimental psychology.* New York: Holt.

CHAPTER 6

Human Memory

A digital computer has a memory in which programs and data are stored. The large amount of information that can be stored for future use gives the digital computer impressive powers. The same can be said of biological memory. Almost all living creatures, including some insects, have the capability of storing what they have learned for future use. Without memory, an organism is a simple creature, at the mercy of stimuli of the moment, but with memory the past can govern behavior as well. The storehouse of past information that can contribute to current behavior is vast for many creatures. Our human memory is the most notable of all.

There were speculations about memory for more than 2000 years, and it was only with the advent of experimental psychology in the past 100 years that the properties of memory were put on an objective footing. Almost all of the research and theory has been on verbal behavior, although other response classes have begun to receive more attention in modern times.

History and Overview of Theories of Verbal Memory

The Distant Past

The ancient Greeks, who were early intellectual ancestors of ours, related memory to the goddess Mnemosyne. Plato (427–347 B.C.), a philosopher among those ancient peoples, said that human memory, which was a gift from Mnemosyne, was like a wax tablet on which experience left its imprint. The imprint can be lost with time, however, so forgetting occurs. The ancient Greeks speculated about the physiological locus of memory. Setting aside notions that memory might be in the heart, Erasistratus (310–250 B.C.) rightly concluded that the brain is the seat of memory and other mental functions.

Origins of Objective Research

Progress was made in the practical uses of memory throughout the ages, as we will see later in this chapter, but the scientific study of memory did not begin until the nineteenth century. Hermann Ebbinghaus (1850–1909) was the founding father of the experimental study of memory (Ebbinghaus, 1885/1964). Ebbinghaus would construct a list of nonsense syllables, which are meaningless consonant–vowel–consonant combinations, and then would learn it himself to a criterion such as two errorless repetitions. The list would be put aside for the duration of the retention interval, which is the interval between the end of learning and recollection, and then would relearn it to the same criterion as before. The time to relearn a list, relative to the time it took to learn it originally, was the measure of forgetting; the more time it took to relearn, the greater the forgetting. Using retention intervals from 19 minutes to 31 days, Ebbinghaus objectively defined a curve of forgetting, which was a decay function of time where rate of forgetting decreased as the retention interval increased. For a long time psychologists thought in terms of *the* curve of forgetting, as Ebbinghaus found, but they no longer think that way. There are many determinants of forgetting and many curves of forgetting.

Short-Term Verbal Retention

Ebbinghaus set the tone of memory research for the next 75 years. He tended to settle the study of memory on verbal behavior, long-term retention, and verbal lists. It was not until experiments by Brown (1958) in Great Britain, and Peterson and Peterson (1959) in the United States, that change occurred (psychologists refer to the Brown–Peterson findings in recognition of their similar pioneering experiments). The study by Peterson and Peterson illustrates the procedures and findings. A single nonsense syllable was presented and then recalled after 3 to 18 seconds. The subject counted backwards by 3's (or 4's) in the retention interval to block any covert rehearsal that might offset whatever forgetting occurred. Forgetting was almost complete in 18 seconds. Psychologists had been accustomed to thinking about forgetting occurring over hours and days, in the Ebbinghaus tradition, and now forgetting was found to occur in seconds. On reflection it was remembered that forgetting over very brief intervals happens every day. You find a number in the telephone book and walk across the room to dial it. The forgetting of a digit or two is common. These data caused psychologists to think in terms of a *short-term memory system,* wherein material is forgotten in seconds. To think in terms of a short-term memory system added complexity to human memory because there had to be a long-term memory system as well. Obviously, verbal material can be remembered for a long time.

Psychologists then turned to other short-term retention situations and findings in the research literature and used them to define the characteristics of short-term memory. How many items, such as random digits or letters of the alphabet, can be recalled immediately after one brief presentation? We all can recall one item, but how about five or ten? The number that can be recalled is called your *memory span,* and it is commonly believed to be seven plus or minus two (Miller, 1956). Memory theorists of the 1960s, thinking of the new short-term memory system, saw memory span as its capacity. Short-term memory had a limited capacity, and the capacity was seven plus or minus two.

Sensory Memories

There was another research development in the 1960s that influenced theoretical thinking about memory. The research was on visual memory (also called iconic memory), and it opened the topic of *sensory memory.* Sperling (1960) and Averbach and Coriell (1961) found that verbal items left a trace in the visual system that lasted about 200 milliseconds after the stimulus had vanished. The visual memory system was different from the Brown–Peterson function, which lasted for seconds.

An implication of the research on visual memory was that all senses have storage capabilities, and subsequent research has supported it (Adams, 1980, Chap. 11). The duration of trace persistence in sensory memory depends on the modality. The persistence of a trace in auditory memory (also called echoic memory) has been estimated as long as 20 seconds (Watkins & Watkins, 1980; Watkins & Todres, 1980).

The Information-Processing Model

If Ebbinghaus, and those who followed in his tradition, could be said to have a theory, it was a strength theory. The repetition of verbal items increased their strength, and the greater the strength, the greater the resistance to forgetting. Forgetting processes eroded the strength of items. Strength theory was not enough to account for the new findings on short-term and sensory retention, as well as established findings on long-term retention, so new theory arrived to take its place. The new theory was an information-processing model, and the most prominent version of it was by Atkinson and Shiffrin (1968).

The *information-processing model* of human verbal memory was a computer analogy. A digital computer has the input and the output of data, a working memory of limited capacity and a major memory of large capacity. Data are transferred from one memory to another, and data are processed, transformed, compared, searched, scanned, and retrieved. When we think about our own memories we find similar processes, so it is easy to assume that the human memory and the digital computer have corresponding processes.

Figure 6-1 shows an information-processing model of verbal memory. A stimulus input creates a trace in sensory memory, and while the trace is active, a search is made of perceptual systems of long-term memory for recognition of the stimulus. Upon recognition the stimulus is transferred to the short-term memory of limited capacity, where forgetting processes proceed to operate on it. Rehearsal will counteract forgetting processes by maintaining the item in short-term memory, and it will transfer the item to long-term memory and give it strength for long-term retention as well. Notice also that an item in long-term memory can be moved into short-term memory for conscious attention and action, such as using it in the ongoing solution of a problem. The latter use of short-term memory has been called *working memory* (Baddeley, 1976, Chapter 8; Baddeley & Hitch, 1974). Working memory has been called a mental scratch pad.

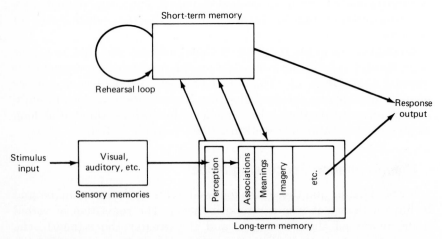

Figure 6-1 An information processing model of human verbal memory.

Long-term memory is the repository of verbal associations and meanings, and nonverbal perceptual systems and imagery. In contrast to short-term memory, the capacity of long-term memory is very large (and unknown). New material can be deposited there by rehearsal, or it can be incorporated into long-term memory by establishing associative, meaningful, or imagery relationships with material that is already stored there. The learning of new material is facilitated by using a memorization strategy that consciously organizes the new material into meaningful verbal or imagery structures. The retention of new material is benefited because the verbal and imagery structures are already secured in long-term memory.

Verbal response output can be from either short-term memory or long-term memory, and either recall or recognition. *Recall* is the production of a response, and *recognition* is an indication that a stimulus has been experienced before.

Levels of Processing Theory

A rival of the information processing model arose in the 1970s, called *levels of processing theory* (Craik & Lockhart, 1972). The basic idea is that there are levels at which a verbal item can be processed, and the deeper the level, the less the forgetting. The first level is shallow, where only the sensory, physical features of the item are processed. Recognition might occur at an intermediate level of processing, and an appreciation of associations and meaning can occur at the deepest level. The item's remembering depends on the depth of processing that was given to it at the time of learning. An experiment, for example, might present words and ask questions about their typography (shallow processing) or their meaning (deep processing). Studies like this have found forgetting inversely related to level of processing, as the model contends (Craik and Tulving, 1975).

A difficulty that critics find with levels of processing theory is that there is no independent definition of levels (Nelson, 1977; Baddeley, 1978). Levels are either intuitively defined by the investigator, or inferred from the amount of forgetting that they intend to predict, which is circular. Neither of these ways of defining levels is scientifically acceptable.

Status of Models of Memory

The information-processing model and the levels of processing model have been the most successful models of modern times. Both are still receiving attention, but there does not seem to be as much enthusiasm as before. Models in science seldom die quickly and decisively. Scientists are reluctant to part with a model because they remember how useful it has been, but they also notice new data that fail to fit the model or problems in the structure of the model (as with levels of processing theory). Enthusiasm wanes, and eventually a new model arrives to replace the old ones. Useful though a good model can be for organizing findings and suggesting ideas, research goes on without it. A creative scientist, with ideas of her own, does nicely without a model.

Strategies of Verbal Memorization

There is a choice in the method of verbal memorization, and success in remembering depends on the method that is chosen and how well it is used. These methods of memorization are called *strategies of verbal memorization,* and there are three of them: rehearsal, verbal elaboration, and nonverbal elaboration.

Rehearsal

Rehearsal is the number of times an item is given rote repetition. Ebbinghaus was the first to conduct an objective investigation of amount of rehearsal and forgetting (Ebbinghaus, 1885/1964). He found that forgetting decreased as amount of rehearsal increased, and the relationship has been a common finding since then.

The inverse relationship between amount of rehearsal and forgetting is illustrated in Figure 6-2, which shows the results of an experiment by Hellyer (1962). A short-term verbal retention paradigm was used, much like that which was used by Peterson and Peterson. A consonant syllable like KXC was presented visually one, two, four, or eight times, and the subject read it aloud each time it was displayed. Retention intervals were either 3, 9, 18, or 27 seconds. Digits were read aloud in the retention interval to prevent rehearsal. The curve for one rehearsal repetition in Figure 6-2 shows rapid forgetting like Peterson and Peterson (1959) found, but retention steadily increased as number of rehearsal repetitions increased.

An implication of levels of processing theory is that an item can be given rehearsal repetitions with such shallow processing that the rehearsal will have virtually no effect on retention. Evidence for this assertion has led to the distinction between *elaborative rehearsal,* which affects retention, and *maintenance rehearsal,* which does not. The Hellyer experiment above is an example of elaborative rehearsal. An experiment on maintenance rehearsal by Glenberg et al. (1977) used a short-term retention paradigm, where the items were four-digit num-

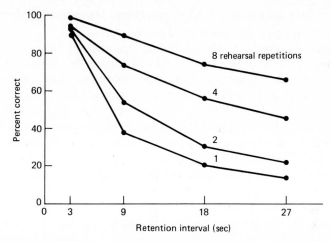

Figure 6-2 Short-term verbal retention as a function of rehearsal repetitions. [Data from Hellyer, S., *Journal of Experimental Psychology* 64, 650 (1962), Table 1. Copyright 1962 by the American Psychological Association. Used by permission of the author.]

bers and words were the filler activity in the retention intervals to prevent rehearsal. At the end of the experiment Glenberg et al. asked for recall of the words instead of the numbers, which, from the subject's point of view, had been of no consequence for the main task of learning and memorizing digits. The number of times that words were repeated increased with length of the retention interval, so their retention should be a function of number of repetitions, according to a strength view. Not so. Recall of the words was unrelated to amount of rehearsal. As provocative as these findings are, the overall pattern of experiments like that of Glenberg et al. is a mix of findings (see Baddeley, 1978, for a summary; Nairne, 1983), and it is fair to say that the concept of maintenance rehearsal, and the conditions that define it, are not yet secure. Whatever the outcome for maintenance rehearsal, rehearsal is clearly a more complex process than some have thought it to be.

Verbal Elaboration

Repeating a verbal item over and over in rehearsal is ordinarily a rote operation, with no conscious effort to associate it with material that resides in long-term memory. Yet, to do so would be a good strategy. The established material in long-term memory is well learned, and to integrate the new with the old should impart the strength of the old to the new and facilitate retention. Rote rehearsal is probably the only effective strategy that a young child has, but adults can have more elaborate strategies and will often use them. One strategy that adults use is called natural language mediation.

A *natural language mediator* is any association for a verbal item that spontaneously comes to the subject during learning. An experimental approach to natural language mediation has used *paired-associate learning*. In paired-associate learning a list of word pairs or nonsense syllable pairs defines the learning task. The first item of a pair is the stimulus term and the second is the response term, and the subject must learn to give the response term when the stimulus term is presented. The learning task is completed when a criterion number correct for the list is met. Thus, the pair BEAR–HAIR requires the subject to learn that BEAR and HAIR go together, and to give HAIR as the response when BEAR is presented. Some subjects will learn this pair by rote rehearsal, but others will have natural language mediators such as "A bear is hairy" or "Bear rhymes with hair." Natural language mediators will vary from pair to pair and subject to subject. BEAR–HAIR is a meaningless word sequence but "A bear is hairy" is not. Natural language mediators occur for nonsense syllables as well. A nonsense syllable pair such as TIL–DEH might elicit the natural language mediator "Until death do us part."

How is a natural language mediator used in recall after it has been formed in original learning? The natural language mediator may be remembered in substantially its original form when the stimulus term is presented, remembered in a distorted form, or not remembered at all. If remembered, there is the problem of decoding it to recover the response term. Decoding HAIR from "A bear is hairy" might be easy, but decoding DEH from "Until death do us part" might not be. A subject must not only identify the word among the five of "Until death do us part" that embodies DEH, but must also identify the particular three letters and their order. The difficulty of doing this reduces the chances of recalling DEH correctly.

An experiment by Montague et al. (1966) on natural language mediation and retention used a long list of 96 nonsense syllables pairs. The list was presented once in learning, and a record was made of each natural language mediator that occurred in learning. If no natural language mediator occurred, it was assumed that the pair was learned by rote. The retention interval was 24 hours. At recall the stimulus term of each pair was presented and the subject was asked to recall the natural language mediator, if one had been formed in learning, and the response term. Correct recall of items that had been learned by rote was 6 percent. Recall of items that had been learned with a natural language mediator depended on whether the mediator was recalled in its original form. If the natural language mediator was recalled in the same or similar form, the percent correct recall was 73 percent. If the natural language mediator was recalled in a different form, the correct recall was 2 percent. Twenty-seven percent of the natural language mediators that were correctly recalled did not yield the response term because they could not be decoded correctly. Compared to rote learning, natural language mediation benefited retention by a factor of 12 *provided* that the mediator was recalled correctly and correctly decoded.

The recall of random digits presented once at a 1-second rate for immediate recall in order is a way of measuring memory span. Typical memory spans are about seven items. That this number can be increased dramatically with natural language mediation is demonstrated in an experiment by Chase and Ericsson (1982). Their principal subject was a college student who trained for 264 sessions over two years. In the early sessions the subject was holding the digits in short-term memory and was averaging seven. As the sessions progressed he began to work out a system based on natural language mediators. The subject was a long-distance runner, so the subgroup of digits 3492 might be coded "near world-record mile time." He also would use dates, 1943 being coded as "near the end of World War II." Ages were used, 896 being "eighty nine point six years old, very old man." The subject's performance at the end of two years was a remarkable 82 digits.

The principle to remember is that the integration of new verbal material with familiar verbal material at the time of learning benefits retention. When familiar associations fail to come to mind, rote rehearsal is an alternative. We will see in the next section that mental imagery is also an alternative.

Nonverbal Elaboration

Natural language mediation is verbal elaboration but mental imagery is nonverbal elaboration, and it is another strategy of memorization. The use of mental imagery is our oldest recorded method of memorization (rote rehearsal is undoubtedly older, but no one found it interesting enough to write about it). The use of mental imagery in recall is said to go back to the ancient Greeks. A roof fell in during a dinner party and Simonides the poet was able to identify the mangled bodies by using mental imagery and remembering where each person had been sitting (Cicero, 1959).

It has been said that if you doubt mental imagery, answer this question without it: How many windows do you have in your house? Most likely you have never walked around your house, deliberately counted the windows, and memorized the number. Instead, you form an image of your house and mentally stroll around it, counting the windows.

The Method of Loci. The best illustration of imagery in memorization is the oldest, called the *method of loci* (or locations). The method of loci comes to us from an unknown Roman teacher of rhetoric who compiled a textbook for his students in about 86–82 B.C. The book is called *Ad Herrenium*. The old meaning of the term "rhetoric" was the art of persuasive speechmaking. Today a speaker uses notes to remember the points of a speech, but in ancient times before writing and paper were common, the points had to be remembered—no small feat for a two-hour speech. *Ad Herrenium* recommended the method of loci, which is summarized as six steps:

1. Imagine a location with a rather large number of distinctive places that you know very well.
2. The places of the location should form a series so that you can progress through them in serial order, forward or backward, from any place in the series.
3. When given a series of items to remember, imagine each one at a place in your location.
4. Give each item a distinctive place in the location. Do not place it near other items.
5. Unusual, bizarre images are the most memorable.
6. For recall, mentally move through your location, mentally see the images that you have placed there, and decode the images to recover the items.

Suppose that a politician is to make a long speech about the plight of the aged. He decides to use his home, with which he is very familiar, as a location. The first point of the speech is about hunger and the aged. He imagines a skinny old man with an empty plate, begging at his front door step. The second point is about crime and the aged. He imagines a nice-looking old woman on the dining room floor covered with blood and a dagger in her heart. The third point is about health care and the aged. He visualizes an old woman in his bedroom, stretched out on his bed, contorted with pain. He proceeds in this way for all points of the speech, putting each image in a distinctive place. At the time of delivery, the speaker mentally strolls through his location, recovers each image, decodes it, and delivers a well-organized speech. The same location could be used for the next speech because there is no interference from one use of a location to the next, it was contended.

Does the method of loci work? It was said that the Roman philosopher Seneca, a teacher of rhetoric, could repeat 2000 names in the order in which they were given. Seneca would have each student of a class of 200 speak a line of poetry. Starting with the last student, he would systematically repeat all the lines (Yates, 1966, p. 16). In modern times Groninger (1971) experimentally evaluated the method of loci. The subjects of an imagery group had the method of loci explained to them and were told to find a location. A subject was then given a pack of 25 cards, each with a word on it, and told to memorize them by the method of loci. The subjects of a control group were given the cards and told to learn them in any way that they wished. All subjects returned for retention tests after 1 week and after 5 weeks. The imagery group recalled 92 percent after 1 week and the control group recalled 64 percent. After 5 weeks the imagery group recalled 80 percent and the control group recalled 36 percent.

Modern research has challenged the assumption of imagery bizarreness which the method of loci makes. With minor disagreement (McDaniel & Einstein, 1986), the evidence seems to be that ordinary images are as good as bizarre ones (Nappe & Wollen, 1973; Hauck et al., 1976; Senter & Hoffman, 1976; Cox & Wollen, 1981; Wollen & Cox, 1981; Kroll et al., 1986). Furthermore, there is evidence (Wollen et al., 1972) that the interaction of imagery elements is the fundamental variable for recall effectiveness, not bizarreness. Common images that are consistent with everyday experience, where elements of the image meaningfully interact, is the best. If the method of loci is used for remembering a grocery list that has steak on it, an interacting, common image of your mother frying steak in the kitchen will have faster formation time (Nappe & Wollen, 1973) and will induce higher retention than will a noninteracting, bizarre image of a steak hanging on the kitchen wall wearing a moustache.

Theories of Forgetting

An item is learned today and there is the capability to recall it. Next week the item cannot be recalled. What has happened to it? Psychology has no single theory to explain forgetting. Instead, there are three theories of forgetting, each with explanatory merit. The three theories are trace decay, interference, and failure to retrieve. All of the theoretical thinking about forgetting is rooted in verbal behavior.

Trace Decay Theory

The oldest theory of forgetting, going back to the ancient Greeks, is that the trace of the item stored in memory decays spontaneously, like the decay of radioactive materials. The reason that forgetting occurs is that the trace becomes too weak to activate the response. Nothing can be done to alter the course of forgetting because trace decay is presumably an inevitable biological process.

Verification of trace decay theory is difficult because a state of empty time is required for the retention interval so that the trace can decay (assuming that is what it does) without the influences of other variables. A conscious human being has other events impinge on her during the retention interval, and these could affect the trace that is supposedly decaying. A motivated human being might attempt covert rehearsal during the retention interval, which would offset decay. Whether complete control of these contaminating events is achieved in an experiment is always questionable. Good experimental attempts have been made to create empty time (Reitman, 1971, 1974), and some support for trace decay theory has been found.

Interference Theory of Forgetting

Trace decay theory is a passive theory that asserts inevitable biological processes, but the interference theory of forgetting is an active theory which contends that forgetting depends on experiences. Forgetting may or may not occur, depending on the kinds of experiences.

There are two sources of interference. One source, called *retroactive interference,* comprises events that occur in the retention interval and act on the criterion material that was previously learned and stored in memory. The action of the interfering material lowers the performance of the criterion material at recall. The other, called *proactive interference,* stems from material that is learned *before* the criterion material is learned and recalled. Previously learned material stored in memory interacts with the learning of criterion material and interferes with recall of the criterion material.

Many laboratory experiments have been done on interference processes, and there is no doubt that interference affects retention. Whether these interference processes that occur in the laboratory are the same ones that cause forgetting in everyday life is not clear. The most convincing studies of interference as an explanation of everyday forgetting are those which have had sleeping or waking in the retention interval. If interference theory is correct, reducing the opportunities for interference by sleeping during the retention interval should reduce forgetting. Support for this deduction goes back to 1924 (Jenkins & Dallenbach, 1924), and it gave a big boost to interference theory. Jenkins and Dallenbach had their subjects learn verbal lists and then either sleep or be awake for retention intervals up to 8 hours. Recall after sleep was more than twice that after being awake. Modern research confirms this old finding (Ekstrand, 1967, 1972; Barrett & Ekstrand, 1972), so the finding is a secure one. Sleep does not eliminate all of the forgetting, so interference may not be the only factor involved.

Failure to Retrieve

Failure to recall a verbal item does not necessarily mean that the item was irretrievably lost from memory. An item may exist in full strength and yet not occur at the retention test because of failure to find it. Failure to retrieve from memory has been compared to a library and its card catalog. The book is somewhere in the library, but it will never be found if its card cannot be found in the card catalog. The failure is in the retrieval operation.

Efforts to retrieve from memory are familiar to us all. Suppose that we are trying to remember a woman's name. We might say "It's Mary. No—that's not it. Mabel? No. It's Marie! No, that's not it either. Marlene! That's it!" Marlene was in memory all the time. Recall was only a matter of finding it. Retrieval efforts are not always successful, but the effort can pay off. Without the effort, Marlene never would have had her name remembered.

Prompting. We often get help in retrieval. Seeing a face reminds us of a name. The first line of a poem reminds us of the entire poem. The use of a stimulus helper in retrieval is called *prompting,* and the reminder stimulus is called the *prompter.*

The most comprehensive study of prompting has been by Wagenaar (1986). Wagenaar's experiment was one of *autobiographical memory,* where he was the single subject of his own experiment, studying his own memory. For six years he recorded one or two of the most memorable events that happened to him every day, for a total of 2402 events. A data sheet for each event had Who, What, Where, and When, as well as a question and an answer about a critical detail of the event. A data sheet might look like this:

Who: Queen Elizabeth
What: I saw the Queen of Great Britain
Where: Buckingham Palace
When: July 10, 1987

Critical Detail:
Question: Who was with her?
Answer: Prince Charles and Lady Diana

At the recall attempt there were five retrieval cues for an event:

1. Who
2. What
3. Where
4. When
5. The question

Each event at recall was organized into a booklet:

Format of the Recall Booklet	*Recall Requirement*
Page 1: Cue 1	Cues 2–4
Page 2: Cues 1 and 2	Cues 3 and 4
Page 3: Cues 1–3	Cue 4
Page 4: Cues 1–4 and the question	Answer to the question

Page 1 might have QUEEN ELIZABETH, with Wagenaar required to remember "I saw her at Buckingham Palace on July 10, 1987." Page 2 would have QUEEN ELIZABETH and I SAW THE QUEEN OF GREAT BRITAIN, with the recall requirement of "At Buckingham Palace on July 10, 1987." On Page 3 would be QUEEN ELIZABETH, I SAW THE QUEEN OF GREAT BRITAIN, and BUCKINGHAM PALACE, with the recall requirement of "July 10, 1987." On Page 4, would be QUEEN ELIZABETH, I SAW THE QUEEN OF GREAT BRITAIN, BUCKINGHAM PALACE, JULY 10, 1987, and the question WHO WAS WITH HER? The recall requirement was "Prince Charles and Lady Diana." Thus, Wagenaar successively tested himself on from one to five retrieval cues for each event. Testing himself on only a few events a day, recall took a year. Each event was tested only once.

The results were that one cue was the least effective for retrieval, two cues were of intermediate effectiveness, and three cues or three cues plus the question the most effective of all. Forgetting occurred for all retrieval conditions at about the same rate. There was no condition of retrieval that eliminated the forgetting. Even the most favorable retrieval condition had about 50 percent forgetting for the longest retention interval. With so much forgetting it cannot be contended that all memories remain at full strength until the proper retrieval cues come along. Unless more provocative retrieval cues can be devised, it must be assumed that some of the memories are lost permanently, perhaps through trace decay or interference. Retrieval is unquestionably a part of recall, but there is probably more to it.

Recognition Memory _____

Recall is the production of a response, but recognition is deciding that a stimulus has been experienced before. You are listening to songs on the radio and one of them is familiar, and you know that you have heard it before. A physician is reading chest X rays and decides that one has the familiar pattern of tuberculosis.

Methods of Studying Recognition

There are two basic methods for studying recognition behavior. Both expose the subject to stimuli and then later, in a recognition test, require the subject to discriminate the old stimuli from new ones that have never been experienced before. The two methods, called the method of single stimuli and the forced-choice method, differ in the conduct of the recognition test.

Method of Single Stimuli. The *method of single stimuli* has old and new stimuli mixed at the recognition test and presented one at a time. The subject evaluates each stimulus and decides whether it is old or new. Signal detection theory, which was discussed in Chapter 3, is usefully applied to data collected by this method, and the rationale is the same as for detecting auditory stimuli. In the extreme case, a subject in a recognition situation can have a very relaxed criterion and designate all stimuli as old. All old stimuli will be correctly classified (hits) but all new stimuli will be incorrectly classified (false alarms). Or in the other extreme case, a subject can have a very strict criterion and designate all stimuli as new. All new stimuli will be correctly classified (correct rejections) but all old stimuli will be incorrectly classified (misses). A criterion can fall between these extremes, and usually does.

Medical scientists knew before signal detection theory came along that the criterion plays a role in the recognition of disease patterns on X ray films (Garland, 1949). The physician with a relaxed criterion will have a high hit rate but a high false alarm rate also. The benefit is that the diseased patients will be detected, but the cost is that normal patients will be labeled as diseased. Conversely, a strict criterion will correctly reject normal patients but miss some of the diseased patients. Chapter 1 discussed errors in the reading of X rays, and the physician's inappropriate criterion is undoubtedly a contributor to them.

Morgan et al. (1973) used signal detection theory to analyze errors in the recognition of pneumoconiosis in chest X rays. Pneumoconiosis is caused by prolonged breathing of dust. Miners can be afflicted with it, for example. The "signal" to be detected is small inflamed nodules in the lung tissue that tend to be opaque and diffusely distributed throughout the lung field. Diagnosis is based on the number of nodules per unit of area, their shape, and their size. This may seem straightforward, but the signal is embedded in "noise" that can obscure the signal and make correct classification difficult. Blood vessels in certain configurations can appear rounded and opaque, similar to the nodules of pneumoconiosis. Other diseases, including some kinds of tuberculosis, can have small opaque nodules also. The veins of older people can become thicker and develop configurations that can be confused with the nodules of pneumoconiosis.

First, Morgan et al. had five expert radiologists agree on the reading of 200 X rays so that there was no doubt who had pneumoconiosis and who did not. The X rays were then submitted to seven other radiologists for interpretation. Their average hit rate was 83 percent and their average false alarm rate was 10 percent. The ROC curve is shown in Figure 6-3. The data points fall along the curve reasonably well, indicating that the readers had similar sensitivity (the average d' was 2.43), but the distribution of data points along the curve shows that the decision criteria (β) vary substantially among the readers. One reader had a strict criterion, with 57 percent hits and 2 percent false

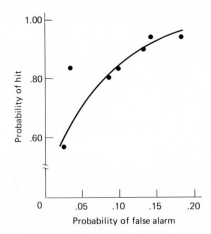

Figure 6-3 Receiver operating characteristic curve for the reading of X-rays. [Data from Morgan, R. H., Donner, M. W., Gayler, B. W, Margulies, S. I., Rao, P. S., and Wheeler, P. S., Decison processes and observer error in the diagnosis of pneumoconiosis by chest roentgenography. *American Journal of Roentgenology* 117, 757–764 (1973), Table V. Copyright 1973 by the American Roentgenology Society. Used by permission of Williams & Wilkins.]

alarms. Another had a lax criterion, with 94 percent hits and 14 percent false alarms. Each, in his own way, was making unacceptable errors.

The Forced-Choice Method. The *forced-choice method* has two or more stimuli presented simultaneously. Only one of the stimuli is old, and the subject's task is to identify the old one from among the alternatives. Students are familiar with the multiple-choice examination, which is a version of the forced-choice method.

The level of recognition that is found depends on the design of the forced-choice test. The greater the number of alternatives, and the greater the similarity of the alternatives, the poorer the recognition (Adams, 1980, p. 323).

The Recognition of Complex Visual Stimuli

Our capabilities and capacity for visual recognition are remarkably good when the stimuli are distinctive and dissimilar. We can recognize hundreds, indeed thousands, of pictures that we have seen once and only briefly. Shepard (1967) presented 612 common pictures to his subjects and then gave them an immediate forced-choice recognition test with two alternatives. Percent correct recognition was 97 percent. Standing (1973) presented 10,000 slides of common pictures over five successive days and then gave a forced-choice recognition test with two alternatives after an average retention interval of 2 days. Recognition was 83 percent correct.

The recognition of pictures is helped by their discriminability and distinctiveness, because recognition is poorer for complex stimuli that are highly similar. Goldstein and Chance (1970) found that correct recognition was 46 percent for ink blots and 33 percent for snow crystals. Recognizing your car in a large parking lot is not always easy.

Variables for Recognition

Familiarity. The basic variable for recognition is sensory experience, or exposure time; it is called *familiarity*. Potter and Levy (1969) presented pictures with exposure times that ranged from 125 milliseconds to 2 seconds. Correct recognition in an immediate test was virtually 100 percent after only 2 seconds of exposure. This finding makes it easier to understand the high level of recognition that was found in an experiment such as that of Standing's (1973) described above. Standing presented each picture for 5 seconds, which is an appreciable degree of familiarity. It is not surprising that correct recognition was so high for so many pictures.

Retention Interval. The high levels of recognition that can be found should not imply no forgetting. In the Shepard (1967) experiment, above, correct recognition of pictures was 97 percent on an immediate test but only 58 percent after 4 months.

The forgetting of recognition is most evident when exposure times are brief. When stimuli are presented for longer periods of time, perhaps repeatedly, recognition is high for long periods of time. Bahrick et al. (1975) used pictures and names from high school class yearbooks and tested for recognition over retention intervals ranging 3 months to 48 years. The recognition of names remained in the vicinity of 90 percent for 14 years and dropped to only 77 percent after 48 years. Recognition of faces remained in the vicinity of 90 percent for 34 years and dropped to 73 percent after 48 years. Familiarity must be the responsible variable. Forgetting occurs readily with exposure times measured in seconds, but high school classmates are seen on a daily basis for three years. Familiarity is very high, and recognition after long retention intervals correspondingly high.

Change in the Stimulus. The criminal who wears a disguise for a crime and then is not recognized at the lineup is capitalizing on the principle that change in the stimulus from exposure to test makes a big difference for recognition. Patterson and Baddeley (1977) showed their subjects pictures of actors and then had either identical or changed appearance at the recognition test. The change was in hair style, addition or removal of a beard, or addition or removal of glasses. Correct recognition was 98 percent when appearance was identical but 45 percent when the appearance was changed.

The Retention of Motor Behavior

There have been psychology textbooks which assure their readers that motor skills are hardly forgotten at all. There is some truth in this. We never seem to forget how to ride a bicycle or to ice skate. From the turn of the century on, the research literature has reported studies where the retention of motor skills is high (Adams, 1987). An example of the high retention of motor skills is an experiment by Fleishman and Parker (1962).

Fleishman and Parker used a tracking task where three dimensions of a motor task had to be controlled at the same time. A programmed dot moved in two dimensions on a cathode ray tube display, and the subject manipulated a control stick in attempts to keep the dot centered in the display as much as possible. Below the tube was a meter

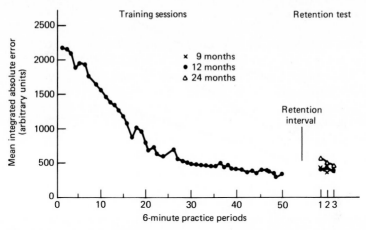

Figure 6-4 The long-term retention of a complex motor skill. [Adapted from Fleishman, E. A., and Parker, J. F., Jr., *Journal of Experimental Psychology* 64, 215–226 (1962), Figure 1. Copyright 1962 by the American Psychological Association. Used by permission of the authors.]

with a needle whose action was programmed, and the subject manipulated a rudder bar with the feet in efforts to keep the needle centered. The score was absolute integrated error, or summation of the time that display elements were off center, disregarding the direction of the error. The more the display elements were centered, the lower the score. Extensive original training was given (17 daily sessions), followed by retention intervals of 9, 14, and 24 months. The results are shown in Figure 6-4. Retention was almost perfect for all retention intervals.

One possible explanation for data like these is that motor responses are intrinsically more resistant than verbal responses to forgetting, maybe because of the large involvement that motor responses have with proprioception. Another explanation is overlearning. We have known since the time of Ebbinghaus (1885/1964) that the greater the degree of learning, the greater the retention, and certainly a great deal of original learning was given in the Fleishman and Parker experiment. The high retention of bicycling and ice skating could easily be explained by overlearning. If the overlearning hypothesis is plausible, a motor response that is given a small amount of practice should show substantial forgetting, perhaps over brief intervals, just like a verbal response. This was the experiment of Adams and Dijkstra (1966).

Adams and Dijkstra studied the recall of a simple linear movement. The subject reached through an opaque curtain and moved a slide along a track until it hit a stop, which defined a length of movement to be remembered. A subject made this move 1, 6, or 15 times, which was the rehearsal variable. Retention intervals ranged from 5 to 120

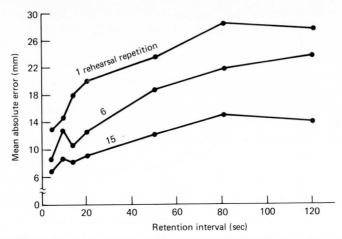

Figure 6-5 The short-term retention of a simple motor response as a function of number of rehearsal repetitions. [Adapted from Adams, J. A., and Dijkstra, S., *Journal of Experimental Psychology* 71, 314–318 (1966), Fig. 1. Copyright 1966 by the American Psychological Association. Used by permission of the authors.]

seconds. Recall was with the stop removed, and the subject attempted to reproduce the length of the movement. The measure of performance was absolute error. The results are shown in Figure 6-5. The short-term forgetting of motor responses was found to be rapid, just like verbal responses. And the more the practice, the less the forgetting, just as with the short-term retention of verbal responses (Figure 6-2).

There is no reason to believe that the forgetting of motor responses is intrinsically different from verbal responses. The many hundreds of hours of practice that we may give a continuous motor skill such as bicycling imparts an enormous resistance to forgetting, so it is not surprising that forgetting is slight, even over long retention intervals. But when amount of practice is small, motor responses are readily forgotten. Small amounts of practice is probably the reason why discrete motor activity such as throwing switches is readily forgotten (Adams, 1967, Chap. 8; Adams, 1977).

Summary

From prescientific interest in memory that reaches back centuries, we have come to the scientific study of memory in the past 100 years. Research on memory, intensifying in the past 25 years or so, has led to new conceptual developments. One development was the information-processing model, based on a digital computer analogy. The similarity between the systems and operations of a digital computer and the functions of human memory can be compelling. A rival model, called levels of processing theory, appeared in the 1970s. The premise of levels of processing theory was that an item can be given shallow processing where it is scarcely thought about, or it can be given deep processing and integrated into knowledge. The deeper the processing, the less the forgetting. Today, memory models are in transition, with no particular model dominating.

There is choice in the method of verbal memorization, and success in remembering depends on the method that is chosen and how well it is used. These methods are called strategies of memorization, and the basic three are rehearsal, verbal elaboration, and nonverbal elaboration. One of the earliest findings of research on human memory is that rehearsal repetitions of an item improve its retention. Verbal elaboration is making the unfamiliar to-be-remembered verbal material familiar by integrating it with established material that is stored in memory. Verbal elaboration improves retention. Nonverbal elaboration is where the to-be-remembered material is integrated with a mental image. Nonverbal elaboration improves retention also.

Trace decay, interference, and failure to retrieve are the three main theories of forgetting; there is no one theory of memory that can explain why material that has been learned cannot be remembered. Trace decay theory says that the trace of the material that is stored in memory decays spontaneously with time. Interference theory asserts the action of interfering experiences that induce the retention loss. Failure to retrieve is a theory which contends that the material is available in memory but that our mental search and retrieval efforts fail to make contact with it and energize it as a response.

Recall is the production of a response, but recognition is deciding that a stimulus has been experienced before. Recognition behavior is studied by exposing subjects to stimuli and then later, in a recognition test, requiring them to discriminate old stimuli from new ones that have never been experienced before. Human powers of recognition can be very good. High levels of retention over long intervals have been found. Prominent variables for recognition are amount of exposure to stimuli or familiarity, similarity of stimuli, length of retention interval, and change in the stimulus from exposure to test.

Almost all of the research on human memory has been done on verbal behavior. Research on motor behavior has been low key over the history of research on memory, but it has increased in recent times. For a long time it was thought that motor skills such as bicycling were exceptionally resistant to forgetting processes. Now it appears that the large amounts of practice commonly given motor skills are responsible for the high retention. A motor skill with a small amount of practice can be forgotten rapidly, just as a verbal response can be.

References

Adams, J. A. (1967). *Human memory.* New York: McGraw-Hill.

Adams, J. A. (1977). Motor learning and retention. In M. H. Marx & M. E. Bunch (Eds.), *Fundamentals and applications of learning.* New York: Macmillan.

Adams, J. A. (1980). *Learning and memory: An introduction.* Revised Edition. Homewood, IL: Dorsey Press.

Adams, J. A. (1987). Historical review and appraisal of research on the learning, retention, and transfer of human motor skills. *Psychological Bulletin, 101,* 41–74.

Adams, J. A., & Dijkstra, S. (1966). Short-term memory for motor responses. *Journal of Experimental Psychology, 71,* 314–318.

Atkinson, R. C., & Shiffrin, R. M. (1968). Human memory: A proposed system and its control processes. In K. W. Spence and J. T. Spence (Eds.), *The psychology of learning and motivation.* Volume 2. New York: Academic Press. Pp. 89–195.

Averbach, E., & Coriell, A. S. (1961). Short-term memory in vision. *Bell System Technical Journal, 40,* 309–328.

Baddeley, A. D. (1976). *The psychology of memory.* New York: Harper & Row.

Baddeley, A. D. (1978). The trouble with levels: A reexamination of Craik and Lockhart's framework for memory research. *Psychological Review, 85,* 129–152.

Baddeley, A. D., & Hitch, G. (1974). Working memory. In G. H. Bower (Ed.), *The psychology of learning and motivation.* Volume 8. New York: Academic Press. Pp. 47–89.

Bahrick, H. P., Bahrick, P. O., & Wittlinger, R. P. (1975). Fifty years of memory for names and faces: A cross-sectional approach. *Journal of Experimental Psychology: General, 104,* 54–75.

Barrett, T. R., & Ekstrand, B. R. (1972). Effect of sleep on memory: III. Controlling for time-of-day effects. *Journal of Experimental Psychology, 96,* 321–327.

Brown, J. (1958). Some tests of the decay theory of immediate memory. *British Journal of Psychology, 10,* 12–21.

Chase, W. G., & Ericsson, K. A. (1982). Skill and working memory. In G. H. Bower (Ed.), *The psychology of learning and motivation.* Volume 16. New York: Academic Press, Pp. 1–58.

Cicero. (1959). *De Oratore* (Books I and II). Revised Edition (E. W. Sutton and H. Rackham translation). Cambridge MA: Harvard University Press.

Cox, S. D., & Wollen, K. A. (1981). Bizarreness and recall. *Bulletin of the Psychonomic Society, 18,* 244–245.

Craik, F. I. M., & Lockhart, R. S. (1972). Levels of processing: A framework for memory research. *Journal of Verbal Learning and Verbal Behavior, 11,* 671–684.

Craik, F. I. M., & Tulving, E. (1975). Depth of processing and the retention of words in episodic memory. *Journal of Experimental Psychology: General, 104,* 268–294.

Ebbinghaus, H. (1885/1964). *A contribution to experimental psychology* (H. A. Ruger and C. E. Bussenius translation). New York: Dover.

Ekstrand, B. R. (1967). Effect of sleep on memory. *Journal of Experimental Psychology, 75,* 64–72.

Ekstrand, B. R. (1972). To sleep, perchance to dream (about why we forget). In C. P. Duncan, L. Sechrest, & A. W. Melton (Eds.), *Human memory: Festschrift in honor of Benton J. Underwood.* New York: Appleton-Century-Crofts. Pp. 59–82.

Fleishman, E. A., & Parker, J. F., Jr. (1962). Factors in the retention and relearning of perceptual-motor skill. *Journal of Experimental Psychology, 64,* 215–226.

Garland, L. H. (1949). On the scientific evaluation of diagnostic procedures. *Radiology, 52,* 309–327.

Glenberg, A., Smith, S. M., & Green, C. (1977). Type I rehearsal: Maintenance and more. *Journal of Verbal Learning and Verbal Behavior, 16,* 339–352.

Goldstein, A. G., & Chance, J. E. (1970). Visual recognition memory for complex configurations. *Perception & Psychophysics, 9,* 237–241.

Groninger, L. D. (1971). Mnemonic imagery and forgetting. *Psychonomic Science, 23,* 161–163.

Hauck, P. D., Walsh, C. C., & Kroll, N. E. A. (1976). Visual imagery mnemonics: Common vs. bizarre mental images. *Bulletin of the Psychonomic Society, 7,* 160–162.

Hellyer, S. (1962). Supplementary report: Frequency of stimulus presentation and short-term decrement in recall. *Journal of Experimental Psychology, 64,* 650.

Jenkins, J. G., & Dallenbach, K. M. (1924). Obliviscence during sleep and waking. *American Journal of Psychology, 35,* 605–612.

Kroll, N. E. A., Schepeler, E. M., & Angin, K. T. (1986). Bizarre imagery: The misremembered mnemonic. *Journal of Experimental Psychology: Learning, Memory, and Cognition, 12,* 42–53.

McDaniel, M. A., & Einstein, G. O. (1986). Bizarre imagery as an effective memory aid: The importance of distinctiveness. *Journal of Experimental Psychology: Learning, Memory, and Cognition, 12,* 54–65.

Miller, G. A. (1956). The magical number seven, plus or minus two. *Psychological Review, 63,* 81–97.

Montague, W. E., Adams, J. A., & Kiess, H. O. (1966). Forgetting and natural language mediation. *Journal of Experimental Psychology, 77,* 7–13.

Morgan, R. H., Donner, M. W., Gayler, B. W., Margulies, S. I., Rao, P. S., & Wheeler, P. S. (1973). Decision processes and observer error in the diagnosis of pneumoconiosis by chest roentgenography. *American Journal of Roentgenology, 117,* 757–764.

Nairne, J. S. (1983). Associative processing during rote rehearsal. *Journal of Experimental Psychology: Learning, Memory, and Cognition, 9,* 3–20.

Nappe, G. W., & Wollen, K. A. (1973). Effects of instructions to form common and bizarre mental images on retention. *Journal of Experimental Psychology, 100,* 6–8.

Nelson, T. O. (1977). Repetition and depth of processing. *Journal of Verbal Learning and Verbal Behavior, 16,* 151–171.

Patterson, K. E., & Baddeley, A. D. (1977). When face recognition fails. *Journal of Experimental Psychology: Human Learning and Memory, 3,* 406–417.

Peterson, L. R., & Peterson, M. J. (1959). Short-term retention of individual verbal items. *Journal of Experimental Psychology, 58,* 193–198.

Potter, M. C., & Levy, E. I. (1969). Recognition memory for a rapid sequence of pictures. *Journal of Experimental Psychology, 81,* 10–15.

Reitman, J. S. (1971). Mechanisms of forgetting in short-term memory. *Cognitive Psychology, 2,* 185–195.

Reitman, J. S. (1974). Without surreptitious rehearsal, information in short-term memory decays. *Journal of Verbal Learning and Verbal Behavior, 13,* 365–377.

Senter, R. J., & Hoffman, R. R. (1976). Bizarreness as a nonessential variable in mnemonic imagery: A confirmation. *Bulletin of the Psychonomic Society, 7,* 163–164.

Shepard, R. N. (1967). Recognition memory for words, sentences, and pictures. *Journal of Verbal Learning and Verbal Behavior, 6,* 156–163.

Sperling, G. (1960). The information available in brief visual presentations. *Psychological Monographs, 74,* (Whole No. 498).

Standing, L. (1973). Learning 10,000 pictures. *Quarterly Journal of Experimental Psychology, 25,* 207–222.

Wagenaar, W. A. (1986). My memory: A study of biographical memory over six years. *Cognitive Psychology, 18,* 225–252.

Watkins, M. J., & Todres, A. K. (1980). Suffix effects manifest and concealed: Further evidence for a 20-second echo. *Journal of Verbal Learning and Verbal Behavior, 19,* 46–53.

Watkins, O. C., & Watkins, M. J. (1980). The modality effect and echoic persistence. *Journal of Experimental Psychology: General, 109,* 251–278.

Wollen, K. A., & Cox, S. D. (1981). Sentence cuing and the effectiveness of bizarre imagery. *Journal of Experimental Psychology: Human Learning and Memory, 7,* 386–392.

Wollen, K. A., Weber, A., & Lowry, D. H . (1972). Bizarreness versus interaction of mental images as determinants of learning. *Cognitive Psychology, 3,* 518–523.

Yates, F. A. (1966). *The art of memory.* Chicago: University of Chicago Press.

Job Layout

PART **III**

Job Layout

CHAPTER 7

Philosophies of Job Design

Human factors engineering has always had an involvement in workplace design, perhaps its largest. The design of displays, controls, and layout of the workplace have always been central efforts of human factors engineering. Independent of human factors engineering and job design, industrial psychologists have been concerned with worker satisfaction and productivity. The activities of these different scientific disciplines blur and lose their identities when it is asked how job design relates to job satisfaction and thus productivity. Productivity may be as much related to satisfaction with job design as with pay, working hours, and fringe benefits. A human factors engineer has the optimization of human behavior in systems as emphasis, and she cannot be insensitive to the interrelation of job design and job satisfaction as a possible determinant of productivity even though the emphasis may be nontraditional for the field. The past 30 years or so have seen a big interest in job design and what it means for productivity.

Historical Background _____

Taylor and Gilbreth were pioneers in work analysis and job design, as we saw in Chapter 1. Even though Taylor and Gilbreth had the individual worker and his job as emphasis, their thinking laid the foundation of the assembly line, where many workers are organized to build a product. The "Taylor system," or "scientific management" as it was called, sought to specify the elements of an individual worker's job so that productivity is optimized. The same ideas and techniques can be applied to many workers on an assembly line. As the task elements of a job for a single worker can be defined and organized for optimal productivity, so the total manufacturing process for a plant can be broken down into specialized jobs, each accomplished by a worker, for optimal efficiency. In his textbook on motion and time study, Barnes (1963, p. 319) gives the advantages of a high degree of job specialization and division of labor:

1. Job specialization enables the worker to learn the job in a short period of time.
2. A short work cycle permits rapid and almost automatic performance with little or no thought required.
3. Less capable people can be employed to perform highly repetitive short-cycle operations, with a lower hourly wage being paid.
4. Less supervision is required because of the routinized character of the work.

Barnes (1963, Chap. 6) presented the steps to be followed in laying out the jobs for a manufacturing plant. The steps correspond to those that the Taylor system articulated for the individual worker, and they indicate how Taylor's principles can be overarching and include plant layout as well:

1. Eliminate all unnecessary work.
2. Combine operations or elements.
3. Change the sequence of operations.
4. Simplify the necessary operations.

The extension of Taylor's principles for the individual worker to large-scale industrial production has been successful. Today these methods for industrial production are called the *engineering approach,* not because they were founded by an engineer but because the underlying philosophy is the same one that governs an engineer's design of a machine. Just as an engineer seeks the essential components for the simplest machine that will minimize error and optimize performance, so the engineering approach will look for the essential elements that can be structured as a job with minimum error and optimal output. The engineering approach recognizes the importance of certain psychological variables, such as the relationship between pay and productivity, but status of the worker is not one of them. The diminished status of the worker has given rise to a new point of view, called the *job enrichment approach,* which is human centered.

Job Enrichment

In a book intended to popularize Taylor's principles of scientific management, Gilbreth (1914) denied that the rigid structure of a job would impair the worker and his efficiency. "Does it make machines out of men?", Gilbreth asked rhetorically (pp. 49–50). No, he replied—no more than a highly skilled golfer is a machine. Would not monotony bring a worker to the brink of madness? No, Gilbreth answered again (pp. 53–54). Some workers are ideally suited for monotonous work. Scientific management will identify workers unsuited for monotonous jobs and find more suitable jobs for them. Wouldn't it be better to train an all-around worker rather than a narrow specialist? No. This is the age of specialization, Gilbreth said (pp. 51–52). Just as the physician and the dentist specialize, so should the industrial worker.

A Pioneering Experiment

As assembly-line factories proliferated in the first half of this century, most notably in the automobile industry, Gilbreth's defense of routinized work was accepted. Doubt surfaced in the 1950s, however.

Walker (1950) wrote a pivotal paper. He said that there had been growing concern about the monotony of repetitive tasks in industry, and remedies such as more frequent rest pauses had had some success in increasing output and worker satisfaction. The problem, however, is more fundamental and lies in job design and the inherent dullness of simple, repetitive work. An obstacle to research on job design is what he called "the sacrosanct character of certain engineering assumptions" (p. 54), namely that the more subdivided the job, the lower plant costs and the greater the output. Walker said that there is a limit to the subdivision of operations because bored workers are not productive. To make his point, Walker reported an experiment that was conducted by the International Business Machines Corporation (IBM). The experiment was started during World War II and continued several years thereafter.

The purpose of the experiment was to evaluate the effects of enriching the jobs of machine operators with variety and interest. Before the experiment the worker placed a part in the machine, started the machine, and the drill or cutting tool went to work. The machine was then stopped and the worker removed the part. The next part was placed in the machine, and the sequence was repeated. The machines were set up by setup men. Other workers sharpened the tools. When the part was completed there were inspectors who checked it for quality. In the experiment, when the job was enlarged, a worker set up the machine for each new batch, sharpened his own tools, made the part as before, and then inspected it for quality.

The outcomes of the experiment were considered impressive. All of the setup men were eliminated and the number of inspectors was reduced, decreasing labor costs. Losses from defects and scrap were reduced, indicating a better quality product. There was less idle time because the workers did not have to wait for setup men and inspectors. The elimination of setup men and inspectors removed an echelon of personnel between the workers and the foremen, raising the status of workers and placing them in a better working relationship with their supervisors. In addition, workers began to receive more money for their work. All these benefits had come from job enlargement.

Definition of Job Enrichment

The IBM experiment is background for an explicit definition of the job enrichment approach. The main assumption of the job enrichment approach is that industrial work is based on a technical system and a social-psychological system, which are two interdependent systems (Trist et al., 1977). The technical system is machines and tools. The design and use of the technical system is dedicated to minimizing costs and maximizing profits. The social-psychological system is not the strength and motor skills of workers but in their *attitudes, beliefs, and feelings*. In the job enrichment approach the designer organizes the technical system so that the social-psychological system is optimized. With the machines and tools organized so that workers are satisfied, job satisfaction will increase and productivity will increase along with it, it is hypothesized.

What is this job satisfaction from which all good things flow? *Job satisfaction is a pleasurable, positive emotional state resulting from the appraisal of one's own job or job experiences* (Locke, 1976, p. 1300). And how do the job enrichment advocates propose to achieve it? Job satisfaction is a complex state of the worker, and a function of many variables, but it is the special hypothesis of job enrichment that restructuring the job in challenging ways can produce a sizable increase in job satisfaction (Conant & Kilbridge, 1965; Hulin, 1971). The IBM experiment telegraphed *the essentials of the job enrichment approach:*

1. The number of tasks in the job should be increased.
2. The variety of tasks in the job should be increased.

3. The worker inspects the quality of her own work and has responsibility for it.
4. The worker chooses her own work methods.
5. The worker chooses her own pace.
6. The worker has responsibility for building the entire product, or a major part of it.

Some versions of job enrichment have workers choosing their own supervisors and making their own work rules.

There is no question that the job enrichment approach is different from the engineering approach. The simplified, low-level jobs fostered by the engineering approach are monotonous and have led to bored, dissatisfied workers who have low productivity, it is contended. The solution is to challenge the worker with job variety and complexity and give him freedom to do the job the way that he wants. The challenge and freedom bring job satisfaction, and the satisfaction brings productivity.

Experimental Studies of Job Enrichment

The attraction of the job enrichment approach is easy to understand. The hypothesis has an intuitive appeal to those who pride themselves in their individualism and unwillingness to be regimented. For scientists concerned with industrial processes, the job enrichment approach is seen as a challenge to the engineering approach, which has dominated the industrial world for decades, so the consequences of the job enrichment hypothesis are far-reaching if the hypothesis is true. The job enrichment hypothesis became increasingly plausible after the IBM experiment, and other experiments followed in the 1960s and 1970s, which was good because enthusiasm can outrun evidence.

Conant and Kilbridge (1965) investigated job enrichment for some operations in the manufacture of home laundry equipment. The water pump for an automatic washer was spotlighted in their analysis. The pump had 27 parts. Pump assembly on the paced line was done by six workers, who required an average of 1.77 minutes for it. Each worker performed an average of six operations. The job enrichment approach had one worker assemble the entire pump, and it took 1.49 minutes to do it. The worker set his own pace in the enlarged job and did his own inspection and quality control. The speed of the assembly line is an item that management negotiates with the labor union and which can be varied, but nevertheless, under the conditions of work that prevailed in the plant, there was clear advantage for the enriched job.

Impressed with this evidence, the company went ahead and enlarged other jobs.

What was the basis of advantage for the enriched job? The workers preferred the enlarged jobs two to one, so increased job satisfaction could be the cause of the increased productivity, as the job enrichment hypothesis says. But increased satisfaction could have been a function of the method of incentive pay that had to be changed for the enlarged job. In contrast to the assembly line, where incentive pay was given for group performance, individual incentive pay was given for the enriched job, so the method of pay and job design were confounded; it is not clear which was the basis of job satisfaction. The main source of advantage for the enlarged job probably came from a reduction in nonproductive work time and balance-delay time, not job satisfaction. Work time on an assembly line has three components: (1) *productive work time,* which is the time spent in performing the operations of the job; (2) *nonproductive work time,* which is the time spent in handling the product and tools and in movement about the workstation; and (3) *balance-delay time,* which is idle time due to the irregular flow of the line. A worker early in the line who does not keep the pace causes those who follow him to wait. The individual in an enlarged job has no balance-delay time because there is no waiting on anyone else. Also, nonproductive work time is reduced.

An experiment by Locke et al. (1976) found that job enrichment affected productivity but not job satisfaction. The subjects of the experiment were clerical workers of a large government agency, and mailroom jobs were manipulated. A control group continued as before but an experimental group had their jobs enriched with greater responsibility, variety, autonomy, and informative feedback about their performances. Results were that the experimental group had higher productivity, less absenteeism, less personnel turnover, and fewer complaints and disciplinary action than did the control group. The level of job satisfaction, however, was about the same for the two groups and remained unchanged over the course of the eight-month experiment. What caused the increase in productivity for the experimental group if an increase in job satisfaction did not? In addition to more job complexity, the job enrichment procedures brought about other changes that had a positive effect on productivity: (1) more efficient use of manpower because the workers could work where they were needed rather than where they were assigned, (2) elimination of unnecessary work procedures, (3) more precise and frequent informative feedback about performance, and (4) competition among some of the subjects of the experimental group (some subgroups of the experimental group acknowledged competing against one another). Job enrichment had beneficial effects that did not have anything to do with job satisfaction.

Criticisms of the Job Enrichment Hypothesis

These studies on job enrichment indicate problems with conceptualization of the job enrichment approach. There are two other problems with the concept:

1. *Job satisfaction and worker productivity are unrelated.* The job enrichment hypothesis says that satisfied workers produce more than unsatisfied ones, but there is no evidence of a positive correlation between job satisfaction and worker performance. Iaffaldano and Muchinsky (1985) analyzed the outcomes of 74 studies of job satisfaction and worker productivity, for many kinds of workers and jobs, and found the correlation to be virtually zero. The positive relationship between job satisfaction and worker performance, which is so obvious to some, is illusory. A happy worker is not necessarily a productive worker, and vice versa.

2. *Industrial workers are basically satisfied.* A basic premise of the job enrichment hypothesis is that industrial workers are dissatisfied. The hypothesis goes on to assert that the dissatisfaction is a cause of low productivity.

The job enrichment hypothesis is weakened if workers are found to be generally satisfied. The Manpower Administration of the U.S. Department of Labor (Quinn et al., 1974) analyzed survey data on job satisfaction from 1958 to 1973. The surveys all asked the same basic question: "All in all, how satisfied are you with your job?" The results are shown in Figure 7–1. The level of satisfaction is high and has remained so for years. The percentage of satisfied workers always tops 80 percent, and in some years tops 90 percent. There is no general evidence that workers in the United States are dissatisfied.

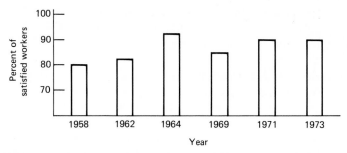

Figure 7-1 Percent of satisfied workers, 1958–1973, based on national surveys. Both men and women were surveyed, except in 1958 and 1964, when only men were surveyed. [Adapted from Quinn et al. (1974).]

How is Figure 7–1 explained? The data seem counterintuitive. Some workers in the surveys undoubtedly had relatively complex jobs and were challenged by them and found them satisfying, but the surveys also included industrial workers in repetitive jobs on assembly lines. Is it possible that workers can be satisfied with dull, monotonous jobs? The answer appears to be "yes." Hulin (1971, p. 165) says that we have misperceived the industrial worker because of *ethnomorphising*, which is the tendency to attribute the values, desires, and aspirations of one's own group to the entire population. Well-educated middle-class engineers, executives, and behavioral scientists who formulate theories of worker behavior such as the job enrichment hypothesis, are driven by a work ethic which says: "Accept challenging jobs, work hard, and you will be rewarded." The work ethic is the route to satisfaction, and those who hold it cannot conceive of anyone being satisfied with a lifetime of monotonous, repetitive work. Yet there are such workers, apparently millions of them in the United States. There are cultural differences in the way that job complexity relates to satisfaction, and a classic study of it was done by Turner and Lawrence (1965).

Turner and Lawrence expected to find job satisfaction related to job complexity. The subjects were 470 workers in 47 jobs in 11 companies. Their important finding was that workers from factories in small towns tended to find satisfaction in complex jobs, which supported the job enrichment hypothesis, but that workers from factories in industrialized cities found satisfaction in repetitive, monotonous jobs, contrary to the hypothesis. City workers were alienated from the work ethic of middle-class America, but small-town workers were not. As Hulin and Blood (1968) say in their commentary on the Turner and Lawrence investigation, there is no reason to expect that workers with fathers and grandfathers who were probably semiskilled laborers like themselves should share the work ethic of middle-class America; they have never benefited from it. Small-town workers come from a segment of the culture where middle-class values remain strong, and the workers share them. The Turner and Lawrence study bears on the data in Figure 7–1. Worker satisfaction in the United States is high, but the reasons for it are not the same for all workers.

Conclusion on Job Enrichment. The positive outcomes of job enrichment do not appear to come from increased job satisfaction. The IBM experiment that was reported by Walker (1950) could have had its results attributable to the higher pay that workers received when their jobs were enlarged, or to improved relationships with their supervisors. Conant and Kilbridge (1965) reduced or eliminated nonproductive work time and balance-delay time, which produced gains in efficiency. Locke et al. (1976) saw this clearly when they said that the gains in productivity that they found were probably attributable to more efficient use of personnel, elimination of unnecessary work procedures, better informative feedback, and worker competition. As Locke et al. concluded, it is the nonenrichment consequences of traditional variables that makes job enrichment seem to work.

The Future

The engineering and job enrichment approaches are based on linear production systems. A *linear production system* is a relatively rigid system, where one product moves through successive stages and emerges finished at the end. The job enrichment approach has problems, as we have seen, and it is not a threat to the engineering approach, but flexible manufacturing systems that modern technologies have made possible will be a threat. Flexible manufacturing has implications for what personnel will do and the skills that they will be required to have.

A *flexible manufacturing system* is an integrated assembly of machines, with the means of automatically transferring items throughout the system, and all under computer control (Kochan, 1986). A flexible manufacturing system in its most sophisticated form is capable of untended production. Being computer-controlled, a flexible manufacturing system is a nonlinear production system that is capable of producing a variety of items. No longer need a factory build one product by moving it through fixed stages. A flexible manufacturing system may, of course, be operated in a linear fashion, but to do so would fail to capitalize on an inherent versatility to produce different items. The power to produce different items allows quick reprogramming of design changes, perhaps in fast reaction to competitive pressures.

There are not many flexible manufacturing systems as yet. In 1980 the number of computer-aided manufacturing systems was estimated to be 25 in the United States, 40 in Japan, 25 in western Europe, and 25 in eastern Europe (Kochan, 1986, p. 34). Notwithstanding, it is the factory of the future, and for good reason. In 1981 the Niigata Engineering Company of Japan inaugurated a flexible manufacturing system for the machining of 30 different types of cylinder heads in batches of 6 to 30. The system runs 21 hours a day, including untended night operation. The number of machines required to produce the parts was reduced from 31 to 6, the number of operators was reduced from 31 to 4, the time during which parts are being produced was increased from 9 hours a day to 21, and the lead time for these parts was reduced from 16 days to 4 (Kochan, 1986, p. 39).

Who are the personnel of a flexible manufacturing system? Jaikumar (1986), in his study of flexible manufacturing systems in the machine tool industry in the United States and Japan, found that flexible manufacturing changes the composition of the work force as well as its size. Engineers now outnumber production workers three to one. Flexible manufacturing systems were largely run by a few well-trained, technically educated individuals. In the Japanese companies with flexible manufacturing systems that Jaikumar studied, 40 percent of the work force was made up of college-educated engineers and all had received training in computer-controlled machines. (The U.S. companies with flexible manufacturing systems that Jaikumar studied had 8 percent of the work force as engineers. Less than 25 percent of them had received training in computer-controlled machines.) Furthermore, the best engineers for flexible manufacturing systems had knowledge in several engineering disciplines and computer science.

The new responsibility for managers of flexible manufacturing systems is not to define tasks for workers, as would be done under the engineering and job enrichment approaches, but to orchestrate the activities of a small cadre of engineers. The manual skills of workers are the main assets of linear production systems, but the assets of flexible manufacturing systems are the intellectual skills of engineers.

Summary

The design of displays, controls, and layout of the workplace have always been central efforts of human factors engineering. Independently of human factors engineering and job design, industrial psychologists have been concerned with worker satisfaction and what it means for productivity. The activities of these different scientific disciplines blur and lose their identities when it is asked how job design relates to job satisfaction and thus productivity. Productivity may be as much related to satisfaction with the job design as pay, working hours, and fringe benefits. The engineering approach and the job enrichment approach are two basic approaches to job design.

The engineering approach looks for the essential work elements of a job that can produce minimum error and optimal output. In recent decades some industrial psychologists have contended that there is more to job design. The argument says that workplaces, such as those of an assembly line defined by the engineering approach, induce low worker satisfaction and low output. A job should be enriched and challenge a worker with complexity, variety, and responsibility, and the higher job satisfaction will produce greater output.

As attractive as the job enrichment approach has been to some, there are convincing criticisms of it. Research on job enrichment has found that benefits to productivity from enlarged jobs operate through effects on traditional variables rather than through job satisfaction. For example, enriched jobs on an assembly line reduce balance-delay time, which is idle time due to uneven flow of the assembly line. Balance-delay time is a traditional variable in industrial engineering. Two other criticisms also undermine premises of the job enrichment approach: (1) workers in the United States are generally satisfied, and (2) there is no relationship between job satisfaction and worker productivity.

The engineering approach is not threatened by the job enrichment approach, but technological advances may cause it eventually to give way to flexible manufacturing systems. A flexible manufacturing system is a computer-controlled grouping of workstations linked by automated material-handling machines. Wholly automated, untended production is possible. Linear production, where one product is evolved through a succession of workstations, is typical of the engineering and job enrichment approaches, but flexible manufacturing is nonlinear because it can be programmed for a variety of products. The key personnel of flexible manufacturing systems are a few broadly trained engineers. The manual skills of workers are the assets of linear production systems, but the assets of flexible manufacturing systems are the intellectual skills of engineers.

References

Barnes, R. M. (1963). *Motion and time study.* Fifth Edition. New York: Wiley.

Conant, E. H., & Kilbridge, M. D. (1965). An interdisciplinary analysis of job enlargement: Technology, costs, and behavioral implications. *Industrial and Labor Relations Review, 18,* 377–395.

Gilbreth, F. B. (1914). *Primer of scientific management.* New York: Van Nostrand.

Hulin, C. L. (1971). Individual differences and job enrichment—the case against general treatments. In J. R. Maher (Ed.), *New perspectives in job enrichment.* New York: Van Nostrand Reinhold. Pp. 159–191.

Hulin, C. L., & Blood, M. R. (1968). Job enlargement, individual differences, and worker responses. *Psychological Bulletin, 69,* 41–55.

Iaffaldano, M. T., & Muchinsky, P. M. (1985). Job satisfaction and job performance: A meta-analysis. *Psychological Bulletin, 97,* 251–273.

Jaikumar, R. (1986). Postindustrial manufacturing. *Harvard Business Review,* November–December, 69–76.

Kochan, D. (Ed.). (1986). *CAM: Developments in computer-integrated manufacturing.* New York: Springer-Verlag.

Locke, E. A. (1976). The nature and causes of job satisfaction. In M. D. Dunnette (Ed.), *Handbook of industrial and organizational psychology*. Skokie, IL: Rand McNally. Pp. 1297–1349.

Locke, E. A., Sirota, D., & Wolfson, A. D. (1976). An experimental case study of the successes and failures of job enrichment in a government agency. *Journal of Applied Psychology, 61,* 701–711.

Quinn, R. P., Staines, G. L., & McCullough, M. R. (1974). Job satisfaction: Is there a trend? Washington, DC: Manpower Administration, U.S. Department of Labor, Manpower Research Monograph 30.

Trist, E. L., Susman, G. I., & Brown, G. R. (1977). An experiment in autonomous working in an American underground coal mine. *Human Relations, 30,* 201–236.

Turner, A. N., & Lawrence, P. R. (1965). *Industrial jobs and the worker: An investigation of response to task attributes*. Boston: Division of Research, Graduate School of Business Administration, Harvard University.

Walker, C. R. (1950). The problem of the repetitive job. *Harvard Business Review, 28,* 54–58.

CHAPTER **8**

Task Analysis

Good job design is a route to low human error in systems. There are two kinds of job design. One kind is to study the performance of a worker in an existing job with the intent of improving it. The human factors engineer has reason to believe that performance can be improved, so information about the worker's performance is collected. The workplace may be redesigned, the tools may be improved, or the organization of the work may be restructured. The other kind of job design is more difficult because the system is yet to be built. The details of a job must be specified through a process of inference, with the goal of anticipating sources of human error and minimizing them. Analysis of an existing job is our first topic.

Analysis of an Existing Job

The analysis of an existing job for the purposes of improving it is the oldest approach to job analysis. The methods originated with the work of Taylor and Gilbreth (Chapter 1), and they continue to be useful because they have proved their value for analyzing manual work and improving it.

Methods from Industrial Engineering

Process Analysis. Barnes (1963) is thorough in his account of the methods of modern industrial engineering as they have evolved since

Taylor and Gilbreth. The details of the methods, developed over decades, are numerous, so only essentials of the thinking and procedures can be presented here.

The motions of a manual worker are systematically observed and classified into 17 elementary motion subdivisions. The subdivision is called the *therblig* (Gilbreth spelled backwards, approximately). Examples of the therbligs are

Search. Directing the eyes.
Select. Reaching for an object with the hand.
Transport empty. Movement of an empty hand.
Transport loaded. Movement of a hand with something in it.
Release load. Releasing object in the hand.

Seventeen therbligs like these are capable of describing a wide range of manual skills. Consider the repetitive task of taking a peg from a box and inserting it in a pegboard. An active right hand and an idle left hand would be a common way to do it. The movements of the right hand would be described as

Transport empty. Reach for the peg.
Select. Select one peg from the box.
Grasp. Close thumb and fingers around the peg.
Transport loaded. Carry peg from box to pegboard.
Position (in transit). Peg is turned to vertical position as it is moved to the pegboard.
Position. Peg is lined up over hole in the pegboard.
Assemble. Peg is inserted into pegboard.
Release load. Fingers are opened and the peg is released.

An example of using this analysis to improve performance would be to prescribe use of the left hand also and have the two hands work together.

Barnes (1963, Chap. 15) has a checklist to guide the improvement of each therblig. For example, there is a list of 18 questions to be asked about "transport empty" and "transport loaded." Can a motion be eliminated? Is the path traveled the best one? Are the correct members of the body used? Can changes in movement direction be eliminated? Are frequently used parts near the points of use? And so on, through the list of 18. Each therblig is examined with probing questions of this kind.

Closely related to checklists are *principles of motion economy* that have been formulated. Barnes (1963, Chaps. 17–19) lists 22 of them. There are principles related to use of the human body, layout of the workplace, and design of tools and equipment. Here are two of each of the three types of principles:

Use of the Human Body
1. The two hands should begin as well as complete their motions at the same time.
2. Eye fixations should be as few and as close together as possible.

Layout of the Workplace
1. Tools, materials, and controls should be located close to the point of use.
2. Materials and tools should be located to permit the best sequence of motion.

Design of Tools and Equipment
1. The hands should be relieved of work that can be more advantageously done by a foot-operated device.
2. Two or more tools should be combined whenever possible (e.g., two-ended wrench).

Additionally, there are refinements and elaborations of these methods that can be useful (Barnes, 1963; Chapanis, 1959, Chap. 2). Therbligs can be given a detailed time analysis with motion picture or videotape records of the movements (called *micromotion analysis*). The movements can be plotted with respect to spatial coordinates, so that their patterning is apparent, or the concurrent movements of several interacting workers can be analyzed.

Link Analysis

Process analysis and micromotion analysis are useful for a standardized, repetitive job; the job can be explicitly specified when it is performed the same way each time. There are jobs, however, that are not done the same way each time. The eyes do not have to look at instruments on a panel in the same sequence each time. Controls do not have to be activated in the same order. When behavior varies, a statistical description of it is required. This might be said of repetitive jobs also, because they are not done in precisely the same way each time, but variation is slight, and slight variation can be ignored without introducing appreciable error.

Link analysis is an evaluation of the transitions that an operator makes between instruments on a display, between controls, or between instruments and controls. Link analysis could be the transitions that are made between a human and different machines, or between different humans. The goal of link analysis is to identify transitions that are inefficient, and rearrange the task elements, whatever they are, for efficiency. Links between task elements, as sequences of use, are expressed statistically as relative frequency measures, and a comparison of them can be the basis of redesign. Accounts of the procedures of link analysis are easy to find (Chapanis, 1959, Chap. 2; Van Cott & Kinkade, 1972, Chap. 10; McCormick, 1976, Chap. 11).

Consider, in hypothetical illustration of link analysis, a system that has four instruments on a panel which an operator must monitor. The four instruments, A, B, C, D, are arranged horizontally A–C–B–D. Assume that instrument A is a basic indicator of system functioning and is watched most of the time. Occasionally, instrument A will signal a deviation in system operation, and then instrument B must be checked. The reading on instrument B could suggest that the deviation is inconsequential, at which time the operator returns to the monitoring of instrument A, or the deviation could be further checked on instrument C. Instrument C could dictate a return to monitoring instrument A, or it could dictate a further check of instrument D. The final decision about the system deviation is made with the instrument D reading, and the deviation could either warrant action or it could be judged unimportant, at which time the operator returns to instrument A. The designer of the instrument panel placed instrument A in an easily viewed position because it is monitored most of the time. Instruments B, C, and D were put in arbitrary nearby positions because they were seldom used.

Let us suppose that a human factors engineer has noticed that some system deviations can have significant consequences for the system and that she has decided to optimize arrangement of the instruments. Eye fixations for an operator were recorded, and the following sample of 100 fixations of the four instruments was obtained:

> AAAAAAAAAAAABAAAAAAABCAAA
> AAAAAAAABCDAAAAAAAAAAAAAA
> ABCDAAAAAAAAAAAAABCAAAAAA
> AAAAABCAAAAAAAAAABCAAAAAA

Table 8-1 has the frequency and the probability of fixating each instrument (with 100 observations the calculation of probability is easy). As is known, instrument A is watched most of the time.

Table 8-2 has the frequencies and probabilities of successive fixations or links: Given fixation n, where is Fixation $n + 1$? Table 8-1 is based on a count of single events, but Table 8-2 is based on successive pairs of events. As probability, these transitional relationships are conditional probabilities.

Figure 8-1 shows the original layout of the instruments, overlaid with the conditional probabilities. The inefficiencies of visual scanning implied in Tables 8-1 and 8-2 are apparent in Figure 8-1. In the original design, instrument C was alongside instrument A, although eye movements were never made from A to C. Instead, the movement was always from A to B, but instrument B was some distance from A and required comparatively long eye travel. Similarly, the travel from instrument D to A was long. Furthermore, the pattern of the scanning lacked order. A revised panel layout is shown in Figure 8-2. Instrument A is placed center, as any instrument requiring steady attention should be, and the other instruments are placed below it and equidistant from it. Distance between instruments is minimized and going back and forth among them is efficient. Visually traveling the sequence A–B–C–D is an easy, quasi-circular scanning pattern.

TABLE 8.1   Frequencies and Probabilities with Which Four Hypothetical Visual Instruments Are Fixated*

	Instrument			
	A	**B**	**C**	**D**
Frequency	85	7	6	2
Probability	.85	.07	.06	.02

*Based on 100 fixations specified in the text.

TABLE 8.2   Conditional Probabilities, and the Frequencies on Which They Are Based, for Successive Fixations of Four Visual Instruments*

	A	**B**	**C**	**D**
	Conditional Probability			
A	.92	.08	0	0
B	.14	0	.86	0
C	.67	0	0	.33
D	1.00	0	0	0
	Frequency			
A	77	7	0	0
B	1	0	6	0
C	4	0	0	2
D	2	0	0	0

*Based on 100 fixations specified in the text.

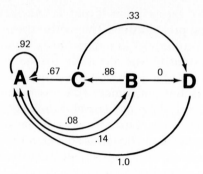

Figure 8-1 Hypothetical example of a link analysis of four instruments before redesign.

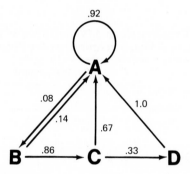

Figure 8-2 Hypothetical example of a link analysis of four instruments after redesign.

Some analysts (Chapanis, 1959, Chap. 2; Van Cott & Kinkade, 1972, Chap. 10; McCormick, 1976, Chap. 11) recommend that importance of the control or instrument be included in the link analysis. If there was weighting implied in the foregoing illustrative example, it was in terms of frequency of use, with instrument A assigned a central panel position. Weighting by frequency of use is justified because frequently used instruments and controls deserve central placement in the workplace (Chapter 9). Importance, however, is not solely a matter of frequency of use. Some instruments and controls may be used infrequently but are important when it becomes necessary to use them. Instruments for system emergencies are of this kind. An automobile horn is a simple example.

A common recommendation is to rank order the transitional frequencies in Table 8-2 according to size. Thus, the frequencies in Table

8-2 could be ranked 1 through 7, according to importance, by experienced users of the system. The rank of frequency could then be added to the rank of importance (Chapanis, 1959, p. 55) for a new index that takes both frequency of use and importance into account and which can be used to design the new layout. As an alternative, McCormick (1976, p. 295) suggests multiplying the ranks. There is no compelling rationale for either approach. The two methods may give the same outcome.

A pioneering application of link analysis was by Fitts et al. (1950), who measured the visual scanning patterns of pilots during the instrument landing of aircraft. (In instrument landing the cues are mostly from inside the cockpit.) Several recommendations for the use of link analysis were made by these investigators. Pilots differed among themselves in their scanning patterns, and it is wise to evaluate the behavior of more than one operator. Eye movements were not the same for different maneuvers, which is to be expected because instruments differ in importance from maneuver to maneuver. The implication is that it would be important to evaluate a range of system operation. Moreover, eye movement patterns may differ from one aircraft to another because of different system requirements.

A link analysis with many connecting elements can involve a large amount of data. Computer programs have been written to ease the burden (Haygood et al., 1964; Cullinane, 1977).

The Critical Incident Technique

Personnel will have memorable incidents in the operation of a system, and these events can be diagnostic for problems that need correction. Day-to-day operation of a system is a routine that is unlikely to have events which are significant for redesign, but unusual events, such as errors, emergencies, accidents, or near-accidents, can be memorable and informative. Recording these memorable events by questionnaire or interview is the *critical incident technique* (Flanagan, 1954).

The best illustration of the critical incident technique for data that can be used in system redesign is found in two articles by Fitts and Jones (1961a,b; reprinted in Sinaiko, 1961). The studies were conducted just after World War II. Several hundred pilots, and former pilots, of the U.S. Air Force were asked about errors that they had made in the operation of aircraft controls or the reading of aircraft instruments. Four hundred sixty errors in the operation of aircraft controls were reported in response to the question:

Describe in detail an error in the operation of a cockpit control (flight control, engine control, toggle switch, selector switch, trim tab, etc.)

which was made by yourself or by another person whom you were watching at the time.

An example response is the pilot who after takeoff turned off his battery and generator switches instead of his landing lights. He became rattled because he thought he was having both an electrical and an engine failure. In confusion he belly-landed the aircraft.

Fifty percent of the errors involving controls were substitution errors where one control was confused with another. Eighteen percent were adjustment errors, where a control was inappropriately moved. The required use of a control was forgotten 18 percent of the time. Other errors had smaller percentages associated with them: moving the control in the wrong direction (6 percent), unintentional activation of the control (5 percent), inability to reach a control (3 percent).

Two-hundred seventy errors were reported in response to the following question on the reading or interpreting of aircraft instruments:

Describe in detail some error which you have made in reading or interpreting an aircraft instrument, detecting a signal, or understanding instructions; or describe such an error made by another individual whom you were watching at the time.

Nine categories of errors in the reading and interpretation of instruments were found. Most notable were errors for instruments that had more than one pointer, such as the altimeter (Chapter 10). The potential difficulties of mistaking altitude are self-evident. Reversal errors were also common, as when the artificial horizon (the instrument that displays aircraft orientation) indicated the aircraft was banked to the right, but the pilot interpreted it to mean that the aircraft was banked to the left.

The critical incident technique can be useful but it takes a number of skilled users of the system to make it so. Naive trainees make errors that derive primarily from inexperience, not necessarily shortcomings in design of the human–machine interface. Aircraft pilots, of the kind that Fitts and Jones used as subjects, usually have thousands of hours of flying time, so the mistakes they make can often be interpreted as design difficulties that experience cannot overcome.

Expert Opinion

The use of human factors experts in job redesign should not be overlooked. It is likely that experts have encountered the problems before and have ideas about solving them. A good example of the use of experts, discussed in Chapter 1, is when the Electric Power Research

Institute in 1975–1976 used the human factors engineers of Lockheed Missiles & Space Company to conduct an analysis of the control rooms of nuclear power plants (Seminara et al., 1977). This analysis was done before the Three Mile Island incident. A similar analysis was conducted by human factors experts after the Three Mile Island incident (Hopkins et al., 1982). Both of these studies were directed at sources of human error.

Analysis of Jobs That Do Not Yet Exist

Seldom is a system completely new—most systems can rely on past experience with similar systems. Exceptions were the first manned space vehicles, which were new in most of their dimensions. Whether a system is partly old or mostly new, the forecasting of job designs is more difficult than analyzing and redesigning existing jobs.

A system is a composite of humans and machines organized to attain one or more goals, and it is usually complex. Various ways of attaining a goal are possible. Various contingencies can arise. The systems analysis for human factors defines the personnel for the system and the details of the jobs that they must perform. Engineers in other disciplines who are working on the system will be doing corresponding systems analyses on subsystems of their own specialties. Electronics engineers will analyze the uses and contingencies of subsystems such as communications, radar, and computers. Mechanical engineers will consider structures, materials, and so on. All analyses must eventually come together and be mutually balanced to define the system that will optimally attain its goals.

There is no standard methodology for systems analysis with respect to job specification, but good general accounts are given by Van Cott and Kinkade (1972, Chap. 1) and Hopkins (1963). Hopkins's analysis for the development of the controls and displays for an astronaut of a hypothetical space vehicle has the essentials of systems analysis for human factors.

Hopkins' Approach

Mission Analysis. Hopkins (1963) begins the analysis of a hypothetical manned space vehicle with a *mission analysis,* which is a broad outline. Narrative accounts were sufficient in his example, although that might not always be so. Hopkins has mission purpose, assumptions and constraints, and major mission requirements as the three parts of mission analysis. The *mission purpose* for the manned space vehicle is to transport personnel, equipment, and supplies from earth to an earth-orbiting space station. After rendezvous with the space station and delivery, the vehicle returns to earth and glides to a landing. The *assumptions and constraints* of the mission are four: (1) the space station is in a polar orbit at an altitude of 500 miles, (2) the space vehicle is launched into a coplanar orbit to an altitude of 300 miles, (3) some change in the plane of the initial orbit for the space vehicle is a normal expectation, and (4) all maneuvers of the space vehicle are accomplished with minimum energy expenditure. There are seven *major mission requirements:*

1. Initiation of powered flight
2. Attainment of planned orbit
3. Rendezvous with the space station
4. Departure from orbit
5. Attainment of aerodynamic flight
6. Arrival in the vicinity of the landing base
7. Controlled landing at the base

These seven requirements define the phases of the mission: pre-launch, boost, rendezvous, deorbit, reentry, terminal navigation, and landing. Each phase is terminated by the accomplishment of a requirement. For example, the rendezvous phase is accomplished by the space vehicle being brought into the orbit of the space station and within range of it.

The mission analysis is followed by specification of essentials to carry out the mission. Consider the rendezvous maneuver. The attitude (orientation) of the vehicle must be kept stabilized, a pitch rate must be maintained so that the longitudinal axis of the vehicle is perpendicular to local vertical and the bottom of the vehicle is parallel to the surface of the earth, the orbit must be determined, and there must be provision for changing the orbit and for translation to the orbit of the space station. Each of these functions suggests other system characteristics in turn—for example, attitude stabilization requires a sensing of vehicular attitude, a means of determining the amount of thrusting required to correct attitude error, and a means of applying the thrust correction.

Allocation of System Functions. At this point, in consultation with other engineers of the project, preliminary assignment of system functions to humans and machines can be made. As we will see in Chapter 16, the allocation of a function to a human or a machine is not always easy, but decisions must be made and thinking about them begins here. Once having made a preliminary allocation, Hopkins devel-

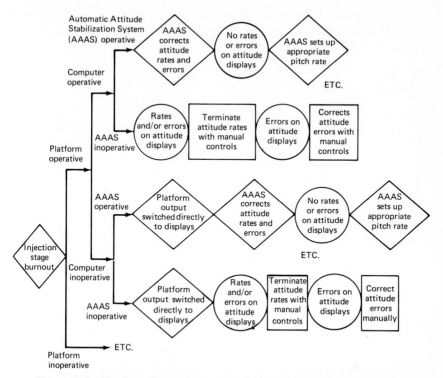

Figure 8-3 Portion of a systems analysis for a hypothetical space vehicle. See text for explanation. [Adapted from Hopkins (1963). In Bennott, et
al. *Human Factors Technology,* 1963, Fig. 34-2. Used by permission of
McGraw-Hill Book Company.]

ops implications for normal and emergency operating conditions of the
system with a version of failure modes and effects analysis (Chapter
2). Diagrams like the one in Figure 8-3 were prepared, where an automatic function was coded as a diamond, information received by an
astronaut as a circle, and action by the astronaut as a rectangle. As
Figure 8-3 suggests, the analysis is progressing toward more and more
information for defining the displays and controls that the astronauts
will require.

Displays and Controls. The analysis so far has shown requirements for displays and for manual controls to complement whatever
automatic controlling there might be. The specifications are very preliminary at this point. Refinements go on and on, as human factors
engineers generate more information in their analyses, weigh the operations of earlier systems of the same general kind if there have been

any, weigh the deliberations of other kinds of engineers who are doing systems analyses of their own, and consider the constraints of cost, weight, reliability, and time.

Timeline Analysis

Timeline analysis is the projected specification of events that pertain to an operator with respect to time, and it is useful for the allocation of functions, the specification of tasks, and the definition of displays and controls. A timeline analysis is greatly detailed, which is a virtue but a burdensome amount of work. Timeline analysis is used selectively.

A prerequisite of a task analysis is that the functions of the humans and the machines, as well as environmental circumstances, are sufficiently understood through other analyses, such as that of Hopkins, so that relevant variables will be included. The frequency of events and their duration are then projected. Figure 8-4 is an example of a timeline analysis with illustrative variables.

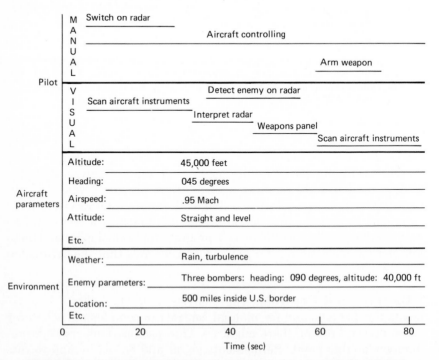

Figure 8-4   Example of a task-timeline analysis for a hypothetical fighter aircraft.

A timeline analysis will make several contributions to the design of a system:

1. It can show the relationship between environmental events and the behavioral requirements that are being imposed on personnel at the same time.
2. The frequency and duration of use of displays and controls may suggest additional controls and displays.
3. The frequency and duration with which displays and controls are used can suggest how these displays and controls should be placed in the workplace. (In Chapter 9 we examine the placement of displays and controls in more detail.)
4. It can be revealing about operator work load. Performance will be degraded when the operator has too much to do per unit of time, and the degradation carries design implications because something must be done about it. In a later section of this chapter we discuss the measurement of work load.

Failure Task Analysis

There can be temptation to emphasize analysis of the normal operations of the system at the expense of events that are rare, but the temptation should be resisted. System emergencies are rare events but often of such importance that system success and survival depend on them.

The failure task analysis methods of Jones and Grober (1961, 1962) were discussed in Chapter 2 within the context of the human as a redundant subsystem to increase system reliability when machine functions fail. Good system design makes the human operator an integral part of the resolution of emergencies, not a bystander. Jones and Grober conducted a failure task analysis for the operator of the *Mercury* manned space vehicle. The failure task analysis is related to the analysis in Figure 8-3 but is more detailed.

The description of system failures in purely hardware terms, such as "internal leakage" or "failed open," is important for hardware engineers, but it is only a beginning for a failure task analysis in behalf of the human operator. Defining the role of the human in emergencies requires an account of the failure in terms of events that impinge on human senses. For each failure there should be documentation of instrument cues such as changes in the readings of basic instruments and emergency warning signals, and changes in ambient cues such as vibrations, odors, body orientation, and the visual scene. Jones and Grober conducted failure task analyses for 400 components of the Mercury manned space vehicle, predicting the consequences of failure of

TABLE 8-3 Format for Prediction of Failures

Failure	Crew Indication	Ground Indication	Corrective Action	Failure Effect
Describes system, component, and mode of failure	Describes characteristics of indicators, lights, and crew perceptions that accompany failure	Describes characteristics of indicators and lights that accompany failure	Describes the recommended actions to be taken by the ground monitor or astronaut based on engineering characteristics of the vehicle for the various phases of flight*	Describes the influence of the failures on the operation of the vehicle

*Subject to approval by operations personnel.

Source: Jones and Grober, Studies of man's integration into the Mercury vehicle. (1962), Fig. 3.0. Reprinted by permission of McDonnell Douglas Corporation.

TABLE 8-4 Sample Failure Pattern

Failure	Crew Indication	Ground Indication	Corrective Action	Failure Effect
Reaction control System Solenoid valve Coil excitation lead Failed to ground	*Attitude indicator:* shows drifting in all axes *Rate indicator:* remains at constant or near constant value *Crew perception:* drift or loss of control in all axes based on window or periscope display	None	*In phase 3 (capsule separation to orbit):* switch to fly-by-wire or manual control to attain and hold retrograde attitude and initiate retrograde as soon as capsule attains retrograde attitude *In phase 6 (retrograde to .05g):* switch to fly-by-wire or manual control for corrections	Partial loss of automatic reaction control system

Source: Jones and Grober, Studies of man's integration into the Mercury vehicle. (1962), Fig. 4.0–3. Reprinted by permission of McDonnell Douglas Corporation.

each from the astronaut's perspective. The format for summarizing a failure is presented in Table 8-3. A specific example from the Jones and Grober analysis is shown in Table 8-4, which documents a failure that degrades the automatic control system, so implications for a manual control system would follow.

Work Load Analysis

The basic assumptions of work load are that (1) tasks are performed with respect to time, and (2) the human operator has a limited capacity for processing information in the time available. The limited capacity is often conceptualized as attention, where only a subset of events can be selected and processed at any one time (Lane, 1982). When events per unit of time exceed the operator's powers to select and process them, overload occurs. An overloaded operator will make errors and omit responses because he has too much to do in the time available to do it. Overload occurs most often in complex tasks with multiple activities (flying an aircraft is an example), but overload can occur with a single activity also. A single but difficult task can overload an operator's capacities because of the intense mental effort that is required.

As with some other topics in human factors engineering, aviation has motivated research and application on work load. The circumstances of flight can overload a pilot, so there has been a strong interest in devising measures of work load as a basis for doing something about heavy work load. The debates in military aviation about one or two crew members in fighter aircraft, and in commercial aviation about two or three crew members in the cockpit, have been questions of work load. Heavy work load may result from poorly designed control and displays that require excessive attention by the operator, so improvement in workplace design can be the issue. Automation can be an answer to some overload questions (Chapter 16), so work load and its measurement become central to system design.

None of today's measures of work load are absolute measures. One cannot look at the value for a measure and say whether the operator was overloaded or not. The measures are relative, useful for comparing two or more situations, such as work load with two different sets of displays and controls, or work load under normal and emergency conditions. Nor does a measure of work load say anything about its determiners. A high value of a work load measure could result from too many things to do in the time available to do it, the intense mental effort required to accomplish a single but difficult task, operation of a system in a difficult environment, or a combination of these conditions.

What are some useful measures of work load? A variety of behavioral and physiological measures have been tried (Hart, 1987; Williges & Wierwille, 1979; Wierwille, 1979). None are completely satisfactory, but two of them are good enough to have practical usefulness. One is task-timeline analysis, and the other is subjective rating scales.

Task-Timeline Analysis and Work Load

Parks (1979) and Stone et al. (1987) have thorough accounts of task-timeline analysis. Examination of the approach used by Stone et al. will illustrate the method. Task-timeline analysis has roots in process analysis and micromotion analysis from industrial engineering, and timeline analysis. Their application was aviation.

The definition of work load used by Stone et al. was the ratio of time required to perform tasks to the time available to do it:

$$\text{work load index} = \frac{\text{time required}}{\text{time available}} \times 100$$

The time available measure begins with an analysis of the flight scenario, where the start and end times of each phase of the flight, and segments within phases, are determined. The difference between start and end time is time available, the denominator of the work load index.

Computing the time required measure begins with drawings of the cockpit configuration and the operating procedures of the aircraft. All of the aircrew tasks and subtasks are specified in chronological order. Working with aircrew personnel familiar with the aircraft, a detailed description of the behavior required to perform each subtask is obtained (e.g., move hand to switch). The time to perform each of these subtasks is then estimated from the human factors literature or from measures obtained previously in recordings made of aircrew behavior in simulators and in flight. The sum of these time values for a segment gives the time required value for the numerator of the work load index.

Summing over the segments can give work load for a phase, and summing over phases can give work load for the entire mission. The work load index computed without leaving the ground was validated against the work load index computed from measures obtained by videotaping aircrew behavior in flight and analyzing the tape by micromotion analysis. The correlation (defined in Chapter 1) between the two versions of the work load index was .81, which is gratifyingly high and shows that the index has predictive power for in-flight criterion behavior. Stone et al. have made several uses of the work load index. The work load for a cockpit with conventional instrumentation was compared with an advanced cockpit having computer-based electronic instrumentation. The change in work load was computed for a simulated emergency where a crew member was incapacitated.

The work load index is useful, but it has several shortcomings:

1. The work load inherent in continuous control tasks is not considered. Whether a pilot's control task is easy or difficult is not reflected in the work load index.
2. The work load in doing two or more things at once is not considered. The work load index is based on subtasks in a series and does not reflect simultaneous tasks such as controlling the aircraft with the right hand, setting a switch with the left hand, and talking on the radio at the same time.
3. It does not consider cognitive activity. The thought processes of an operator, and the load they create, are not considered. Rouse (1979, p. 256) said that being occupied 10 percent of the time with brain surgery is a greater work load than being occupied 90 percent of the time with vacationing. The task-timeline analysis would consider neither the mental load for brain surgery nor vacationing.

Rouse's observation highlights the view that work load has a subjective component and is accompanied by mental effort. Rating scales, to which we now turn, include the subjective side of work load.

Rating Scales

As with task-timeline analysis, rating scales are having their usefulness in aviation. A precedent exists in aviation for the use of rating scales. Two aerodynamically sound aircraft can have different handling qualities, where handling qualities are defined as "those qualities or characteristics of an aircraft that govern the ease and precision with which a pilot is able to perform the tasks required in support of an aircraft role" (Cooper & Harper, 1969, p. 2). Along with ratings of handling qualities, a test pilot would be asked to make specific comments on the characteristics of a new aircraft. Ratings of handling qualities are multidimensional because they embody the stability of the aircraft, the cockpit interface, the environmental circumstances of the flight, work load, and emotional stress.

Recent research on rating scales for work load has used several subscales aimed at multidimensionality. Hart (1987) reports a rating scale developed at NASA–Ames Research Center that provides an overall work load score based on the weighted average of ratings on six subscales: mental demands, physical demands, temporal demands, own performance, effort, and frustration. Each scale is a 20-centimeter line divided in 20 equal intervals, anchored by bipolar descriptors at each end (e.g., extremely low, extremely high). A pairwise comparison among the six scales by each rater gives the importance of each scale for the rater, and these importance ratings are transformed into weights for the ratings. The six ratings are each multiplied by their weights, summed, and divided by the sum of the weights to give an overall work load score. Hart (1987) reports that these ratings have been recorded inflight, although administering them during a post-flight debriefing with a videotape reminder of the pilot's activities during each segment of the mission works as well.

Rating scales have been criticized on several grounds (Williges & Wierwille, 1979): (1) ratings change as a function of adaptation through learning or motivational changes, (2) the operator may confuse mental and physical work load, and (3) all of the work load may not be consciously available to the operator for rating. Rating scales will fail to reflect performance under different conditions of work load to the extent that these criticisms are true. Criticisms notwithstanding, rating scales have been sensitive indicators of work load. Hicks and Wierwille (1979) used an automobile simulator and varied the driver's work load with simulated wind gusts. Measures from a rating scale discriminated the three levels of work load. Wierwille and Connor (1983) established three levels of work load by manipulating the difficulty of motor control in a simulated flying task, and rating scales discriminated among them. Wierwille et al. (1985) used the difficulty level of navigational problems during simulated flying to establish three levels of work load, and rating scales discriminated them once again. Whatever the methodological problems for rating scales might prove to be, they appear to be a good instrument for measuring workload, perhaps the best at present. Williges and Wierwille (1979), in their review of behavioral work load measures, conclude that substantial research support exists for methods of subjective work load assessment such as rating scales (p. 569). Rating scales lack the diagnostic detail of task-timeline analysis, but they have the virtue of including the mental effort required of a task in their scores.

Performance Measurement

When a task exists, and the performance of an operator can be observed, why not measure performance and assess the effects of work load directly? Why bother with indirect measures of work load such as those obtained from rating scales? Effects on performance, after all, are the primary interest.

It would seem self-evident that performance should be poor to the extent that work load is high. Not necessarily. Performance measures in the task can be a sensitive indicator of work load (e.g., Hicks & Wierwille, 1979), but they may not always be. The argument is made (Williges & Wierwille, 1979, pp. 564–568) that an operator may hold performance on a primary task constant by putting out an extra effort as work load is increased by adding secondary tasks. Subjective assessment of work load with rating scales is thought to get at extra effort expenditures of this kind that might not be revealed in measures of performance.

Summary

Task analysis is the study of job requirements for the purposes of improving performance in an existing job or specifying the requirements of a job in a new system.

Analysis of an Existing Job

The pioneers of industrial engineering devised the methods of analyzing an existing job with the intent of improving it. The method of process analysis classifies the motions of an operator into elementary motion categories. Principles of motion economy are then used to judge the motions and to specify improvements for them. Analysis of motion picture or videotape records of operator's movements, called micromotion analysis, may be used.

Link analysis is a method of analyzing operator behavior in an existing job with the intent of improving it. Link analysis is an evaluation of the transitions that an operator makes between elements of a machine, between machines, or between humans of the system, in a situation where the behavior is less standardized than the repetitive jobs that receive process and micromotion analyses. Inefficient transitions are identified, and the job is redesigned to eliminate them.

The critical incident technique, as another approach to improving an existing job, interrogates system operators for atypical events such as errors, emergencies, accidents, or near-accidents. Information on problem areas that is obtained in this way is informative for job redesign.

Analysis of Jobs That Do Not Yet Exist

The jobs for a new system must be forecast. Starting with the broad outline of mission requirements for the system, human factors engineers increasingly refine their forecast of jobs and their characteristics as system development progresses. One method that can be used is a detailed task-timeline analysis of important segments of a job, where events are estimated with respect to a time baseline. Among the uses of timeline analysis are ideas about controls and displays and their placement, and inferring about work load. The implications of system emergencies are examined by failure task analysis, which suggests controls and displays that an operator will need for the resolution of emergencies.

Work Load Assessment

One of the responsibilities of a task analyst is to assess operator work load and take steps to minimize overload that can degrade system performance. How to measure work load is a modern research issue. At present, a relatively easy and useful method is subjective work load assessment, where the operator uses rating scales to indicate the work load demands of a task.

References _____

Barnes, R. M. (1963). *Motion and time study*. Fifth Edition. New York: Wiley.

Chapanis, A. (1959). *Research techniques in human engineering*. Baltimore: Johns Hopkins University Press.

Cooper, G. E., & Harper, R. P., Jr. (1969). The use of pilot rating in the evaluation of aircraft handling qualities. Advisory Group for Aerospace Research and Development of the North Atlantic Treaty Organization, AGARD Report 567, April.

Cullinane, T. P. (1977). Minimizing cost and effort in performing a link analysis. *Human Factors, 19,* 151–156.

Fitts, P. M., & Jones, R. E. (1961a). Analysis of factors contributing to 460 "pilot-error" experiences in operating aircraft controls. In H. W. Sinaiko (Ed.), *Selected papers on human factors in the design and use of control systems.* New York: Dover. Pp. 332–358.

Fitts, P. M., & Jones, R. E. (1961b). Psychological aspects of instrument display. I: Analysis of 270 "pilot-error" experiences in reading and interpreting aircraft instruments. In H. W. Sinaiko (Ed.), *Selected papers on human factors in the design and use of control systems.* New York: Dover. Pp. 359–396.

Fitts, P. M., Jones, R. E., & Milton, J. L. (1950). Eye movements of aircraft pilots during instrument landing approaches. *Aeronautical Engineering Review, 9,* February, 22–29.

Flanagan, J. C. (1954). The critical incident technique. *Psychological Bulletin, 51,* 327–358.

Hart, S. G. (1987). Measurement of pilot workload. In A. H. Roscoe (Ed.), *The practical assessment of pilot workload*. Advisory Group for Aerospace Research and Development, North Atlantic Treaty Organization, AGARDograph 282, June. Pp. 116–122.

Haygood, R. C., Teel, K. S., & Greening, C. P. (1964). Link analysis by computer. *Human Factors, 6,* 63–70.

Hicks, T. G., & Wierwille, W. W. (1979). Comparison of five mental workload assessment procedures in a moving-base driving simulator. *Human Factors, 21,* 129–143.

Hopkins, C. O., Synder, H. L., Price, H. E., Hornick, R. J., Mackie, R. R., Smillie, R. J., & Sugarman, R. C. (1982). Critical human factors issues in nuclear power regulation and a recommended comprehensive human factors long-range plan. Washington, DC: U.S. Nuclear Regulatory Commission, Technical Report NUREG/CR-2833 (Volumes 1–3), August.

Jones, E. R., & Grober, D. T. (1961). Studies of man's integration into the Mercury vehicle. St. Louis, MO: McDonnell Aircraft Corporation, Report 8276, June 15.

Jones, E. R., & Grober, D. T. (1962). The failure task analysis. St. Louis, MO: McDonnell Aircraft Corporation, June 21.

Lane, D. M. (1982). Limited capacity, attention allocation, and productivity. In W. C. Howell & E. A. Fleishman (Eds.), *Human performance and productivity: Information processing and decision making*. Volume 2. Hillsdale, NJ: Lawrence Erlbaum. Pp. 121–156.

McCormick, E. J. (1976). *Human factors in engineering and design*. Fourth Edition. New York: McGraw-Hill.

Parks, D. L. (1979). Current workload methods and emerging challenges. In N. Moray (Ed.), *Mental workload: Its theory and measurement*. New York: Plenum Press. Pp. 387–416.

Rouse, W. B. (1979). Approaches to mental workload. In N. Moray (Ed.), *Mental workload: Its theory and measurement*. New York: Plenum Press. Pp. 255–262.

Seminara, J. L., Gonzalez, W. R., & Parsons, S. O. (1977). Human factors review of nuclear power plant control room design. Palo Alto, CA: Electric Power Research Institute, Final Report, Project 501, Report EPRI NP-309, March.

Stone, G., Gulick, R. K., & Gabriel, R. F. (1987). Use of task-timeline analysis to assess crew workload. In A. H. Roscoe (Ed.), *The practical assessment of workload*. Advisory Group for Aerospace Research and Development, North Atlantic Treaty Organization, AGARDograph 282, April. Pp. 15–31.

Van Cott, H. P., & Kinkade, R. G. (Eds.). (1972). *Human engineering guide to equipment design*. Washington, DC: U.S. Government Printing Office.

Wierwille, W. W. (1979). Physiological measures of aircrew mental workload. *Human Factors, 21,* 575–593.

Wierwille, W. W., & Connor, S. A. (1983). Evaluation of 20 workload measures using a psychomotor task in a moving-base aircraft simulator. *Human Factors, 25,* 1–16.

Wierwille, W. W., Rahimi, M., & Casali, J. G. (1985). Evaluation of 16 measures of mental workload using a simulated flight task emphasizing mediational activity. *Human Factors, 27,* 489–502.

Williges, R. C., & Wierwille, W. W. (1979). Behavioral measures of aircrew mental workload. *Human Factors, 21,* 549–574.

CHAPTER 9

Layout of the Workplace

A main source of human error that degrades system performance is layout of the workplace. A justification for human factors engineering is the low performance and discomfort of workers whose workplaces are poorly designed–work surfaces that are too high or too low, uncomfortable chairs that produced aching backs, controls that cannot be easily reached, or instruments that are inappropriately placed. In later chapters we examine the design of individual instruments and controls for response to them. This chapter is about design of the overall workplace.

Anthropometry

There would be no need for anthropometric data in workplace design if we were all the same size and one workplace would fit us all. Unfortunately, from the standpoint of workplace design, humans are remarkably varied, and the workplace must accommodate the variability. The accommodation is through anthropometry. *Anthropometry* is the science that measures the human body and its biomechanical characteristics.

Consider these case histories which demonstrate failures to consider anthropometry: Twenty percent of drivers of a particular truck model could not operate the clutch or the footbrake without lifting their knees against the steering wheel. In the same vehicle 100 percent of the drivers could not reach the emergency hand brake without twisting the body out of normal position (Damon et al., 1966, p. 38). Gregoire (1972) conducted an anthropometric evaluation of emergency controls in the cockpits of seven U.S. Navy jet aircraft. Sixty-eight percent of the emergency controls could not be reached by the aviators who were in the lowest 5 percent of body size. Ten percent of the controls could not be reached by aviators who were the top 5 percent in body size. A number of the controls were 5 to 10 inches out of reach. One can only surmise the difficulties, if not the dangers, that poor designs like these inflict on pilots.

Methods of Anthropometric Measurement

The two kinds of anthropometric measurement are static and dynamic.

Static Measurement. Static measures are from a passive subject. Most anthropometric measures are static. Typical static measures are measures of body size, such as height, weight, length of the forearm, and size of the foot.

The methods of static measurement are easy. Define a population of people. When the population is small, everyone can be measured. The astronauts in the United States comprise a small population. When the population is large, such as American men 18 years and older, a sample of them must be taken, and the larger the sample the better.

The description of a measure is statistical because of the variability within it. Table 9-1 is an example of data from several dimensions of the human body out of the hundreds that anthropometrists have defined. Anthropometrists are careful to give each measure an explicit

TABLE 9-1   Example Anthropometric Data from Three U.S. Populations

		Percentile					
	Date	5th	10th	25th	50th	75th	90th
Nude Height (inches)							
Air Force women	1968	60.0	60.7	62.1	63.8	65.4	66.9
Air Force enlisted men	1965	64.5	65.5	67.1	68.7	70.4	72.1
Airline stewardesses	1971	62.5	63.0	64.1	65.4	66.8	68.1
Nude Weight (pounds)							
Air Force women	1968	102.3	106.9	115.5	126.1	137.1	148.6
Air Force enlisted men	1965	124.8	131.4	144.0	159.8	177.3	194.7
Airline stewardesses	1971	102.2	104.9	110.0	116.4	123.4	129.8

Source: Churchill et al. (1978b). *Anthropometric source book: Volume II: A handbook of anthropometric data.* Houston, TX: Spacecraft Design Division, Lyndon B. Johnson Space Center.

definition. For example, arm span is the distance between the tips of the right and left middle fingers when the arms are fully extended laterally. Hand circumference is measured around the knuckles. These data were drawn from an anthropometric data bank, comprised of surveys of 61 military and civilian populations, maintained by the Aerospace Medical Research Laboratories of the U.S. Air Force (Churchill et al., 1978b). Table 9-1 shows that the 5th percentile for height of Air Force women is 60.0 inches. Five percent of Air Force women are shorter than 60.0 inches. The 90th percentile is 66.9 inches. Ten percent are taller than 67.8 inches.

Dynamic Measurement. To be strictly correct, dynamic measurement is better identified with the discipline of biomechanics than anthropometry. The intent is the same, however–to provide descriptive measures of properties of the human body. Biomechanics considers the body as a biophysical system with strength, speed, endurance, and capabilities for response to such physical forces as impact, acceleration, and vibration. Static measures are fundamental for defining the workplace, but dynamic measures may be required also. A control might have to be positioned for maximum exertion of force, for example. Damon et al. (1966, Chaps. 3 and 4) have a good account of dynamic measures.

Strength is a good example of a dynamic measure, and the procedures of data collection and presentation are about the same as for static measures. A measure of strength is obtained from a dynamometer, which measures how hard you can squeeze with your hand, and Table 9-2 gives values for the right and left hands of two populations. How long a foot pedal can be depressed, or how fast a control stick can be deflected, are dynamic measures that are collected and presented in about the same way.

TABLE 9.2 Example Anthropometric Data: Hand Strength as Measured by Dynamometer Squeeze

| Subjects and Hand | Date | Maximum Squeeze (pounds) | |
		5th Percentile	95th Percentile
Normal men	1950		
Right hand		74	142
Left hand		65	124
Commercial bus and truck drivers	1955		
Right hand		91	151
Left hand		86	140

Source: Damon et al. (1966), Table 92. *The human body in equipment design.* Cambridge, MA: Harvard University Press. Reprinted by permission of Harvard University Press.

Treating the body as a physical object and understanding its tolerances is a complex collection of methods and measures. Individuals have been put in centrifuges to see the kinds of acceleration forces that can be tolerated, for example. Knowledge of how the body can tolerate injury comes from accident data and research on cadavers. Mathematical models of the human body have been used. Data like these are useful for designing safer workplaces. Surviving an airplane crash is not easy, but the design of some cockpits is better than others for protecting the pilot from impact. Similarly, an appropriate design of automobile interiors will improve a driver's chance of surviving a crash.

How to Use Anthropometric Measures

Variables to Be Remembered. There are several cautions to keep in mind when anthropometric measures are applied:

1. *Use data from a population that corresponds to the users of the workplace.* The smallest people on earth are the pygmy people of central Africa, who average about 4.5 feet tall. The tallest people are the Northern Nilotes of southern Sudan, also Africa, who average about 6 feet tall (Roberts, 1975, p. 13). These extremes are cited to emphasize the wide variation in the human species, and the variation must be remembered when anthropometric data are being used for workplace design.

   The question to ask is: Who will use the workplace? Orientals are smaller than Westerners. The average height of the Thailand armed forces is 5.3 feet, but the average height of U.S. Army ground troops is 5.7 feet (White, 1975, Table 2). It behooves Japanese automobile manufacturers to remember these differences between Easterners and Westerners when they produce cars for export (which they do). In an anthropometric survey of U.S. Army personnel, men were larger than women in all dimensions except hips (White, 1979). Old people are shorter than young adults (Annis, 1978). The literature of anthropometry testifies to populations with many differences. Good source books are publications of the National Aeronautics and Space Administration (Churchill et al., 1978a,b), and Damon et al. (1966).

2. *The time of the survey makes a difference.* The characteristics of the human body are not static. The sons of Harvard men were taller (by 1.3 inches) and heavier (by 10 pounds) than their fathers (Damon et al., 1966, p. 50). Men of the northern forces in the U.S. Civil War (1861–1865) averaged 5.6 feet in height and 136 pounds in weight, but in 1966 the men of the U.S. army averaged 5.7 feet and 159.1 pounds (Annis, 1978, p. II-53). U.S.

women averaged 5.3 feet in height and 133.5 pounds in 1903–1904, and 5.4 feet and 140.4 pounds in 1933 (Annis, 1978, p. II-53). Nutrition could be the cause. Anthropometric measures were made of changes occurring in 2000 Soviet adults during a two-year famine following World War I. Height decreased 1.5 inches, probably because of vertebral shrinkage (Annis, 1978, p. II-8). Not everyone agrees that these trends over time, particularly modern times, are of practical significance. White (1979) compared anthropometric data on U.S. Army men and women from 1946 with data of 1977. Increases were found for some measures, but they were small. If the nutritional argument is valid, it could mean that the nutritional value of the U.S. diet has stabilized in modern times.

3. *Anthropometric measures are almost always made on nudes.* Measurement of the nude body is an understandable approach to data on body size, but the measures underestimate the physical characteristics of the human body in a job. Failure to consider the functioning operator has resulted in hatchways through which personnel with heavy winter clothing cannot readily pass, and controls that gloved hands cannot easily operate. The problem is even worse for personnel who carry such equipment as parachutes or weapons. Ordinary street clothes can add 4 to 6 pounds to weight and 1 inch to height (shoes), and a helmeted soldier, without backpack or weapon, can have 23 pounds added to weight and almost three inches to height. Recommended increments for clothing and equipment are available (Damon et al., 1966, p. 52; Hertzberg, 1972, p. 478).

Application of Anthropometric Data. The two main approaches to designing the spatial dimensions of the workplace are designing for extremes and making features of the workplace adjustable. A third approach, design for the average individual, deserves limited consideration.

Designing for extremes means that the design is referenced with respect to one or both extremes of a body measure, and it is the method that is used when any member of an unselected population is a potential user of the workplace. An automobile must be designed to accommodate an unselected population because anyone can buy one, so the headroom of cars is designed such that all can use them except the tallest members of the using population. The approach is the same with doorways: Only the very tallest are inconvenienced.

Sometimes an approach that excludes both extremes is required. A clothing manufacturer might decide against the unprofitable practice of manufacturing clothes for the few people who are in the upper and lower 5 percent of body size. Those few must resort to tailor-made clothes.

Precision in design is gained if the using population can be selected and the extremes eliminated. A using agency that can reject applicants who are extreme in body dimensions can do a better job of workplace design for the narrower range of applicants that are accepted. Aircraft cockpits can be designed to better accommodate those admitted to flight training because applicants who are too tall or too short are rejected.

Making the workplace adjustable is the best solution. The adjustable car seat is a familiar example. Within wide limits, short, medium, or long legs can comfortably reach the brake and gas pedals. There is a limit to which the world can be made adjustable, however. Adjustable features of a workplace are more complicated and expensive than fixed features. Notwithstanding, having essential features of the workplace adjustable for optimum accommodation of, say, the 5th to the 95th percentile of users, is highly desirable. Conventional office furniture has few options for adjustment. Usually, a desk is not adjustable at all, and office chairs have only a few adjustments. Using telephone directory assistance operators working at video display terminals as subjects, Shute and Starr (1984) compared conventional desks and chairs with desks and chairs of advanced design that allowed many adjustments. The advanced furniture significantly reduced the discomfort of the operators. Operators were found to use different postures throughout the day to increase their comfort, and the advanced furniture allowed adaptation to whatever posture was chosen.

Design for the average individual has never had credibility because the average individual is a myth. No one is average in all dimensions. It would not be unusual for an individual to be in the 80th percentile in height, 30th percentile in arm length, and 50th percentile in weight. Nevertheless, there is some standardization that implies use of an average and it seems to work well enough. There is no realistic way to make counters in American kitchens adjustable, so they are standardized at 3 feet. Undoubtedly, there are very short people who must stand on stools to prepare their dinner, and tall people with short arms who cannot easily reach the counter top; nevertheless, the inconvenience does not appear to be large.

Test and Evaluation. The use of anthropometric data to specify dimensions of the workplace is only a first step. Additional steps should be taken:

1. Build an inexpensive mockup of the workplace. Include enough of the workplace elements so that visibility of display features, reachability of controls, and comfort can be assessed.

2. Select a representative sample of subjects who cover the anthropometric range of eventual users.
3. Dress the subjects in the clothing and equipment of eventual users. Have the subjects go through the motions of the users, and have them do it for realistic time periods so that comfort can be evaluated.
4. Make systematic observations of efficiency, comfort, visibility, and safety if pertinent, such as ease of exit in an emergency.
5. Revise the workplace design, as required. If extensive changes in the design are mandated by the test, a second test to validate the changes is desirable.

Layout of Visual Displays and Controls

The spatial layout of the workplace is structured with the help of anthropometry, and within that layout the controls and the visual displays must be arranged. The task analysis (Chapter 8) is a major source of information about the controls and displays that are required and the comparative importance of each, and now a decision must be made about their placement in the workplace. The controls and displays should be arranged according to three priorities:

1. *First priority.* Give optimal, central placement to visual displays and controls that are used most frequently.
2. *Second priority.* Not central placement, but accessible placement to visual displays and controls that are important but which are seldom used. Visual displays and controls for emergencies are the best examples. Emergency displays and controls are used infrequently—with luck not at all, but when they are required they must be accessible.
3. *Third priority.* Visual displays and controls that are used for routine operations, such as calibration or starting and stopping the system, can be given a low priority because the demands on the operator are low when they are used. Placing them in a less than optimum position in the workplace ordinarily will not carry penalties.

Having decided on general placement of groups of displays and controls, how should they be arranged? There are four principles that will cover most cases:

1. *Arrange displays to facilitate visual monitoring.* The task analysis can indicate the sequential order in which instruments should be scanned. The eye is a highly directional sensor. The eye

has high efficiency in its middle two degrees, and it does a lot of moving to place visual information in the middle, where it can best be seen. The displays should be arranged to minimize eye movements so that visual information is obtained speedily and with the least effort.

Arranging a group of instruments horizontally for left to right scanning is a good way to do it because it corresponds to the reading habits of Western societies and capitalizes on past experience. Arranging the group in a vertical array for top-to-bottom scanning can be justified also, but it is probably less desirable than right–left scanning.

The ease and comfort of visual monitoring is facilitated when the instruments are in the operator's preferred line of sight. Without it the operator is required to assume an uncomfortable head position, which in turn could force an uncomfortable body posture. Human factors handbooks have been recommending a preferred line of sight of 10 to 15 degrees below the horizontal (e.g., Van Cott & Kinkade, 1972, p. 400), but recent research has a new recommendation. The preferred line of sight for an operator sitting erect is about 30 degrees below horizontal (Hill & Kroemer, 1986).

2. *Controls that are used sequentially should be placed together and organized for successive activation.* This principle corresponds to (1) the scanning of visual instruments. When switch 1 has to be thrown, followed by switch 2 and the positioning of a lever, the ordering of the controls should be switch 1–switch 2–lever.

3. *Displays with a common function should be grouped together.* The instruments associated with the same subsystem should be grouped together (e.g., electrical subsystem, communications, etc.). Whenever an operator needs a particular kind of information, he will know where to look for it.

4. *Controls with a common function should be grouped together.* The rationale is the same as (3). Whenever an operator needs a particular control, he will know where to reach for it.

Not a principle of workplace layout perhaps, but good practice nevertheless, is to label all displays and controls clearly. The labels should be consistently placed, such as all below the element or all above it.

Summary

Anthropometry is the science of body measurement, and dimensions of the workplace should be defined by its data. Static measures such as height and weight are from a passive subject. Dynamic measures such as strength of a limb and endurance are from an active subject. The measures are made on a defined population and described statistically. Workplaces are ordinarily designed to neglect users with extremes in pertinent bodily dimensions, such as the too tall or the too short, making the workplace convenient for the majority. Having features of the workplace adjustable, for example, tables whose tops can be tilted and chairs whose heights can be adjusted, is the best solution to design of a comfortable, efficient workplace for a wide range of users. Adjustable workplaces are more expensive.

First priority for the placement of displays and controls in the workplace should be given to displays and controls that are used most frequently. Second priority is displays and controls which are important but seldom used, such as emergency displays and controls. Third priority is displays and controls for easy routines when operator work load is low, such as starting and stopping the system.

There are four principles that govern the arrangement of displays and controls once their general placement has been decided: (1) arrange displays to facilitate visual monitoring; (2) controls that are used sequentially should be together and organized for successive activation; (3) displays with a common function should be grouped together; and (4) controls with a common function should be grouped together.

References

Annis, J. F. (1978). Variability in human body size. In Churchill, E., Churchill, T., Downing, K., Erskine, P., Laubach, L. L., & McConville, J. T. (Eds.), *Anthropometric source book*. Volume II. *A handbook of anthropometric data*. Houston, TX: Spacecraft Design Division, Lyndon B. Johnson Space Center. Pp. II-1 to II-63.

Churchill, E., Churchill, T., Downing, K., Erskine, P., Laubach, L. L., & McConville, J. T. (Eds.). (1978a). *Anthropometric source book*. Volume I. *Anthropometry for designers*. Houston, TX: Spacecraft Design Division, Lyndon B. Johnson Space Center.

Churchill, E., Churchill, T., Downing, K., Erskine, P., Laubach, L. L., & McConville, J. T. (Eds.). (1978b). *Anthropometric source book*. *Volume II. A handbook of anthropometric data*. Houston, TX: Spacecraft Design Division, Lyndon B. Johnson Space Center.

Damon, A., Stoudt, H. W., & McFarland, R. A. (1966). *The human body in equipment design*. Cambridge, MA: Harvard University Press.

Gregoire, H. G. (1972). "Can you reach the controls?" A cockpit anthropometric survey. In H. W. Hendriks & L. R. Chason (Eds.), *Proceedings, 3rd annual symposium, psychology in the Air Force*. Colorado Springs, CO: United States Air Force Academy.

Hertzberg, H. T. E. (1972). Engineering anthropology. In H. P. Van Cott & R. G. Kinkade (Eds.), *Human engineering guide to equipment design.* Washington, DC: U.S. Government Printing Office. Pp. 467–584.

Hill, S. G., & Kroemer, K. H. (1986). Preferred declination of the line of sight. *Human Factors, 28,* 127–134.

Roberts, D. F. (1975). Population differences in dimensions, their genetic basis and their relevance to practical problems of design. In A. Chapanis (Ed.), *Ethnic variables in human factors engineering.* Baltimore: Johns Hopkins University Press. Pp. 11–29.

Shute, S. J., & Starr, S. J. (1984). Effects of adjustable furniture on VDT users. *Human Factors, 26,* 157–170.

Van Cott, H. P., & Kinkade, R. G. (1972). *Human engineering guide to equipment design.* Revised Edition. Washington, DC: U.S. Government Printing Office.

White, R. M. (1975). Anthropometric measurements on selected populations of the world. In A. Chapanis (Ed.), *Ethnic variables in human factors engineering.* Baltimore: Johns Hopkins University Press. Pp. 31–46.

White, R. M. (1979). The anthropometry of United States Army men and women: 1946–1977. *Human Factors, 21,* 473–482.

The Presentation of Information

PART IV

The Presentation of Information

CHAPTER **10**

The Visual Display of Information

Displaying information visually is the most prominent method. The visual system, and the visual perceptual system based on it, are remarkable for their powers of perceiving detail, size, pattern, color, texture, depth, and spatial distribution of stimuli. The demand that is placed on vision in human–machine systems because of its excellence can overload an operator, so the good design and organization of instruments that present visual information become a crucial part of workplace specification. Principles of grouping visual instruments in the workplace were presented in Chapter 9. This chapter covers the design of individual instruments.

There are two basic kinds of instruments for the display of visual information. One kind is the *symbolic instrument,* and the other is the *pictorial instrument.* All instruments are coded representations of relationships and values in the real world, but some are more abstract than others. Symbolic instruments, such as the commonplace dial or a digital counter, have the most abstract coding. Distance, for example, might be encoded with a pointer moving with respect to a fixed background scale that has numerical values on it. Distance, as we perceive it, is unrelated to a pointer moving against a background. A pictorial instrument is more realistic, resembling relationships of the world to some degree, but its fidelity need not be like a photograph. Only essential relationships need to be presented. Angular relationships would be the same as the real world. Distance might be portrayed by size differences, or perspective where lines come together like straight railroad tracks going to the horizon. Like a road map, many details would be omitted. In this chapter we examine these two ways of presenting visual information.

Symbolic Instruments

Types of Symbolic Instruments

The three basic kinds of symbolic instruments for presenting visual information are shown in Figure 10-1. There are, of course, variations

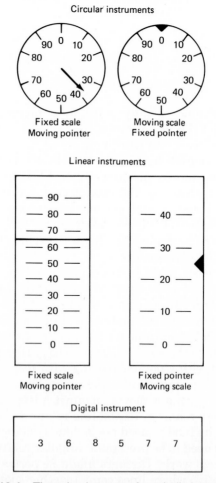

Figure 10-1   Three basic types of symbolic instruments.

in these ways of presenting visual information symbolically, such as a thermometer-like scale, but the three kinds of instruments shown in Figure 10-1 are the ones typically found in human–machine systems.

The most common way of presenting a continuous visual input symbolically is the circular instrument, and there are two kinds. A pointer can move with respect to a fixed scale with numerical values on it, or conversely, the pointer can be fixed and the scale move. Linear instruments are of the same two kinds. Figure 10-1 shows the linear instruments oriented vertically, but they can be oriented horizontally also. The digital presentation is the simplest of all. The numerals can be presented by an electromechanical counter, but today it is common to derive them from a digital computer and present them on an electronic display. The graphics capability of a digital computer could put circular and linear instruments on an electronic display as well. The Appendix to this chapter discusses the strengths and weaknesses of various modern electronic displays.

Which Symbolic Instrument Is Best?

The question has no single answer because it depends on the kind and quality of information that the operator requires. Task analysis, discussed in Chapter 8, will tell the kind of information that an operator needs from a visual source and what will be done with it. There are three uses of an instrument, and an operator can require any one or a combination of them:

1. *Quantitative reading.* Obtaining the numerical value of the variable.
2. *Qualitative reading.* Judging the approximate value, trend, direction, or rate of change of the variable.
3. *Checking reading.* Determining that a change has occurred from the desired value.

Table 10-1 rates symbolic instruments for the three kinds of readings.

With appropriate attention to instrument detail, which will be discussed below, all circular and linear instruments could be designed to give a satisfactory quantitative reading, but none can surpass a digital presentation. It is easy enough to look at the tip of a pointer and obtain a number, but nothing beats reading a number when you need a number. This intuitively reasonable fact is borne out by research. Goolkasian (1984, Experiment 1) compared the vocal reaction time for reading a digital clock with that of reading an analog clock (a circular instrument with a moving pointer and fixed scale). Reaction time to the dig-

TABLE 10.1 Three Basic Kinds of Symbolic Instruments Compared for Their Capabilities in the Three Kinds of Information That Might Be Required from Them

Type of Symbolic Instrument	Behavioral Requirement		
	Quantitative Reading	Qualitative Reading	Checkreading
Circular			
Moving pointer, fixed scale	+	+	+
Fixed pointer, moving scale	+	−	−
Linear			
Moving pointer, fixed scale	+	+	+
Fixed pointer, moving scale	+	−	−
Digital	+ +	−	−

ital presentation was almost twice as fast. Reading errors were much lower for the digital clock (1 percent) than for the analog clock (8 percent).

Qualitative reading and checkreading are both better when there is a moving pointer against a fixed scale, whether the instrument is circular or linear. It would seem that behavioral processes are simpler with the moving pointer-fixed scale instrument than with the other kinds of symbolic instruments. The spatial position of the pointer becomes fixed in perceptual memory, and changes in it are readily detected and interpreted. When the pointer is fixed and the scale moves the detection of change involves remembering the number that has been at the tip of the pointer, reading the number that is at the pointer on each glance, and mentally comparing the two and deciding whether change has taken place. The same reading and mental comparison of numbers are required for a digital presentation. These extra mental operations take more time and are opportunities for error.

Matheny (1961, p. 99) reports an incident that addresses the value of the moving pointer-fixed scale instrument for qualitative reading. A jet bomber lost its canopy in flight, and the pilot elected to stay with the aircraft and attempt to return it to base even though the wind blast was blurring his vision so severely that he was unable to make numerical readings of his instruments. Receiving heading instructions from the ground, the pilot was nevertheless able to discern the approximate position of the compass needle and maintain his aircraft on the heading to his base. Qualitative readings were responsible for returning the aircraft and its crew safely. It is unlikely that the pilot would have been able to read the heading if it had been presented digitally.

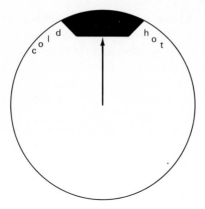

Figure 10-2 Type of symbolic instrument that would suffice if only check reading were required.

Similarly, the approximate positions of the hands of an analog wrist-watch can sometimes be read in a darkened movie theater where the numerals of a digital wristwatch cannot. Some individuals whose near vision is poor report that they cannot read the numerals of a digital wristwatch without their glasses but can see the positions of the hands of an analog wristwatch without glasses, blurred though they may be, and tell the time.

If task analysis shows that checkreading is all that is required for a variable, an instrument could be explicitly designed for it. Suppose there is no problem with a heating system as long as it maintains a specific temperature range, but that action is required whenever the system is too cold or too hot. Figure 10-2 shows an instrument design that would do the checkreading job. An appreciable deviation of the needle from the 12 o'clock position would be a call for action. A research comparison for the 9 o'clock and 12 o'clock positions of the needle has found no difference between them (Mital & Ramanan, 1986).

Design Considerations

Length of Scale. Length of scale is often deserving of no special consideration, going as it might from 0 to 100. When, however, the scale is long, such as 0 to 100,000, difficulties are created and the se-lection of an instrument is a problem. The symbolic instruments in Figure 10-1 are not equal in their capabilities for presenting a long scale of values. Circular instruments are limited in the length and res-olution of the numerical scale that they can display, and the same is true of a linear instrument with moving pointer and fixed scale, but a linear instrument with fixed pointer and moving scale has no such restrictions. The scale could be on a tape that would move by the rectangular opening, and it can be long and finely graduated. Like the linear instrument with fixed pointer and moving scale, a digital presentation has the advantage of a long scale of values with fine graduations. Indeed, the graduations can be the finest of all, limited only by the number of digits to the right of the decimal point.

The difficulty of selecting a symbolic instrument for a long scale of values such as 0 to 100,000 is evident when the prominent choices are examined. There is no problem when gross discriminations will do, such as readings within 5000 of the true value. The moving pointer–fixed scale instruments shown in Figure 10-3 are easy ways to do it. The 100,000 units are presented in a small scale space, but if high accuracy is needed, the instruments in Figure 10-3 will not do. The problem comes when high accuracies, perhaps within 50 units or less, are required. Granting the operator some powers of interpolation between scale divisions, there is no way that an accuracy of 50 units or less can be guaranteed from the instruments in Figure 10-3. Figure 10-4 is a solution in the form of a linear instrument with a fixed pointer and a moving scale. The scale can be any length and so of any degree of accuracy. The scale shown in Figure 10-4 has divisions of 100 units,

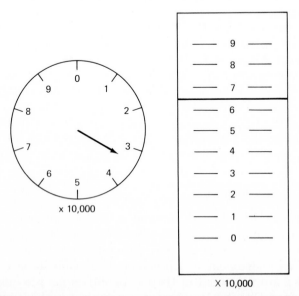

Figure 10-3 Instruments with moving pointer and fixed scale, having gross numerical divisions. The gross divisions will suffice when fine accuracy is not required.

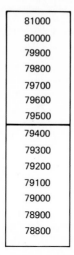

81000
80000
79900
79800
79700
79600
79500
79400
79300
79200
79100
79000
78900
78800

Figure 10-4   A linear instrument with a fixed pointer and a moving scale, such as a tape that moves behind the opening, can solve the problem of fine accuracy when a long scale of numerical values is required.

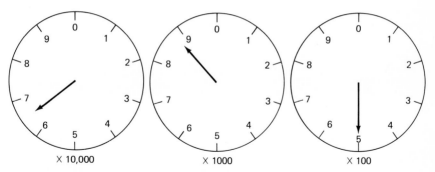

Figure 10-5   A long scale of numerical values, where fine accuracy is required, could be handled with three circular dials—one for units of 10,000 (left), one for units of 1000 (center), and one for units of 100 (right).

but the divisions could be even smaller. Table 10-1 shows that a linear instrument with fixed pointer and moving scale is not very good for qualitative reading and check reading, so if task analysis shows a requirement for one or both of them, the instrument in Figure 10-4 will be deficient.

Figure 10-5 shows another instrument that could accommodate a long scale. The 100, the 1000, and the 10,000 units of the reading are each on a separate dial. The operator must visually scan them and mentally add the values. Good accuracy is potentially possible, although it could be undermined by fallible mental addition processes.

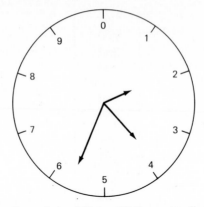

Figure 10-6 The three dials of Figure 10-5 can, in effect, be superimposed, putting the three needles on a single instrument. The short pointer is for units of 10,000, the intermediate-size pointer is for units of 1000, and the long pointer is for units of 100.

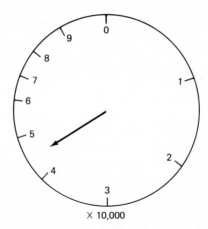

Figure 10-7 If fine accuracy is required at the low end of the scale but not the high end, a long scale of values can be accommodated with a nonlinear scale.

Moreover, both mental addition and visual scanning take time; reading would be slow. Also, the three dials occupy a lot of panel space. Panel space is limited for some workplaces.

Figure 10-6 is a solution that overlays, in effect, the three dials in Figure 10-5, by putting the three pointers on one instrument. The small pointer is for 10,000 units, the intermediate pointer for 1000 units, and the long pointer for 100 units. This instrument is an improvement. Less panel space is required, and some of the visual scanning has been eliminated, although not entirely because three pointers must each be fixated. Mental addition is still a requirement.

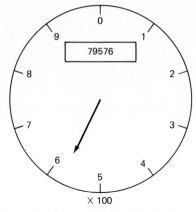

Figure 10-8 The pointer is for values at the low end of the scale. The digital window presents values throughout the full range of the scale.

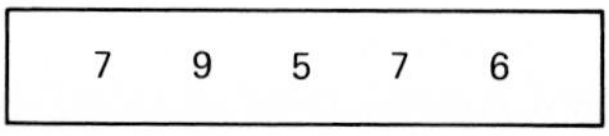

7	9	5	7	6

Figure 10-9 A digital display can provide any length of scale, with any degree of accuracy.

Most scales on symbolic instruments are linear, with equal distance between scale divisions, but Figure 10-7 shows a nonlinear scale. It is a solution to the problem of a long scale of values if accuracy is required for only part of the scale. A logarithmic function describes the scale divisions in Figure 10-7, where scale separations are large for small numerical values and small for large numerical values. This instrument will work for a system where accuracy is required at the low end of the scale and where gross reading suffices for the high end of the scale.

Figure 10-8 is an instrument that combines analog and digital features, where units at the low end of the scale are read by the pointer. This instrument is sometimes called the *drum-pointer* instrument. Why bother with the pointer when the entire reading, in all its accuracy, can be read from the digital presentation? Complete reliance on a digital presentation is certainly one way to do it, as Figure 10-9 shows, but for certain applications the instrument in Figure 10-8 has advantages because the pointer can convey information that is conveyed poorly by digits. The rate of change in a variable is perceptually more apparent from the rate of pointer change than from the rate of digit change. Furthermore, the moving pointer allows qualitative reading and check reading, and digits are poor for it.

Aviation has come close to trying all the alternatives in Figures 10-3 to 10-9 in search of a symbolic instrument that would serve as the altimeter (Rolfe, 1969, 1971; Schum et al., 1963). The search, which is probably not finished, has been motivated by the obvious importance of altitude for a pilot. An altimeter must give accurate readings at low altitudes. Some accuracy can be sacrificed at high altitudes, although good accuracy is useful there also. Qualitative reading and check reading, in addition to quantitative reading, are required. The altimeter configuration that has had the most widespread and longest use is the instrument shown in Figure 10-6, and it was thought to be accurate at all altitudes. Moreover, the instrument has some capability for qualitative reading and check reading. The problem with the three-needle altimeter is that it is prone to reading errors as large as 1000 and even 10,000 feet (Fitts & Jones, 1947), and it is considered to be a source of aircraft accidents (e.g., Tweedie, 1960). There are plenty of three-needle altimeters still in use, but in the past 20 years or so the drum-pointer altimeter shown in Figure 10-8 has become increasingly common. The increasing presence of digital computers aboard aircraft leads to the expectation of digital altimeters because of the ease with which digits are obtained from a digital computer and presented on a display. Beyond what is known in general about digital presentation, the particular human factors strengths and weaknesses of the digital altimeter remain to be clarified, although some research has been done (e.g., Rolfe, 1969, 1971).

Circular versus Linear Instruments. All else equal, what shape of instrument is best? Analysts (Adams, 1967; Semple et al., 1971) have reviewed the research and found no reason to favor one over the other. Semple et al. (1971, p. 218) said: ". . . it can only be concluded that 20 years of research have produced no definitive answer to the question of whether circular scales are any different from linear scales in terms of the amount of time required to read them or the frequency of errors which result from such readings."

Instrument Size. Instrument size depends on viewing distance and the availability of panel space. A viewing distance of 28 inches is commonly recommended, and for this distance it is common to find circular instruments of 2.5 to 3.5 inches diameter. Linear instruments are in the neighborhood of 6 inches long.

Number Systems and Interpolation. The research that has been done shows that scale progressions that are numbered 1, 2, 3, . . ., or 10, 20, 30, . . ., or 100, 200, 300, . . . are better than 2, 4, 6, . . ., or 1, 3, 5, . . . (Chambers, 1956; Barber & Garner, 1951). Whitehurst (1982) found that a numerical progression of 0, 100, 200, and 300 with graduation marks of 10 or 20 between numerals gave faster reading time and fewer errors than a progression of 0, 80, 160, or 240 with graduation marks of 8 or 16.

How many graduation marks should there be between numerals? Too many marks will clutter and impede the reading. There is justification for minimizing the graduation marks between numerals and relying on the operator's powers of interpolation. Grether and Williams (1947), on the basis of their research on circular dials, recommend .7 inches between marks, which would appear to rely on interpolation. Cohen and Follert (1970) placed graduation marks at various points on a linear scale anchored by 0 and 100 and found accurate interpolation of fifths and even tenths of the interval.

Other. The foregoing are big variables for symbolic instruments, but there are other situational variables that need to be considered, such as illumination and contrast of the instrument with respect to its background. Information on these variables is easy to find. Characteristics of symbolic instruments is one of the older research interests of human factors engineering, and reliable reference materials are available (Chambers, 1956; Chapanis, 1965; Grether & Baker, 1972; Semple et al., 1971). In addition, the relationship between instruments and controls must be considered. Which way should the pointer move when the control is moved? Should the pointer of a circular instrument move to the left when the control is moved to the left, or vice versa? In Part V we discuss control–display relations.

Pictorial Instruments

Situational displays and contact-analog displays are the two main kinds of pictorial instruments. *Situational displays* give a masterful view of system–environment relationships; their use is in decision making. A situational display might be maplike but with less clutter than a map, showing where the operator and his system are in relation to the environment. Situational displays have been called *horizontal displays* because they so often portray the world spread out around the system. *Contact-analog displays,* which have been called *vertical displays,* present a forward view from a vehicle and are intended for vehicular control. The operator would see a pictorial representation of the world in front of his vehicle, much as you see the scene in front of you when you drive a car. The term "contact" is from aviation, meaning in visual contact with the external world instead of symbolic contact through instruments. Figure 10-10 shows how a hypothetical, idealized aircraft cockpit might be organized with a situational display and

CONTACT ANALOG AND SITUATIONAL DISPLAYS

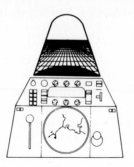

TYPICAL AIRCRAFT INSTRUMENT PANEL

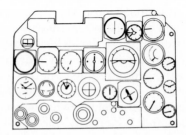

Figure 10-10 The bottom half of the figure shows a conventional aircraft panel of symbolic instruments. The top half of the figure shows the union of a situational display and a contact-analog display that some day might replace symbolic instruments in the cockpit. [Adapted from Adams (1967), Fig. 9-3. From *Contemporary Approaches to Psychology,* edited by Harry Helson and William Bevan, © 1967 by D. Van Nostrand Company, Inc. Reprinted by permission of Wadsworth, Inc.]

a contact-analog display, with the situational display used for navigation and the contact-analog display used for moment-to-moment guidance of the aircraft. The lower half of Figure 10-10 shows a conventional aircraft cockpit with symbolic instruments which pictorial displays might, at least in part, replace. Applications of pictorial displays are not limited to aviation. A police patrol car of the future might have a situational display which would show the relationship of the car's position to crimes in the city, perhaps with priorities assigned to the crimes as an aid in decision making.

Situational Displays

Are situational displays better than corresponding symbolic instruments? Figure 10-11 shows a hypothetical air traffic control situation

SYMBOLIC DISPLAY

AIRCRAFT NUMBER:	0 0 1	0 0 2
AIRCRAFT HEADING (DEGREES):	0 4 5	2 5 5
AIRCRAFT SPEED (KNOTS):	1 5 0	3 0 0
AIRCRAFT RANGE (NAUTICAL MILES):	0 5 0	1 2 5
BEARING FROM STATION (DEGREES):	0 0 0	3 0 0

PICTORIAL DISPLAY

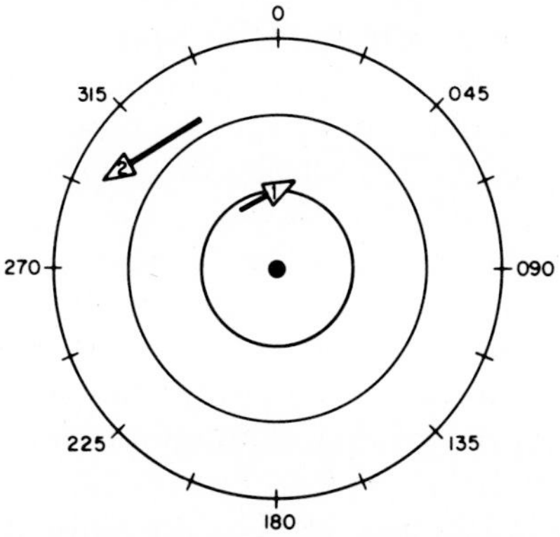

Figure 10-11 Two hypothetical displays for a controller in an air traffic control system. Information about the two aircraft could be presented symbolically or pictorially. [Adapted from Adams (1967), Fig. 9-2. From *Contemporary Approaches to Psychology,* edited by Harry Helson and William Bevan © 1967 by D. Van Nostrand Company, Inc. Reprinted by permission of Wadsworth, Inc.]

for two aircraft. Assume that the air traffic control system has assigned each aircraft a number and must keep track of each aircraft's heading, speed, range, and bearing from the controller's station. The information for the two aircraft could be presented digitally, as in the top of Figure 10-11, or pictorially, as in the lower part of Figure 10-11. Which is better? Pioneering research on the topic was done by Williams and Roscoe (1949) and Roscoe et al. (1950), and their studies indicate the kind of research that encourages innovative change.

A very high frequency omnidirectional radio range is a navigational aid that provides the pilot with continuous information about his angular position with respect to any station that he selects, and the pilot uses the aid to choose the compass heading to or from the station. Distance-measuring equipment would be used to locate the aircraft's position along the selected course. Roscoe and his associates compared the conventional symbolic cockpit instruments for this navigational aid with a new display, not unlike that in the bottom half of Figure 10-11, that had aircraft and station symbols on it. Installing the two kinds of instrumentation in aircraft and having pilots use them both in in-flight navigation problems might seem to be the best experimental approach. There is nothing wrong with the approach, but it is not the best for a beginning because it is too expensive, without a hint that it will materialize into something important. A wiser research plan is to start in well-controlled situations, find indications of important trends, refine the ideas, and then move on to more realistic and expensive approaches.

Roscoe and his associates began with simple test booklets that had drawings of several kinds of navigational displays (Williams & Roscoe, 1949). Some of the displays were symbolic and some were pictorial. Response was selection from multiple-choice answers about the direction and heading required to fly the aircraft on a given track. Aircraft pilots were used as subjects. Pictorial displays had half the solution time of symbolic instruments, and many fewer errors were made on them. Roscoe et al. (1950), in a follow-up laboratory study, used a flight simulator, which is a device that imitates the performance of an aircraft without leaving the ground. Simulators are used in engineering research for design decision, as Roscoe et al. did, and for aircrew training, as we will see in Chapter 18. The conventional symbolic instruments were compared with the pictorial display that was found to be most promising in the test booklet experiment. The main results are shown in Table 10-2. The pilots had an easy navigational time of it with the pictorial display, but not so with symbolic instruments. The difficulties of symbolic instruments are portrayed in Figure 10-12, which shows recordings made of the best and the worst flight paths during the experiment. Eventual adoption of pictorial displays in some operational aircraft resulted from these research beginnings (Platner, 1968).

Why is a pictorial display better than a corresponding set of symbolic instruments? Three reasons are hypothesized:

1. The pictorial display represents instrument integration, where several sources of information are combined into a coherent, perceptual whole, which is easier to scan and interpret (Roscoe, 1968; Roscoe et al., 1981).
2. Work load is reduced. When instruments are not integrated, they must be scanned visually, then the information from each remembered and finally, integrated and interpreted. The apprehension of information is more direct with a pictorial display, with less need for each of these behavioral processes. As a consequence of reduced work load, the operator has more resources to do other things concurrently (Chapter 8; Lane, 1982), which is an impor-

TABLE 10.2   Data from an Experimental Comparison of Symbolic Instruments and a Situational Display for Air Navigation*

Incorrect Problem Solution

   Pictorial display: none unsolved
   Symbolic display: five completely unsolved

Percent Excess Distance Flown on Correct Solutions

Problem No.	Pictorial Display	Symbolic Display
1	0.2	2.5
2	6.6	28.4
3	3.3	27.2
4	0.1	0.3

$$\% = \frac{\text{excess distance}}{\text{correct distance}} \times 100$$

Number of Unnecessary Turns (Greater than 15 Degrees from Track Desired)

Problem No.	Pictorial Display	Symbolic Display
1	0	4
2	0	2
3	0	3
4	1	1
	1	10

*A flight simulator was used. Four experienced pilots were each given eight navigational problems.

Source: Roscoe et al. (1950). Comparative evaluation of pictorial and symbolic VOR navigation displays in the 1-CA-1Link Trainer. Washington, D.C.: Civil Aeronatucis Administration, Research Report No. 92. October.

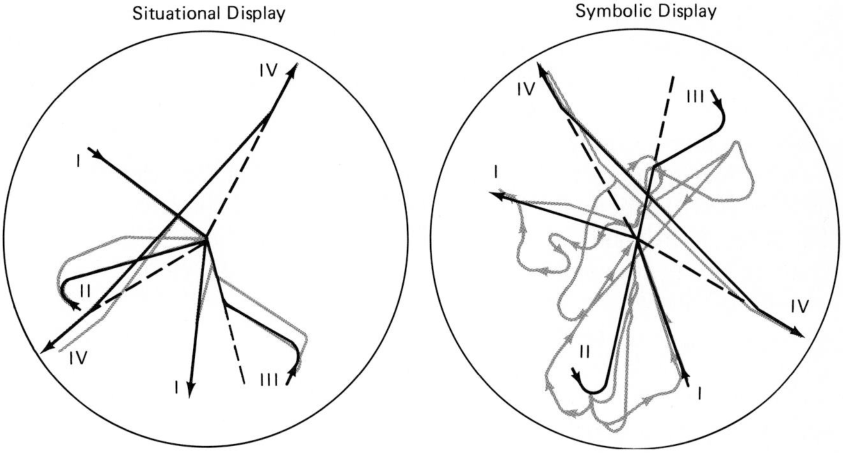

Situational Display   Symbolic Display

Figure 10-12   Best and worst flight path recordings in an experimental comparison of symbolic and pictorial displays for four terminal area navigation problems. [Adapted from Roscoe (1968) from Airborne display for flight and navigation, Fig. 3.10, 321–332. Copyright 1968 by the Human Factors Society, Inc. and reproduced by permission.]

tant consideration for busy operators such as pilots. Research on contact-analog displays, discussed in the next section, provides direct evidence on this point.

3. The adult human has had a vast amount of perceptual experience in the real world, and the pictorial display, being quasi-realistic, capitalizes on it. The operator does well on the pictorial display because it is similar to the world in which we live. In psychological terms, there is "positive transfer" from the world to the pictorial display (Chapter 18).

Contact-Analog Displays

Contact-analog displays for vehicular control, which have received development within the context of aircraft and submarines, have not had the smooth transition from laboratory research to operational system that situational displays have had. An early reason for hesitancy was that dynamic image generation for contact-analog displays required the graphics capability of a digital computer, and substantive thinking about these displays began in the 1960s, before computers were widespread. Other reasons, and they are influential today, relate to display hardware and the state of our human factors knowledge. It would help to have a thin, flat-panel display for workplaces within limited space, such as in aircraft, and we appear only to be on the threshold of useful flat-panel displays for contact-analog presentation (possibilities for flat-panel displays are discussed in the Appendix to this chapter). If flat-panel displays for contact-analog presentation are a forthcoming possibility, the remaining barrier is the state of our knowledge about the human factors effectiveness of contact-analog displays for vehicular control. Human factors research on contact-analog displays is a mix of findings that fall short of a mandate.

One conceptualization of a contact-analog display for aircraft control (Figure 10-10) relied on grid lines that represented the earth's surface. The size of the grid's rectangles was to inform about altitude; their angular orientations, about banking and heading; their rate of increase and decrease, about rate of climb and dive; and the speed with which they go by, about groundspeed. How well the grid lines communicated this information to the pilot was a research question. The gist of the findings was that the contact-analog display communicated all this information poorly, with an accuracy insufficient for flight (Fox, no date; Sidorsky, 1958; Fox et al., 1959; Abbott & Dougherty, 1964; for a summary, see Adams, 1967, pp. 350–355). Perhaps degree of detail in the earth representation was at fault, but Matheny and Hardt (1959a,b) and Elam et al. (1962) were unable to find an appreciable effect of more or less detail. Nor did corresponding research on contact-analog displays for submarine control have much luck in showing that display content and detail can affect operator performance (Blair & Plath, 1962, p. 68).

Since the beginning of instrument flight, we have known that symbolic instruments can be designed well enough for accurate aircraft control. How well does the contact-analog display stand up in a direct comparison with symbolic instruments for vehicular control? Blair and Plath (1962) made the comparison for submarine control. A submarine simulator was the research device. A subject's task was to use an assigned display for the execution of various maneuvers. Figure 10-13 shows the displays that were used. The pictorial display shown in Figure 10-13 is more complex than the one in Figure 10-10. The environment of the submarine has two surfaces. The roadway, which is a command signal that tells the operator which way to steer, represented advanced thinking about the contact-analog display. The contact-analog presentation in the upper right of Figure 10-13 shows how the dis-

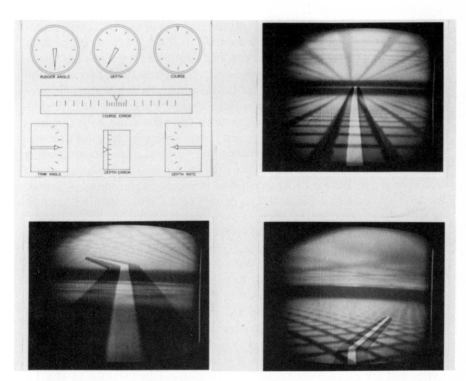

Figure 10-13   Displays that Blair and Plath (1962) used in an experimental comparison of symbolic instruments and a contact-analog display. [Adapted from Adams (1967), Fig. 9-4. From *Contemporary Approaches to Psychology,* edited by Harry Helson and William Bevan © 1967 by D. Van Nostrand Company. Reprinted by permission of Wadsworth, Inc.]

play looked when the operator was on the prescribed course; any deviation from this presentation must be corrected. The bottom half of Figure 10-13 shows a heading error indicated by the road turning. Depth error was indicated by the size of the grid squares and steer up or steer down by roadway commands. With problems that required changes in course and depth, the experimental comparison of conventional symbolic instruments (upper left of Figure 10-13) with the contact analog display found that subjects were unmistakably superior with symbolic instruments.

The Blair and Plath findings might have begun a slow death for the contact-analog display, but matters are not so simple. Subsequent research demonstrated that complex trade-offs are involved. Dougherty et al. (1964) compared symbolic flight instruments with the contact-analog display under different conditions of work load. The task was control of a helicopter simulator in straight and level flying. While attempting to maintain the aircraft straight and level the pilot was occupied with a secondary task of reading digits 0 through 9 aloud from a numerical readout device (a Nixie tube, .5 inch in diameter). The experimental question involved the capabilities of the two instrument displays for communicating deviations from the prescribed flight path while reading digits. The rate of reading digits was the work load variable, and it was either 27, 40, 67, or 120 per minute. The faster the digits, the less time available for the display, so the greater the premium on efficient transmission of information. Figure 10-14 is an artist's representation of the simulator with two pilots and both displays showing, although in the experiment there was only one pilot and one display at any one time. Notice that the earth in the contact-analog display has a pathway imposed on the grid lines. The pathway was a perceptual reference like the grid lines; it was not a command pathway, as used by Blair and Plath (1962), which told the operator which way to steer.

Errors from the required flight requirements were computed for each flight variable and combined to give the results of the Dougherty et al. study, shown in Figure 10-15. With imposed work load, the contact-analog display was better than the symbolic instruments (less error), and the advantage increased as work load increased. With less time available as work load increased, the pilot was able to extract, in brief glances, more information from the contact-analog display than the symbolic instruments. Koch and Dougherty (1965) found the contact-analog display effective for recovery from unusual helicopter orientations and for emergency maneuvers, which again testifies to the capabilities of the contact-analog display as an integrated instrument for communicating complex information. Koch and Dougherty (1965) did not, however, compare the contact analog display with symbolic instruments as Dougherty et al. (1964) had done.

Figure 10-14 Artist's sketch of the helicopter simulator used for the experimental comparison of a panel of symbolic instruments and a contact-analog display, which are shown. [Adapted from Dougherty et al. (1964), Fig. 1. © 1964 Bell® Helicopter Textron Inc. Used by permission of Bell Helicopter Textron Inc.]

Conclusions. The success of situational displays does not translate to success for contact-analog displays; pictorial displays are not uniformly advantageous. What can be said for the contact-analog display? It is not poor for vehicular control, but neither is it as good as symbolic instruments. In the configurations that have been tested, the contact-analog display is probably not good enough to be used in systems such

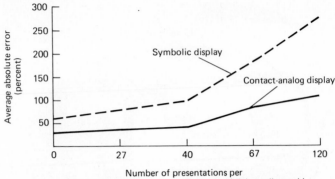

Figure 10-15 Comparison of symbolic instruments and a contact-analog display for helicopter control under different conditions of work load. [Adapted from Dougherty et al. (1964), Fig. 6. © 1964 Bell® Helicopter Textron Inc. Used by permission of Bell Helicopter Textron Inc.]

as aircraft and submarines. Nevertheless, the contact-analog display has greater power than symbolic instruments to communicate complex information in brief glances. Maybe designs of the contact-analog display will emerge with coding that will make it as good, or better, than symbolic instruments for vehicular control, but alternatively, maybe instrument integration is the chief merit of contact-analog displays. A display, good at fast communication of system–environment relationships, particularly in moments of heavy work load, could have its uses.

The contact-analog display represents a significant episode in the research saga on pictorial displays. Research on contact-analog displays has produced tantalizing hints of the boundaries of design and use, but not enough knowledge for practical application. The ease with which computers can generate pictorial presentations should be a stimulus for new research directions. Perhaps an addition of symbolic information to the contact-analog display is the way to go. Color coding will prove useful, although the principles of color coding for dynamic pictorial displays are unclear (Stokes et al., 1985, Chap. 7). Figure 10-16 is an example of a contemporary developmental effort that uses both symbolic information and color on a contact-analog display for a fighter aircraft (Ropelewski, 1986). The ground terrain is coded green, and terrain at or above flight level is coded brown. The "beams" in the rear, coded in red and yellow, are ground-based weapon threats. The "ribbon" symbol in the center is the recommended flight path for minimum risk. The numerical values are airspeed (left), heading (top), and

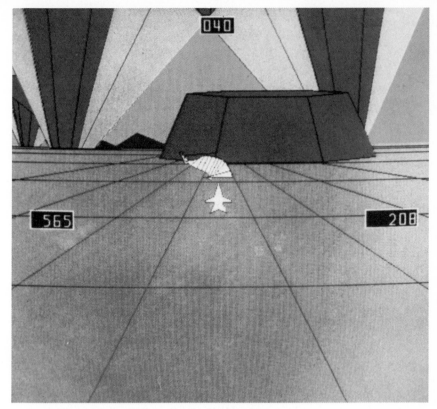

Figure 10-16 Modern contact-analog display undergoing research and development, which is in color and combined with symbolic information. See text for explanation. [Source: Ropelewski (1986), upper figure on p. 51. Reprinted by permission of *Aviation Week & Space Technology*.]

altitude (right). It remains to be seen whether research will establish that modern displays like these are better than their predecessors.

Pictorial and Symbolic Information Combined: The Head-Up Display

Nothing is more pictorial than our perception of the real world, and the superimposition of symbolic information on it is a composite display that is neither wholly pictorial nor wholly symbolic. A pilot must time-share his visual scanning between the symbolic instruments in the cockpit and the visual world outside. If some of the symbolic infor-

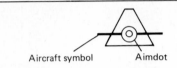

Aircraft symbol   Aimdot

Figure 10-17   Modern head-up display that is used in the DC-9 Super 80 aircraft. [Adapted from Lowe and Ornelas (1983), Fig. 7. Used by permission of the American Institute of Aeronautics and Astronautics.]

mation could be projected on the windshield and overlaid on the visual scene, visual scanning requirements would be reduced. Some modern aircraft do this with a display known as the *head-up display*. The name is apt because a pilot with a head-up display can keep his head up more and not down in the cockpit so often, scanning symbolic instruments. Figure 10-17 shows landing with the head-up display in the DC-9 Super 80 aircraft (Lowe & Ornelas, 1983). The pilot "flies" the aircraft symbol to the touchdown zone at the end of the runway. If the pilot is landing in bad weather and is relying on an electronic instrument landing system, an aimdot is generated at the touchdown zone of the runway, and the pilot aligns to it just as he would do to the touchdown zone of the runway if he could see it. As the point of touchdown comes close, the aimdot moves up and the pilot follows it and, in so doing, levels off and has a soft landing. Head-up displays for other aircraft and applications can include information such as numerical values for airspeed, heading, and altitude. Military aircraft can have information for weapons delivery. There is exploratory investigation of a head-up display for the automobile by the automobile industry (Holusha, 1986). The speedometer is the most likely candidate for projection on the windshield because the driver looks at it so often. Whether an expensive head-up display is justified for the automobile is a research and cost issue that the automobile industry must resolve. At best, the head-up display will keep the eyes oriented more toward the outside world and reduce accidents. At worst, the head-up display will be no more than an expensive add-on gadget for the innocent and the rich.

The head-up display is instructive for human factors engineering because it shows how a human factors problem can be created in the efforts to solve one. The head-up display has three behavioral premises:

1. Eye movements between the instrument panel and the external world will be reduced.
2. Focused at infinity, the symbology of the head-up display and the external world, which are superimposed, receive simultaneous attention and interpretation.

3. The resting eye is focused on infinity where the images are focused.

Item 1 is not debatable, but items 2 and 3 deserve additional discussion. First, item 2.

Evidence that pertains to item 2 is from the laboratory. Neisser and Becklen (1975) had two action sequences superimposed on a television screen and required the subjects to detect and respond to events of one of them. The subjects' central attentive mechanisms were very good in singling out one of the sequences and responding to it. If, however, the subjects had to respond simultaneously to events in both sequences, performance was degraded. When attention was directed to events in one sequence, events in the other sequence were missed. The implication for a pilot with a head-up display is that he cannot apprehend the display symbology and the external world simultaneously but must shift attention back and forth, and when he is attending to one, he is not seeing what is happening in the other.

Modern research indicates that the assumption of item 3 is probably wrong. The imaging of the symbology on the aircraft windshield is at optical infinity, as is the visual world that it overlays, and it has been assumed that the pilot need not change the focus of his eyes as he goes back and forth from the world to the symbols. Focus is with the lens of the eye, just behind the pupillary opening. The curvature of the lens determines how the rays of light are refracted (bent). Therefore, it determines the focus of the image on the retina at the rear of the eye where the neural receptors are located which transmit information to the brain. The shape and focus of the lens are controlled by the ciliary muscles, and it is called the eye's accommodation (Chapter 4). Contracted ciliary muscles produce a spherically shaped lens that brings near objects into focus. When the ciliary muscles relax and the lens is relatively flat, it has been believed that the accommodation associated with relaxed eye muscles is for distant objects at optical infinity. Contrary evidence has been accumulating for some time, and it was Leibowitz and his associates (Leibowitz & Owens, 1975; Owens & Leibowitz, 1976; Owens, 1979) and Roscoe (Roscoe, 1982a,b; Roscoe & Hull, 1982) who propelled the topic into prominence in modern times. Leibowitz and Owens (1975) established that under low levels of illumination, whether in the laboratory with artificial stimuli or outdoors with naturalistic scenes, the relaxed lens is focussed at an intermediate, not a far distance. Accommodation of the eye under these circumstances is called the *dark focus,* which averages at about arm's length for young adults (Leibowitz & Owens, 1975). Uniform visual fields such as empty sky, fog, or snowstorms, without compelling stimuli, would produce dark focus responses also, they hypothesized. When stimuli are adequate, however, accommodation tends to correspond to stimulus distance. As Roscoe said (1982a, p. 973): "Eye accommodation is a tug-of-war between the stimulus and the dark focus, with the stimulus normally pulling just hard enough to be seen and recognized." When, however, a near stimulus intervenes between the eye and a distant stimulus, the eye is biased toward its dark focus by the near stimulus (Owens, 1979). The latter finding has implications for head-up displays.

The idea of the head-up display is that computed symbology on the windshield is focused at infinity, the same as the visual scene. The relaxed eye is simultaneously accommodated to the symbology and the visual scene because they are both at infinity, and both of them will be visually clear. Apparently, this is not so. Owens (1979), in his research, found that accommodation of the eye was drawn by a near stimulus, which in the case of the head-up display is the computed symbology. Iavecchia (1985) used an optometer to measure the accommodation of the eye while her subjects viewed scenes through a head-up display in a simulated situation. The focus of the eye, while simultaneously viewing the head-up display and the visual scene, was not the same as viewing the visual scene alone. She found that the head-up display produced a shift toward the dark focus level of an individual. The symbology appears to be a compelling stimulus that releases the focus to lapse inward and away from distant objects. A laboratory study by Norman and Ehrlich (1986) has confirmed that displays like the head-up display draw the accommodation of the eye in from infinity and reduce the users' ability to detect and recognize small targets at infinity. A pilot landing his aircraft in poor weather with the head-up display would not have the runway in focus when he came out of clouds. A military pilot using a head-up display in dropping a bomb on a target would not have the target in focus when he is attending to symbology on the windshield. Reasoning similarly, Roscoe (Roscoe, 1982a,b; Roscoe & Hull, 1982) contends that the pulling power of near stimuli, such as aircraft window posts in the cockpit, accompanied by the comparative absence of pulling power of the sky, hold a pilot's eyes near their dark focus position and cause air collisions because approaching aircraft are out of focus and are not seen in time.

A benefit of the head-up display is hypothesized to be a reduction of visual scanning, but these problems with the head-up display could offset the benefit. A study by Soliday and Milligan (1968) shows that the net advantage of a head-up display is small under some circumstances. They compared the head-up display with conventional symbolic instruments for terrain following, a particularly demanding kind of low-altitude, high-speed flight that should bring out the advantages of a head-up display if they exist. A simulator for the F-4 fighter-bomber aircraft, which included simulation of the external world, was used. The pilot subjects preferred the head-up display to symbolic in-struments, but performance measurement showed only a small 5 per-cent advantage for altitude error with the head-up display. Heading error was actually worse by 34 percent with the head-up display. The *g* forces required in the maneuvering, which say something about the characteristics of the maneuvers, did not differ appreciably for the two displays. It cannot be said that a head-up display creates more prob-lems than it solves, but the human factors advantages and disadvan-tages may not combine to net a big gain.

Warning Signals

Visual warning signals, which are common, occupy space on the dis-play panel and provision must be made for them. Auditory warning signals, even though they do not occupy space, enter the picture be-cause they are an option when too many visual warning signals clutter the panel and create a visual work load problem for the operator.

Visual Warning Signals

How the eyes and head move, and the efficiency with which they function in detecting spatially arrayed signals on a display panel, are useful background for the topic of visual warning signals. An impor-tant feature of the eye is that only a relatively small region, called the fovea, is capable of detailed perception (Chapter 4). The foveal region is the center 2 degrees of the eye. Outside the fovea is the peripheral region. How well we see in peripheral vision depends on such variables as stimulus size, brightness, and contrast, but there is no question that it is poor. The readability of a digital display drops to about 50 percent at 40 degrees from the fovea, for example (Robinson, 1979). The reason that we move our eyes so much is that we constantly strive to keep stimuli of interest in the capable cone of foveal vision. The range of eye movements is about 45 degrees, so with peripheral vision the extent of useful perception is in the neighborhood of 80 to 90 degrees. Head movements do not wait on the eye to reach their maximum deflection of 45 degrees before they occur, however. Head movements may occur for stimuli lying 20 degrees from center, and they most certainly occur for stimuli beyond 45 degrees (Robinson, 1979, p. 351). How long does it take the eye to react to a signal when the operator is monitoring central instruments and manipulating controls? The reaction time of the eye under these circumstances can be as much as 700 milliseconds when the operator does not have advance warning of where the signal will occur. Once the reaction has begun, it will take additional time for the eyes and head to move to the signal, and the time will be a function of angle of signal displacement and signal brightness (Robinson, 1979, Figure 3). Ideally, the detection of a visual signal in peripheral vision could take about 1 second or so. Under the practical circumstances of system operation it could take longer. Moreover, all of this assumes an alert operator. An indifferent or a fatigued operator could take longer, or possibly miss the signal entirely.

How should visual warning signals be placed on the operator's display panel? They cannot be placed centrally because the instruments that are frequently used in routine system operation are located there; there is competition for the central space of a panel. They must be placed off center, therefore, but how far off center? Within 30 to 40 degrees from center would guarantee that they would be seen in peripheral vision, provided that the signals were sufficiently bright. Flashing, two to three times per second, would increase the attention-getting value of a signal. And how many warning signals can be accommodated in this 30- to 40-degree cone? Therein lies a problem because there may be insufficient space for all the warning signals and the labeling that they require. One solution (Grether & Baker, 1972, p. 78) is to use a centrally placed master warning light. When the master warning light comes on the operator turns to a detailed panel of warning indicators, well off to the side where there is more space, for specific information. The detailed panel could be meaningfully structured according to subsystems so that interpretation is easy.

What color should a warning light be? Red is standardized as a danger signal in the United States, and it is clearly discriminable in a 20- to 30-degree cone around a center fixation point (Coren et al., 1984, pp. 193–194). Moreover, reaction time to red is fast. Reynolds et al. (1972, Experiment 1) compared reaction times to red, green, white, and yellow signal lights on four different background colors and with dim or bright ambient illumination. Red consistently gave the fastest reaction time. White gave the slowest reaction time.

Colors can be combined with shapes to give a large number of informative signals. A backlighted red triangle could mean engine fire, but a blue triangle could mean engine temperature higher than desirable but not necessarily dangerous. We all are familiar with the shape coding of traffic signs, and as we all know, it takes study time to learn the meanings of shapes; training of operators would be required. Furthermore, the meaning of the shapes can be partly forgotten, which is a requirement for retraining. There will not be a forgetting problem if the operator has frequent experience with the shapes, but warnings can be rare, so forgetting can be an expectation. Not dealing with visual signals, but pertinent nevertheless, is an experiment by Patterson and Milroy (1980) on the learning and retention of 10 auditory warning signals used in cockpits of seven commercial aircraft. A ringing bell meant "fire," and a siren meant "disconnected autopilot," for example. After fully learning the meaning of the 10 signals the subjects returned a week later for a retention test. On average they remembered the meaning of only seven of them.

One final caution: There are several kinds of color blindness, and small percentages of the population have one or another of them (Chapter 4). If all individuals with any kind of color deficiency are counted, 8 percent of males and .05 percent of females are color blind (Coren et al., 1984, p. 191). Warning signals that depend on the discrimination of colors can run into trouble with unscreened operators.

Auditory Warning Signals

The eye is a highly directional sensor, but the ear is not. A great advantage of an auditory warning signal is that an operator can be oriented in any direction and hear it. The ambulance siren and the fire alarm are unmistakable, and we hear them no matter where we are looking.

Sometimes auditory warning signals appear to be a good thing gone wild. Modern commercial aircraft have 14 to 17 auditory warning signals, and there are aircraft with as many as 40 (Thorburn, 1971; Cooper, 1977, p. 14). Moreover, standardization among aircraft is lacking. Bells, chimes, tones, clackers, and warblers in aircraft are becoming what one analyst called "a growing cacaphony of zoo noises" (Hunter, 1969). These signals, intended to be informative, have a potential for confusion. Also, as discussed above, there is the problem of forgetting what they mean.

A final caution: Operators with impaired hearing could have problems with auditory warning signals.

Modern Solutions

Both visual and auditory warning signals are made to work in many systems, but none is without problems. What do modern designers see as answers?

Voice Warning Systems. One answer, and research and experience may prove it a good one, is the voice warning system. Instead of a red light or chimes coming on, the operator hears a calm but attention-getting voice deliver a brief message. Instead of warning the operator indirectly with signals such as red lights or chimes that require translation, a voice warns the operator directly, leaving nothing for translation.

The voice warning system was first installed in aircraft, and thinking about it continues for aircraft (Cooper, 1977), although it is easy to see possibilities for other kinds of systems. The U.S. Air Force installed a voice warning system in the B-58 bomber in the early 1960s (Cooper, 1977; Hunter, 1969; Thorburn, 1971), and the pilots liked it. Since then other branches of the U.S. armed forces have investigated it, and limited use of it is being made today in commercial airliners (Kantowitz & Sorkin, 1983, pp. 241–243). The adoption of voice warning is slow, as any appreciable departure from the practices of decades would be, but acceptance of the idea is gaining.

Analysts have worried about potential problems with voice warning systems, none of them insurmountable.

1. The warning voice will interfere with other audio transmissions. A pilot during takeoff or landing can tolerate interference from only the most dangerous emergencies. A solution would be to let the pilot turn the voice warning system off but allow the system to override the cutoff with the most critical messages, such as engine fire. This means an automatic decision mechanism to evaluate the severity of the problem and prioritize messages.

2. The voice warning system is unable to handle simultaneous, multiple failures. Two warning lights can come on simultaneously, or two auditory warning signals can sound simultaneously (not without problems of its own), but a voice warning system has only one voice and can report only one event at a time. Here, again, an automatic decision mechanism would be required to prioritize messages.

3. Operators hear various voices in the course of their work and the voice of the warning system must be distinctive among them. Psychological research can solve a problem of this kind. Just as research can show one tone discriminable from all others, or one control knob discriminable from all others, so one voice could be shown fully discriminable. The pioneering voice warning system on the B-58 aircraft was an audiotape system, but future systems will undoubtedly use synthetic speech. Synthetic speech could be electronically fine-tuned to be distinctive from all other voices in the system.

4. Like any kind of auditory warning signal, a warning voice that continues until the emergency is resolved can be annoying and distracting. There should be a way of turning the voice off once

the message is understood, with a provision that the voice could come on again if a new emergency arose.

5. What should the voice say? Should it only call attention to the emergency, or should it also tell the operator what to do? The next section, on computer-generated menus, which faces the same issue, includes this topic.

Computer-Generated Menus. The cockpits of modern advanced aircraft have fewer and fewer electromechanical instruments that look like the cousins of clocks. Instead, parameters for control, navigation, engine and electrical subsystems, weapons (in the case of military aircraft), and so on, are processed by onboard digital computers, and their status is displayed in the cockpit on cathode ray tubes. Similarly, warning information is now being presented visually on cathode ray tubes instead of with visual or auditory signals. Furthermore, the pilot is not only told about the problem on such displays but may be told what to do about it as well. Ropelewski (1982) reported a test flight in the A310 aircraft manufactured by France's Airbus Industrie. The pilot has two cathode ray tubes for aircraft management centered directly in front of him, as does the copilot. Between the two of them are two cathode ray tubes for monitoring the state of the aircraft, and it is here that warning messages appear. A fuel pump failure on an engine was simulated. An advisory warning message appeared on the left display, while a schematic diagram of the fuel system appeared on the right display. The three steps to resolve the emergency were presented under the message on the left, printed in blue. As the pilot executed each of the steps correctly the display turned from blue to white, and when all steps had been completed the pilot pressed a button that cleared the screens. A priority decision-making logic is built into the A310 warning subsystem so that most critical emergencies appear first. Very critical emergencies such as engine fire are displayed in red.

Whatever the merits of other ways of presenting warning messages, it would seem that computer-generated visual messages are here to stay. An innovative feature of the A310 system is that the pilot is not only told what is wrong but is told what to do about it, which is a procedure that could be used by voice warning systems also. Instructing the operator what to do about an emergency is an approach with merit and also a potential problem. The merit is that the operator does not have to rely on fallible memory. The problem is that procedures to resolve an emergency can depend on the situation and not always be the same. An aircraft emergency at 40,000 feet with plenty of time and space available may be a minor difficulty, but not at 100 feet. Conceivably the warning subsystem could automatically sense altitude and present actions appropriate for 100 feet, but it would be unrealistic to expect the system to sense and interpret the contexts of all emergencies. The best way to view these recommendations for action is as advisory messages. Let the pilot, on the basis of his intelligence and training, take the advice or not, as he sees fit.

Final Comment

This chapter has drawn heavily on aviation for the context of its discussions, and it is not surprising. Flight is complex and high-speed; there is a premium on efficient presentation of visual information if missions are to be successful and safe. Aviation has been a test bed for thinking about instrument design, and lessons have been learned for all systems.

Summary

Symbolic instruments and pictorial instruments are the two main ways of presenting visual information in the workplace. Symbolic instruments, like a pointer moving against a fixed scale with numerals on it, have the most abstract coding of relationships and values in the real world. Pictorial instruments, although not photographic in quality, represent the objects and relationships of the world more realistically.

Symbolic instruments have three uses: (1) quantitative reading where a numerical value must be obtained, (2) qualitative reading where an approximation must be obtained, and (3) check reading where change from a desired value is determined. Task analysis will indicate which of these functions are required for an instrument, and the instrument can then be chosen accordingly. Common types of symbolic instruments are fixed scale with moving pointer, fixed pointer with moving scale, and digital.

Situational displays and contact-analog displays are the two main types of pictorial displays. Situational displays give a large view of the system–environment relationships and are used in decision making. Contact-analog displays present a forward view from a vehicle and are intended for vehicular control. Research has shown that situational displays improve system performance, but contact-analog displays have not had corresponding success.

Advanced aircraft will often have a head-up display where symbolic information is projected at eye level so that the pilot can jointly attend to the symbolic information and the outside world. The name of the display is apt because the pilot can keep his head up more and not down in the cockpit so often, scanning instruments. Modern research has found human factors advantages and disadvantages for the head-up display.

Visual and auditory warning signals were examined. The trend in warning signals appears to be away from warning lights and sounds and toward voice warning systems and computer-generated menus of information and advice.

Appendix: Display Technologies

It is overstatement to say that the familiar electromechanical instruments, like the clocklike face with the pointer that moves on it, are disappearing. Instruments like these will be with us for a long time, partly because entrenched items do not yield easily to new technology, and partly because they can be reliable and low cost. Notwithstanding, the increase in electronic modes of presentation has been dramatic in modern times, and it is worthwhile to summarize their characteristics and capabilities here (for more details, see Sherr, 1979; Snyder, 1980).

The ubiquitous cathode ray tube display, which is represented in our television sets and the visual display units of digital computers, deserves to be contrasted with the prominent technologies for flat-panel displays. Some workplaces are compact and can profitably use flat-panel displays that occupy rather little space.

Cathode Ray Tube (CRT) Display

The CRT is bulky because an electron gun (the cathode) is positioned behind a glass panel. The glass has a phosphor coating and the gun fires a beam of electrons at it. The phosphor particles are illuminated momentarily when the electrons strike them. The electron beam is deflected and moved in a pattern, and the patterned activation of the phosphor is the image that the viewer on the other side of the glass panel sees. A monochromatic display has only one kind of phosphor element. In the case of color, the phosphor has elements of three dots— the three primary colors red, blue, and green—and they are activated by one or three electron guns, depending on the type of CRT.

Bulky though the CRT may be, its strong virtues explain why it is used so widely. The CRT can be small or large, it has excellent color capability, its resolution is high, and its cost is low. The CRT can display video or radar, alphanumerics, or computer-drawn graphics.

Flat-Panel Displays

Flat-Panel CRT Display. Small flat-panel CRTs have been developed that are a scant 2 inches thick. One version (Snyder, 1980, p. 83) has a thin cathode whose electrons are controlled by a number of switching plates that lie between the cathode and the glass panel. The on and off switching of the plates determines whether the electron beam passes through and activates the phosphor on the glass panel or not. Activating the phosphor, whether it is monochromatic or multicolor, gives an image like a conventional CRT.

The flat-panel CRT appears to have the qualities for alphanumerics, graphics, and video, but so far it has been limited to small sizes and is costly.

Light-Emitting Diode (LED). We are all familiar with LED displays in digital wristwatches and calculators. The basic element of an LED is a solid-state semiconductor which undergoes visible change when it is electrically activated. Activating a pattern of the elements can give an image, such as an alphanumeric character. LEDs are available in either red, orange, yellow, or green.

An LED display is low in cost and excellent for alphanumeric presentations. The many elements for a matrix to be addressed for graphics and video greatly increases the cost of an LED display. It is unlikely that an LED display will be competitive with other flat-panel technologies for graphics and video for some time.

Electroluminescence (EL). As the term implies, electroluminescence is light produced by the application of electricity (the LED display is a specialized kind of EL display). A typical EL device will have an electrical conductor, a layer of phosphor, and then a metallic electrode deposited on a sheet of glass. Application of an electrical current will cause the phosphor to glow. By addressing various elements constructed in this way, patterns can be formed. The color of the light depends on the type of phosphor.

EL appears suitable for alphanumerics, graphics, and monochromatic video. EL displays can be small or large, but they are costly.

Plasma Displays. The basic element of a plasma display is a small gas-filled unit. The gas is ionized and glows when electrically activated. The addressing of a number of these units on a display creates a pattern for the viewer.

A plasma display is excellent for alphanumeric readout and has some graphics and video capability, but it is costly. Color capability is limited so far.

Liquid Crystal Display (LCD). The LCD is common in watches and calculators. A liquid crystal material is sandwiched between two transparent sheets. Application of voltage causes realignment of the crystal molecules and a change in its visual properties. Elements formed in this way are addressed to form a pattern that makes an image for the viewer. The LCD is excellent for alphanumeric readout, but it has a slow response time which limits its application for graphics and video. Moreover, the contrast changes with viewing angle. The complexities of addressing the elements makes large displays expensive. Color quality is low. Resolution is ordinary.

References

Abbott, B. A., & Dougherty, D. J. (1964). Contact analog simulator evaluations: Altitude and groundspeed judgments. Fort Worth, TX: Bell Helicopter Corporation, Report D228-421-015. March.

Adams, J. A. (1967). Engineering psychology. In H. Helson & W. Bevan (Eds.), *Contemporary approaches to psychology*. Princeton, NJ: Van Nostrand, Pp. 345–383.

Barber, J. L., & Garner, W. R. (1951). The effect of scale numbering on scale-reading accuracy and speed. *Journal of Experimental Psychology, 41,* 298–309.

Blair, W. C., & Plath, D. W. (1962). Submarine control with the combined instrument panel and a contact analog-roadway display. *Journal of Engineering Psychology, 1,* 68–81.

Chambers, E. G. (1956). Design and use of scales, dials and indicators. Cambridge, England: Applied Psychology Research Unit, Medical Research Council, Report 254/56, October.

Chapanis, A. (1965). *Man–machine engineering*. Belmont, CA: Wadsworth.

Cohen, E., & Follert, R. L. (1970). Accuracy of interpolation between scale graduations. *Human Factors, 12,* 481–483.

Cooper, G. E. (1977). A survey of the status of and philosophies relating to cockpit warning systems. Moffett Field, CA: National Aeronautics and Space Administration, NASA Ames Research Center, Contractor Report NASA CR 152071, June.

Coren, S., Porac, C., & Ward, L. M. (1984). *Sensation and perception*. Second Edition. New York: Academic Press.

Dougherty, D. J., Emery, J. H., & Curtin, J. G. (1964). Comparison of perceptual work load in flying standard instrumentation and the contact analog vertical display. Fort Worth, TX: Bell Helicopter Corporation, Report D228-421-019, December.

Elam, C. B., Emery, J., & Matheny, W. G. (1962). Tracking performance as affected by the position of the attitude display. Fort Worth, TX: Bell Helicopter Corporation, Report D228-421-010, March.

Fitts, P. M., & Jones, R. E. (1947/1961). Psychological aspects of instrument display. I: Analysis of 270 "pilot-error" experiences in reading and interpreting aircraft instruments. Reprinted in H. W. Sinaiko (Ed.), *Selected papers on human factors in the design and use of control systems.* New York: Dover, Pp. 359–396.

Fox, G. A. (no date). Detection of angular change in a grid line display. Fort Worth, TX: Bell Helicopter Corporation, Report D228-420-001.

Fox, G. A., Hardt, H. D., & Matheny, W. G. (1959). Detection of small changes in the size of the squares in a grid line display. Fort Worth, TX: Bell Helicopter Corporation, Report D228-420-002, February.

Goolkasian, P. (1984). Picture–digit differences in processing clock times. *American Journal of Psychology, 97,* 259–283.

Grether, W. F., & Baker, C. A. (1972). Visual presentation of information. In H. P. Van Cott & R. G. Kinkade (Eds.), *Human engineering guide to equipment design.* Washington, D.C.: U.S. Government Printing Office. Pp. 41–121.

Grether, W. F., & Williams, A. C., Jr. (1947). Speed and accuracy of dial reading as a function of dial diameter and angular separation of scale divisions. In P. M. Fitts (Ed.), *Psychological research on equipment design.* Washington, DC: U.S. Government Printing Office, Army Air Forces Aviation Psychology Program Research Reports, Report 19. Pp. 101–149.

Holusha, J. (1986). 'Speedometer' at eye level. *New York Times,* April 3.

Hunter, G. (1969). Interest rising in voice warnings. *Aviation Week & Space Technology,* March 3.

Iavecchia, J. H. (1985). Response biases with virtual imaging displays. Warminster, PA: Naval Air Development Center, Aircraft and Crew Systems Technology Directorate, Technical Memorandum ACSTD-TM-2287, June 25.

Kantowitz, B. H., & Sorkin, R. D. (1983). *Human factors: Understanding people–system relationships.* New York: Wiley.

Koch, C., & Dougherty, D. J. (1965). Contact analog simulator evaluations: Unusual attitude maneuvers. Fort Worth, TX: Bell Helicopter Corporation Report D228-421-504, December.

Lane, D. M. (1982). Limited capacity, attention allocation, and productivity. In W. C. Howell & E. A. Fleishman (Eds.), *Human performance and productivity: Information processing and decision making.* Volume 2. Hillsdale, NJ: Lawrence Erlbaum, Pp. 121–156.

Leibowitz, H. W., & Owens, D. A. (1975). Night myopia and the intermediate dark focus of accommodation. *Journal of the Optical Society of America, 65,* 1121–1128.

Lowe, J. R., & Ornelas, J. R. (1983). Applications of head-up displays in commercial transport aircraft. *Journal of Guidance, Control, and Dynamics, 6,* 77–83.

Matheny, W. G. (1961). Human operator performance under non-normal environmental operating conditions. In S. B. Sells & C. A. Berry (Eds.), *Human factors in jet and space travel.* New York: Ronald Press. Pp. 78–111.

Matheny, W. G., & Hardt, H. D. (1959a). Further study in the display of spatial orientation information. Fort Worth, TX: Bell Helicopter Corporation, Report D228-421-002, August.

Matheny, W. G., & Hardt, H. D. (1959b). The display of spatial orientation information. Fort Worth, TX: Bell Helicopter Corporation, Report D228-421-001, August.

Mital, A., & Ramanan, S. (1986). Results of the simulation of a qualitative information display. *Human Factors, 28,* 341–346.

Neisser, U., & Becklen, R. (1975). Selective looking: Attending to visually specified events. *Cognitive Psychology, 7,* 480–494.

Norman, J., & Ehrlich, S. (1986). Visual accommodation and virtual image displays: Target detection and recognition. *Human Factors, 28,* 135–151.

Owens, D. A. (1979). The Mandelbaum effect: Evidence for an accommodative bias toward intermediate viewing distances. *Journal of the Optical Society of America, 69,* 646–652.

Owens, D. A., & Leibowitz, H. W. (1976). Oculomotor adjustments in darkness and the specific distance tendency. *Perception & Psychophysics, 20,* 2–9.

Patterson, R. D., & Milroy, R. (1980). Auditory warnings on civil aircraft: The learning and retention of warnings. Cambridge, England: Applied Psychology Unit, Medical Research Council, Civil Aviation Authority, Contract 7D/S/0142, Stage II, Final Report, February.

Plattner, C. M. (1968). Hughes tests area navigation unit. *Aviation Week & Space Technology, August 26.*

Reynolds, R. E., White, R. M., Jr., & Hilgendorf, R. L. (1972). Detection and recognition of colored signal lights. *Human Factors, 14,* 227–236.

Robinson, G. H. (1979). Dynamics of the eye and head during movement between displays: A qualitative and quantitative guide for designers. *Human Factors, 21,* 343–352.

Rolfe, J. M. (1969). Human factors and the display of height information. *Applied Ergonomics, 1,* 16–24.

Rolfe, J. M. (1971). Numerical display problems. *Applied Ergonomics, 2,* 7–11.

Ropelewski, R. R. (1982). Advanced technology in Airbus A310. *Aviation Week & Space Technology,* October 11.

Ropelewski, R. R. (1986) New technologies offer quantum leap in future fighter capabilities. *Aviation Week & Space Technology,* June 23.

Roscoe, S. N. (1968). Airborne displays for flight and navigation. *Human Factors, 10,* 321–332.

Roscoe, S. N . (1982a). Landing airplanes, detecting traffic, and the dark focus. *Aviation, Space, and Environmental Medicine, 53,* 970–976.

Roscoe, S. N. (1982b). Neglected human factors. In R. Hurst & L. R. Hurst (Eds.), *Pilot error: The human factors.* New York: Aaronson. Pp. 131–151.

Roscoe, S. N., & Hull, J. C. (1982). Cockpit visibility and contrail detection. *Human Factors, 24,* 659–672.

Roscoe, S. N., Smith, J. F., Johnson, B. E., Dittman, P. E., & Williams, A. C., Jr. (1950). Comparative evaluation of pictorial and symbolic VOR navigation displays in the 1-CA-1 Link trainer. Washington, DC: Civil Aeronautics Administration, Research Report 92, October.

Roscoe, S. N., Corl, L., & Jensen, R. S. (1981). Flight display dynamics revisited. *Human Factors, 23,* 341–353.

Schum, D. A., Robertson, J. R., & Matheny, W. G. (1963). Altimeter display and hardware development 1903–1960. Wright-Patterson Air Force Base, OH: Flight Control Laboratory, Aeronautical Systems Division, Air Force Systems Command, Technical Documentary Report ASD-TDR-63-288, May.

Semple, C. A., Jr., Heapy, R. J., & Conway, E. J., Jr. (1971). Analysis of human factors data for electronic flight display systems. Wright-Patterson Air Force Base, OH: Air Force Flight Dynamics Laboratory, Air Force Systems Command, Technical Report AFFDL-TR-70-174, May.

Sherr, S. (1979). *Electronic displays.* New York: Wiley.

Sidorsky, R. C. (1958). Absolute judgments of static perspective transformations. *Journal of Experimental Psychology, 56,* 380–384.

Synder, H. L. (1980). Human visual performance and flat panel display image quality. Arlington, VA: Office of Naval Research, Department of the Navy, Technical Report HFL-80-1/ONR-80-1, July.

Soliday, S. M., & Milligan, J. R. (1968). Terrain-following with a head-up display. *Human Factors, 10,* 117–126.

Stokes, A., Wickens, C., & Kite, K. (1985). A review of the research on aviation display technology. Savoy, IL: Aviation Research Laboratory, Institute of Aviation, Final Contract Report, Contract 84-Pri-G/W-1539 with General Motors Research Laboratory, Transportation Research Department, December.

Thorburn, D. E. (1971). Voice warning systems—a cockpit improvement. In H. W. Hendrick (Ed.), *2nd annual symposium, psychology in the Air Force.* Colorado Springs, CO: United States Air Force Academy.

Tweedie, P. G. (1960). Pilots misread altimeters in BOAC crash. *Aviation Week,* August 1.

Whitehurst, H. O. (1982). Screening designs used to estimate the relative effects of display factors on dial reading. *Human Factors, 24,* 301–310.

Williams, A. C., Jr., & Roscoe, S. N. (1949). Evaluation of aircraft instrument displays for use with the omni-directional range. Washington, DC: Civil Aeronautics Administration, Research Report 84, March.

CHAPTER 11

Speech Communication

In this chapter we examine speech as another way of presenting information to an operator. System operators talk with one another and transmit information. Today, operators are being presented information with synthetic speech. How well operators understand what is being said by human or artificial sources is basic to their performance and their contribution to system effectiveness.

The Characteristics of Speech

The Physiological Essentials of Speech

Sometimes the physiological functioning and mechanisms of our bodily organs are vaguely known to us, but not so for speech. Most of us are familiar with the physiological elements of speech production.

The *lungs* provide the power for speech. Air is forced from the lungs up the *trachea,* or windpipe. The air passes through the vocal organ, called the *larynx*. The larynx has two muscular folds called the *vocal folds,* or vocal cords. The space between the folds is called the *glottis*. When vocal folds are relaxed the air passes freely into the *pharynx,* or throat, up into the mouth, but when folds are tense they restrict the passage of air. The air pressure will force the folds apart, emitting a puff of air, and then the folds close again. The vocal folds open and close, often very rapidly as a function of the changing air pressure, and this vibration of them is fundamental to the production of sound. The quality of speech is influenced by the size of the vocal folds and the tension on them. A soprano singer has smaller vocal folds than a bass singer. The pitch of sound increases as tension on the vocal folds increases.

Above the larynx is the *oral tract,* made up of the pharynx and the mouth, and the *nasal tract,* or nasal passages. These cavities importantly influence the characteristics of vocal sounds. The *articulators* are the parts of the oral tract that are used in the production of sound, and they are evident as the lips, tongue, teeth, and the hard and soft palate. The articulators change the sound by changing the size and shape of the oral cavity. Consonants and vowels are the two main types of speech sounds in the English language, and they are distinguished primarily by the roles of articulators. The vowels tend to be uttered with an open mouth, with a prominent role played by the tongue. Vowels may have the lips rounded, as in the letter *o,* or not rounded, as in the letter *e.* The mouth tends to be more constricted with consonants, with the other articulators combining to restrict passage of the air in various ways for the different consonants. The vocal folds always vibrate for vowels, but they may or may not vibrate for consonants and so are said to be voiced or voiceless. The consonant *z* is voiced; the consonant *s* is not.

Phonemes

The vowel–consonant distinction is not the most fundamental way that speech sounds can be classified. The phoneme is the smallest unit of sound in the language, and there are about 40 of them (there is a minor controversy about the number) in standard North American English (neglecting regional dialects). All of our syllables and words are comprised of them. The numbers and kinds of phonemes differ from language to language. Examples of phonemes in standard North American English are the *p* in *pill,* the *v* in *vat,* and the *th* in *thigh.* The phonemes can be further reduced to the physiological level and related to voicing and the articulators that are used in their production. The outcome is 13 two-valued *features* which describe all the phonemes (Clark & Clark, 1977, pp. 177–191).

The Intelligibility of Speech

The Measurement of Speech Intelligibility

An objective way of understanding how well speech is understood is *intelligibility tests.* A new human–machine system, which has speech as a variable, can have intelligibility testing as a part of its early evaluation. If intelligibility is low, changes might be required in microphones, loudspeakers, headsets, soundproofing, training of the users, and so on. The speech dimension of a system must be given attention like any system variable; speech cannot be left to common sense because it is commonplace.

Intelligibility Test Methods. Intelligibility testing has talkers and listeners who are representative of users of the system and who are motivated to do well. After training in the use of system equipment a talker will read the material to a listener. The score is percent of items correctly heard.

The type of test materials used may be realistic. If a system is being evaluated, realistic system interchanges between personnel might be used. Often, however, standardized test materials are used, particularly in laboratories. There are four basic kinds of standardized test materials (Kryter, 1972). One is the *nonsense syllable test.* A nonsense syllable is a meaningless verbal unit such as *faz* or *zaf.* The talker reads from a list of nonsense syllables, embedding a syllable in a carrier sentence such as "You will write _____ now.", and the listener writes the best estimate of what was heard. Nonsense syllables are difficult to understand, so they are sensitive to small amounts of noise and distortion in the audio system. Another is the *phonetically balanced word list,* where the occurrences of sounds are proportional to those in everyday speech. Examples are *smile* and *fern.* The phonetically balanced word lists are considered less sensitive than nonsense syllables, but they inform about the intelligibility of the various sounds. Phonetically balanced word lists require less training than nonsense syllables before testing can begin because they are meaningful. The *modified rhyme test* has the speaker say a word like *hot,* and the listener must choose on a multiple-choice answer sheet from among *hot, tot, lot, not, got,* and *pot.* This test is sensitive to consonant phonemes, in this case the first letter of the word. The *sentence test* has such meaningful and realistic items as *Men are but infants to a smart girl.*

The Articulation Index. Speech intelligibility tests are used because of their self-evident value: If you want to know about the intelligibility of speech in a situation, administer an intelligibility test. Self-evident though they may be, intelligibility tests have their difficulties. Intelligibility tests are expensive and time consuming, they require a number of subjects and a skilled experimenter, and scores on them are not easily replicated because of the local characteristics of the laboratories, the equipment, the experimenter, the speakers, and the subjects.

Acoustic engineers have developed a technique for the prediction of intelligibility test scores from physical measures of speech and noise, sidestepping intelligibility testing. An index has been devised, called the *articulation index*. The articulation index predicts the scores of intelligibility testing rather well (Kryter, 1962b), probably because the physical characteristics of the speech stimulus are such a powerful variable for speech perception. The physical stimulus is not the whole story, as we will see later in this chapter; nevertheless, the articulation index does a commendable job. Calculation of the articulation index is not easy (Hawley, 1977, Part III; Kryter, 1962a, 1972, 1985; Webster, 1978); it is best left to specialists in the speech and hearing sciences.

Noise and the Degradation of Speech

Extraneous sounds are the enemy of speech intelligibility. The background noise of engines, industrial operations, radio static, and other voices, lower the intelligibility of speech. A good demonstration of the effects of noise on speech perception is a study by Miller (1947). White noise of different bandwidths and intensities was used to mask an intelligibility test of monosyllabic words. The results are shown in Figure

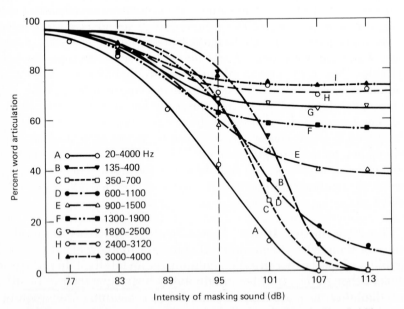

Figure 11-1 Masking of words by white noise of different bandwidths. [Adapted from Miller (1947), Fig. 4.]

11-1. The louder the noise, the more effective the masking. Wide-band noise (20 to 4000 hertz) produced better masking than narrow bands. Among the narrow bands, particularly at the high intensity levels, low-frequency bands were the most effective. The latter finding is expected because most of the energy of the speech spectrum is below 1000 hertz.

Variables that Offset the Degrading Effects of Noise

The damaging effects of noise on speech perception would be easy to understand if only the noise, the speech signal, and the peripheral ear were involved. There is more to it, however—the interpretative mind. A listener can use her knowledge about incoming speech and infer something about the coherent signal that is partly buried in noise and indistinct. The distinction is made between *bottom-end processing* and *top-end processing* (Klatt, 1977). Sometimes the distinction is called *bottom-up* and *top-down* (Salasoo & Pisoni, 1985). Bottom-end processing is use of the physical auditory stimulus by the auditory system. Top-end processing is the interpretative mind using knowledge to infer about the auditory input. The two kinds of processing combine to give us auditory perception.

Experiments that are discussed in this section refer to signal-to-noise ratio as a way of specifying the intensity of speech relative to intensity of noise. *Signal-to-noise ratio* is defined as the ratio of speech intensity to noise intensity in decibels. The values of signal-to-noise ratio range from negative to positive. Negative values mean a relatively high level of noise with respect to speech, and positive values, vice versa.

Size of the Vocabulary. Restricted vocabularies in system operations produce no-nonsense communications for the sake of efficiency (pilots and air traffic controllers do not chat breezily about their children). As a bonus, restricted vocabularies have increased intelligibility. Miller et al. (1951) administered intelligibility testing to their subjects with vocabularies of 2, 4, 8, 16, 32, and 256 single-syllable words. The subjects were informed of the vocabularies. Amount of noise was systematically varied. The results are shown in Figure 11-2. The number of words correctly heard is high when the vocabulary is small. Even when a word is not heard very well, a good job can be done of inferring what it might be because the number of alternative possibilities for it are small. As the number of alternatives increase, inference becomes more difficult.

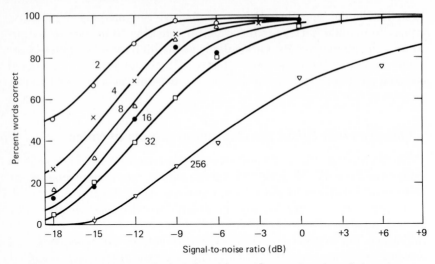

Figure 11-2 Masking of words by white noise as a function of size of vocabulary. [Adapted from Miller et al. (1951), Fig. 2.]

Language Constraints. The limited requirements of a system can constraint vocabulary and improve the perception of speech, but the structure of natural language itself also constrains vocabulary and improves the perception of speech. A noun will follow *the* in the English language. It is likely that a noun at the beginning of a sentence will be followed by a verb. When the sentence *The dog ate a bone* is presented under noisy conditions, and the listener hears only *dog* and *bone,* it is reasonable to infer that something like *chewed* or *smelled* or *ate* lies between. Miller et al. (1951) presented sentences with five key words (the other words were such as *a* and *the*). The sentence presentations were compared with the individual presentation of key words. Various signal-to-noise ratios were used. The results are shown in Figure 11-3. Reduction in vocabulary size by language constraints clearly improved the perception of speech. Another experiment by Miller et al. (1951) can be interpreted in the same way. For various signal-to-noise ratios, the digits 0 through 9, words of sentences, and nonsense syllables were compared for intelligibility. Digits are a highly restricted vocabulary, the words of sentences are constrained by language but the constraints are less than digits, and nonsense syllables are the least constrained of all. The order of constraint determined the accuracy of the perception. Digits were heard the best, words of sentences next, then nonsense syllables.

Salasoo and Pisoni (1985) used a different experimental approach and came to a similar conclusion about language context and speech

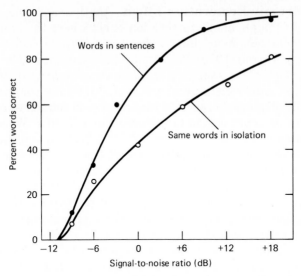

Figure 11-3 Structure of language improves the perception of speech in noise. [Adapted from Miller et al. (1951), Fig. 3.]

processing. The speaker presented subjects either simple, meaningful sentences such as *The stray cat gave birth to kittens,* anomalous sentences (sentences that use meaningful words in a nonsensical arrangement) such as *The end home held the press,* and the key words of these sentences individually. On the first trial the key words were replaced with noise. On each consecutive trial 50 milliseconds of noise was removed and replaced by the corresponding section of the word, and as trials progressed the subjects heard more and more of the word. By the final trial all the noise had been removed and all the words were presented. Their hypothesis was that the context of familiar language structure for meaningful sentences would make words more easily recognizable with a greater proportion of their words blanked with noise, but anomalous sentences and individual words, which lack language structure, would require more word and less noise to be understood. The hypothesis was supported. These findings of Salasoo and Pisoni affirm the role of top-end processing in speech perception because our knowledge of language contributes to speech recognition, beyond the acoustic input. Salasoo and Pisoni concluded, however, that top-end processing does not contribute as much to speech perception as bottom-end processing. For all that the mind contributes, there is no escaping the power of the impinging stimulus. Without bottom-end processing there is no top-end processing.

The Phonemic Restoration Effect. Another example of top-end processing is the phonemic restoration effect where the auditory perceptual system can "hear" inputs that are not there. The pioneering study was by Warren (1970). Warren used the sentence: *"The state governors met with their respective legislatures convening in the capital city."* The first *s* in "legislatures" was replaced with a cough or a tone, and virtually all subjects reported that all speech sounds were present; the auditory perceptual system restored the deleted *s*. Warren called the phenomenon the *phonemic restoration effect.* In continuation research, Warren and Obusek (1971) found that the deleted *s* was restored when coughs, tones, and buzzes were used, but not silence. Reasoning in behalf of top-end processing, Samuel (1981) hypothesized that restoration of deleted segments would be better if the deletions were in longer words, more frequent words, or sentences because of top-end processing and the expectations that a listener would have about the missing elements. His results supported the premise.

Warren and Obusek (1971, p. 361) say that the phonemic restoration effect explains the *dinner party problem.* A listener is trying to hear an after-dinner speaker amidst clattering dishes, laughter, and nearby

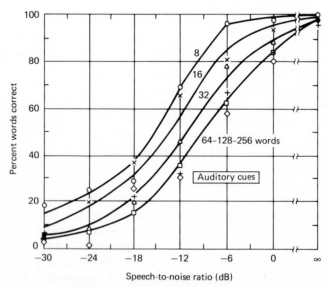

Figure 11-4 Intelligibility of words in noise with auditory cues only, as is ordinarily done in intelligibility testing. These curves are the reference for interpreting the data in Figure 11-5, where the listener could see the speaker. The curve parameter is vocabulary size. [Adapted from Sumby, W. H., and Pollock, I., *Journal of the Acoustical Society of America* 26, 212–215 (1954). Reprinted by permission of the American Institute of Physics.]

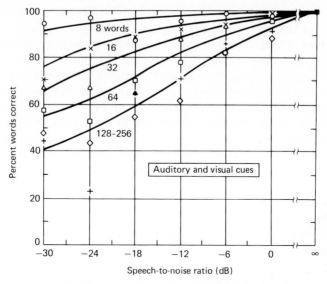

Figure 11-5 Intelligibility of speech where the listener could see as well as hear the speaker. These curves are to be compared with Figure 11-4, where the listener could only hear the speaker. The curve parameter is vocabulary size. [Adapted from Sumby, W. H., and Pollock, I., *Journal of the Acoustical Society of America* 26, 212–215 (1954). Reprinted by permission of the American Institute of Physics.]

conversations that mask entire phonemes and groups of phonemes. The phonemic restoration effect helps fill the gaps and makes the speech understandable.

Visual Cues. Speech perception is larger than speech. Our facial expressions, the patterning of our lip movements, and the movements of our bodies when we speak are other ways that we express ourselves, and we use them to supplement speech when we communicate. Sumby and Pollack (1954) presented vocabularies of two-syllable words, consisting of 8, 16, 32, 64, and 128 words. Various signal-to-noise ratios were used. The subjects either faced the speaker or were turned from him. Figure 11-4 is for auditory speech alone, and Figure 11-5 is for speech accompanied by visual cues. The visual cues make a large contribution to speech perception. Sumby and Pollack concluded that the contribution of visual cues is largest when noise levels are high and perception from the auditory source is poor.

Simultaneous Messages

Trying to understand a speaker amidst a babble of nearby voices is a common occurrence for us all. This has been called the *cocktail party problem,* although the problem is hardly restricted to social occasions. An operator in a system can be required to understand a voice over a loudspeaker or the voice of a co-worker in the presence of other voices. Simultaneous voices can mask a message like noise can do, but there is a difference. Noise is not coherently related to speech and does not sound like it, but simultaneous voices can sound like the voice that is receiving attention and can interfere as well as mask. The similarity among voices in quality and content can be confusing.

A pioneering study of simultaneous messages was by Broadbent (1953a). Two messages, in different voices, were presented simultaneously over a loudspeaker. Each message had a different call sign, and the listener was required to respond to only one of them. Broadbent's subjects were able to answer less than half the messages directed to them. From initial findings like these, investigators went forward to determine variables that influence the understanding of one message among others.

Physical Variables Influencing Reception of One of Two Simultaneous Messages. The physical separation of two simultaneous voices has been found to be a basic variable for the sorting out of simultaneous messages. Poulton (1953) had two loudspeakers separated by 180 degrees horizontally, opposite the left and right ears, or two of them on top of one another with 6 degrees of vertical separation. When the subjects had to respond to one of the two simultaneous messages the omissions and information heard wrongly were reduced by half when the sources were widely separated. Spieth et al. (1954) manipulated the amount of separation of two loudspeakers by having either 10, 20, 90, or 180 degrees horizontal separation between them. The greater the separation, the better the reception of one of two simultaneous messages.

Reception is better when simultaneous voices are distinctive. Broadbent (1953b) found that reception of a message was better when the two simultaneous voices were different speakers than when they were the same speaker. Differences in voices can be enhanced artificially by using audio filters which eliminate a portion of the speech spectrum (Spieth et al., 1954; Egan et al., 1954). Spieth et al. (1954) made two voices more distinctive by distorting them with a 1600-hertz high-pass filter in one message channel that passed only frequencies above 1600 hertz, and a 1600-hertz low-pass filter in the other message channel that passed only frequencies below 1600 hertz. Message reception was better when the voices were filtered than when they were not.

Perception in the Irrelevant Message Channel. The auditory system is capable of selective attention, given sufficient separation of the sources or distinctiveness of the simultaneous voices. Does this mean that attentiveness is complete, and that when attention is directed to one message channel nothing is perceived in the other? The answer is "yes, mostly," but with qualification. There is little doubt that a great deal is missed in the irrelevant message channel. Cherry (1953) presented one message in the left ear of a headset and a different message in the right ear. The subject had to attend to the message in one ear and say it aloud, which is called *shadowing*. Cherry found that his subjects failed to perceive that the voice speaking the irrelevant message in the other ear was speaking in a foreign language. Even when the irrelevant message was in a subject's own language, there was no perception of individual words or the meaning of the message.

The qualification is that some kinds of information get through. The study that established this qualification was by Moray (1959). Like Cherry (1953), Moray found that virtually nothing was perceived in the irrelevant channel. The exception was that one's own name in the irrelevant channel was heard. Broadbent (1971, p. 140), who is the leading analyst of this topic, concludes: "One can sum up these experiments as showing that the selection of one ear shuts out most of the information on the other. But if some does break through, the items which do so are not random, but are those whose content is of certain kinds." Moray concluded that the selective blocking of information in the irrelevant channel, where everything was screened out except the personally meaningful name, implies a central attentional mechanism in the auditory system.

Synthetic Speech

The day of synthetic speech is here. Machines talk, not in the rigid sense of an audio tape message but in a flexible way where a digital computer swiftly assembles words or their phonemic elements into speech. The potential of synthetic speech is large. There will be minor uses of synthetic speech, like a voice in your car reminding you to fasten your seat belt, but among the major uses will be an automatic reading machine which will read the printed page aloud to the blind, or voice warning systems that will tell an operator of an emergency (Chapter 10).

Synthetic speech has the same problems of intelligibility as natural speech. Synthetic voices often sound unnatural, like a robot in a science fiction movie. Certainly, this unnaturalness will disappear as further technical developments are made, but in the meantime there is interest in the intelligibility of synthetic speech. How does it compare with natural speech? The intelligibility of synthetic speech is evaluated in the same way as natural speech.

Pisoni and his colleagues at Indiana University, who have had an extensive research program on the intelligibility of speech, used a voice synthesizer called MITalk. MITalk uses a digital computer and was originally designed for converting any English text into natural-sounding speech as a reading machine for the blind. The design of MITalk is instructive on how a good speech synthesizer works (Pisoni & Hunnicutt, 1980). MITalk first looks up a word in a 12,000-word dictionary stored in the computer. Parts of speech and pronunciation are indicated for words that are found. Prosodic variation is the stress and timing of speech, which is important for naturalness, and MITalk looks up prosodic information on the duration, timing, and pause parameters for the word. Finally, taking all this information into account, 20 parameters for each phoneme are calculated and the word is translated into sound. If the word was not found in the dictionary, MITalk makes a letter-to-sound translation of the word. MITalk is an advanced system. Voice synthesizers are usually not so sophisticated.

Table 11-1 has the results of research by Pisoni (1982a,b) that compared the intelligibility of natural and synthetic speech. The various intelligibility tests that he used have been discussed earlier in this chapter. Synthetic speech, generated by a sophisticated voice synthesizer, compares favorably with natural speech. Anomalous sentences were an exception but of no particular importance in this context.

Pisoni (1982b) found evidences of improvement with training in the use of synthetic speech, and Schwab et al. (1985) tested the hypothesis that lesser speech synthesizers than MITalk can have the perception

TABLE 11-1   Intelligibility of Natural Speech and Synthetic Speech Compared*

Type of Test	Natural Speech	Synthetic Speech
Modified rhyme test	99.4	93.1
Meaningful sentences	99.2	93.2
Prose passages	67.8	70.3
Anomolous sentences	97.7	78.7

*The speech synthesizer was of high quality. The entries are percent correct.

Source: Pisoni (1982a). Speech technology: The evolution of computers that speak . . . and listen. Distinguished Faculty Research Lecture. Bloomington, IN: Office of Research and Graduate Development, Indiana University. And Pisoni (1982b). Perception of speech: The human listener as a cognitive interface. *Speech Technology, 1,* 10–23.

of their speech improved with training. The training was with informative feedback, where the listener could correct errors that were made. Schwab found significant improvement over 10 days of training, with much of the improvement being retained in a follow-up test six months later. Egan (1948) and Miller et al. (1951), in related findings, found improvements with experience in the intelligibility of speech in noise. Improvements in the intelligibility of indistinct speech through learning is analogous to living among foreigners who speak heavily accented English. Understanding is difficult at first, but becomes easier with experience.

Summary ───

The visual mode is the most prominent way of presenting information to a system operator, but speech is important also. As visual information must be clearly seen, so speech must be intelligible. There are several kinds of intelligibility tests that measure how well speech can be understood.

A degrading influence on speech intelligibility is noise. Loud noise and wide-band noise are effective masking agents. There are variables that offset the degrading effects of noise. When the vocabulary is small, the intelligibility of speech is higher because the listener can infer some of what is being said. The structure of language constrains the vocabulary and improves the perception of speech. Inferences about missing segments of words improve speech perception. Seeing the speaker, and watching movements of the lips and body, improve intelligibility.

Synthetic speech is with us, and we can expect to find increasing uses of it in systems. There is the same concern with the intelligibility of synthetic speech as with natural speech. The research evidence is that the intelligibility of synthetic speech, produced by a high-quality speech synthesizer, compares favorably with natural speech. Training improves the intelligibility of synthetic speech.

References ───────────────────────────────────────

Broadbent, D. E. (1953a). Listening to one or two synchronous messages. *Journal of Experimental Psychology, 44,* 51–55.

Broadbent, D. E. (1953b). Failures of attention in selective listening. *Journal of Experimental Psychology, 44,* 428–433.

Broadbent, D. E. (1971). *Decision and stress.* New York: Academic Press.

Cherry, E. C. (1953). Some experiments on the recognition of speech, with one and with two ears. *Journal of the Acoustical Society of America, 25,* 975–979.

Clark, H. H., & Clark, E. V. (1977). *Psychology and language.* New York: Harcourt Brace Jovanovich.

Egan, J. P. (1948). Articulation testing methods. *Laryngoscope, 58,* 955–991.

Egan, J. P., Carterette, E. C., & Thwing, E. J. (1954). Some factors affecting multi-channel listening. *Journal of the Acoustical Society of America, 26,* 774–782.

Hawley, M. E. (Ed.), (1977). Speech intelligibility and speaker recognition. In *Benchmark papers in acoustics.* Volume 11. Stroudsburg, PA: Dowden, Hutchinson & Ross.

Klatt, D. H. (1977). Review of the ARPA Speech Understanding Project. *Journal of the Acoustical Society of America, 62,* 1345–1366.

Kryter, K. D. (1962a). Methods for the calculation and use of the articulation index. *Journal of the Acoustical Society of America, 34,* 1689–1697.

Kryter, K. D. (1962b). Validation of the articulation index. *Journal of the Acoustical Society of America, 34,* 1698–1702.

Kryter, K. D. (1972). Speech communication. In H. P. Van Cott & R. G. Kinkade (Eds.), *Human engineering guide to equipment design.* Washington, DC: U.S. Government Printing Office. Pp. 161–226.

Kryter, K. D. (1985). *The effects of noise on man.* Second Edition. New York: Academic Press.

Miller, G. A. (1947). The masking of speech. *Psychological Bulletin, 44,* 105–129.

Miller, G. A., Heise, G. A., & Lichten, W. (1951). The intelligibility of speech as a function of the context of the test materials. *Journal of Experimental Psychology, 41,* 329–335.

Moray, N. (1959). Attention in dichotic listening: Affective cues and the influence of instructions. *Quarterly Journal of Experimental Psychology, 11,* 56–60.

Pisoni, D. B. (1982a). Speech technology: The evolution of computers that speak . . . and listen. Distinguished Faculty Research Lecture. Bloomington, IN: Office of Research and Graduate Development, Indiana University.

Pisoni, D. B. (1982b). Perception of speech: The human listener as a cognitive interface. *Speech Technology, 1,* 10–23.

Pisoni, D. B., & Hunnicutt, S. (1980). Perceptual evaluation of MITalk: The MIT unrestricted text-to-speech system. *Proceedings of the International Conference on Acoustics, Speech and Signal Processing.* New York: IEEE Press.

Poulton, E. C. (1953). Two-channel listening. *Journal of Experimental Psychology, 46,* 91–96.

Salasoo, A., & Pisoni, D. B. (1985). Interaction of knowledge sources in spoken word identification. *Journal of Memory and Language, 24,* 210–231.

Samuel, A. G. (1981). Phomenic restoration: Insights from a new methodology. *Journal of Experimental Psychology: General, 110,* 474–494.

Schwab, E. C., Nusbaum, H. C., & Pisoni, D. B. (1985). Some effects of training on the perception of synthetic speech. *Human Factors, 27,* 395–408.

Spieth, W., Curtis, J. F., & Webster, J. C. (1954). Responding to one of two simultaneous messages. *Journal of the Acoustical Society of America, 26,* 391–396.

Sumby, W. H., & Pollack, I. (1954). Visual contribution to speech intelligibility in noise. *Journal of the Acoustical Society of America, 26,* 212–215.

Warren, R. M. (1970). Perceptual restoration of missing speech sounds. *Science, 167,* 392–393.

Warren, R. M., & Obusek, C. J. (1971). Speech perception and phonemic restoration. *Perception & Psychophysics, 9,* 358–362.

Webster, J. C. (1978). Speech interference aspects of noise. In D. M. Libscomb (Ed.), *Noise and audiology.* Baltimore: University Park Press. Pp. 193–228.

PART V

Manual Control

CHAPTER **12**

Tracking Behavior: The Continuous Manual Control of a System

Tracking is familiar to us all. Driving a car, flying an airplane, or steering a bicycle are examples. Continuous control of a vehicle is the common example of tracking, but there are others. Watching a dial and manipulating a control to hold a fluctuating temperature on 1000 degrees in an industrial plant is an example of tracking also. A formal definition of tracking will be given in a moment, but these examples show that an intuitive grasp of tracking is easy.

Manual steering of a vehicle, like other manual functions, are giving way to automation, it is said. The human being will no longer be needed for control. Machines will do the controlling, with the human operator becoming an executive, doing the higher decision making which the machine does not do. There are, indeed, automatic tracking devices that contribute to the control of human–machine systems. The automatic pilot of the aircraft is a well-known example. Automatic recovery techniques for hazardous flight conditions, where the aircraft is programmed on a safe course if the pilot should pass out from excessive *g* forces during extreme maneuvers, is a more advanced conception of automatic control that is forthcoming (Klass, 1986). Notwithstanding, the prophecy that the human controller is being displaced by machines is not being fulfilled at a rapid rate. We need do no more than look around us to see that airplanes, automobiles, tractors, submarines,

249

and tanks are almost always controlled manually. There is a trend toward automatic control, but it is weaker than many suggest. Even when the trend becomes strong, and automatic control is routine for vehicles, we will see the operator involved with tracking, and the reasons are two:

1. In Chapter 2 we discussed principles of system reliability. It was said there that the operator can contribute to system reliability by being a redundant subsystem. The complexity of human–machine systems edges them toward unreliability, and a human operator that takes over when an automatic subsystem fails can be instrumental in keeping system reliability high. An automatic subsystem that is reliable enough to use means that the operator's manual skills will be needed infrequently, but when the automatic subsystem fails the manual skills must be available and proficient, as good as if the system was manually controlled all the time.

2. An automatic control subsystem must have the values of its variables specified; it must be told what it is to do for the system. There will always be unusual system sequences that cannot be preprogrammed which will require an operator to evaluate the situation perceptually and do something about it manually.

This chapter is about visual tracking. The visual emphasis is not arbitrary. Most tracking situations of the world are visual–motor, as is the research that has been done on them.

Tracking Defined

Tracking has an externally programmed input signal that defines a system requirement for the operator, and the requirement is met by manipulating a control that generates an output. The input minus the output is error, and the operator's requirement is to null the error. Tracking is error nulling. The displayed error difference between the input and the output is a continuous response requirement. The error elicits a corrective movement whose consequences for the system are displayed, which in turn may create a new response requirement, and so on. The continuous presentation of error and response to it makes tracking a closed-loop task. Closed-loop regulation was defined in Chapter 5, but as a reminder, a system is closed-loop when the thing controlled affects the agent that does the controlling. In tracking, the response to the error that is displayed changes the error quantity and thus the display.

Steering a car is a tracking task that is familiar to us. The road is the input that establishes the system requirement, the position of the car in the lane is the output established by manipulation of the steering wheel, and the driver responds continually to error in vehicle alignment in efforts to maintain the car's correct position on the road. Controlling an aircraft or a submarine is, in principle, the same although the input is often from instruments. Tracking was an early interest of human factors engineering because of its importance for so many human–machine systems.

Laboratory tracking tasks are analogs of real-life tracking tasks, and use of them in research has given us most of the scientific understanding that we have of variables that determine tracking behavior. A typical laboratory task will have a cathode ray tube for the display. The display might be simple, such as a moving dot for the input and a circle for the output which the subject tries to keep on the dot by manipulating a control. The input dot is programmed to move in a pattern. The control might be a pivoted stick, a handle, or a knob to be rotated.

The Measurement of Tracking Behavior

The measurement of tracking behavior is almost always some function of error. Figure 12-1 shows a segment of a continuous input and the response that might be made to it. The difference between input

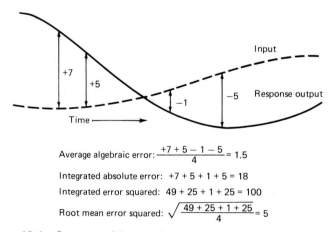

Average algebraic error: $\dfrac{+7 + 5 - 1 - 5}{4} = 1.5$

Integrated absolute error: $+7 + 5 + 1 + 5 = 18$

Integrated error squared: $49 + 25 + 1 + 25 = 100$

Root mean error squared: $\sqrt{\dfrac{49 + 25 + 1 + 25}{4}} = 5$

Figure 12-1 Segment of the continuous input and output of a tracking task, and four commonly used measures of the error difference between them.

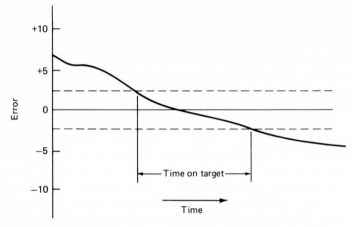

Figure 12-2 Measurement of time on target. The heavy line is the function for error difference between input and output. Time on target is recorded whenever the error function falls within a defined target zone, indicated by the dashed lines.

and output at any moment is error, and it is this difference score that is the index of operator proficiency and the basis of measurement. Measurement at four points is illustrated in Figure 12-1, and four commonly used measures based on them are shown. Average algebraic error retains algebraic sign, or direction of the error, and so is an index of directional bias in responding. The value of 1.5 in Figure 12-1 illustrates a bias toward positive error. Often, however, the analyst is interested in an overall index of error, not the directional bias, so the algebraic sign is dropped and one of the other three measures is computed.

Time on target is another useful measure, and it is illustrated in Figure 12-2. A scoring zone around zero error is defined, and time is cumulated whenever the operator maintains error within the zone. An engineer evaluating a new control system for an automobile might define the scoring zone as centerline of the car within plus/minus 1 foot of the center of the lane, and record the time that test drivers maintain the car within this tolerance band. Similarly, time on target might be used to document the position of an aircraft with respect to the runway centerline during landing.

Control System Dynamics

Control dynamics is defined as the relationship of the system's output to the control action of the operator. System output can be simple and proportional to the operator's response, but it need not be. A number of transformations are possible. Three of the most prominent ones are shown in Figure 12-3. At the top of the figure is a block diagram of an operator controlling a human–machine system that would describe tracking. The operator manipulates a control in response to error information on the display; the control action changes the state of the system, which is information that is fed back to the display and which, in turn, may require further control action. The complexity of the control–system relationship is called *control order*. The simplest control order is *position control*, which is also called zero-order control. The relationship between control action and system output is linear and proportional. Figure 12-3 assumes a simple step input signal and shows that the position control requires a correspondingly simple response to it. Moving the control 1 inch to the left might move the system 1 foot to the left, 2 inches to the left moves it 2 feet to the left, and

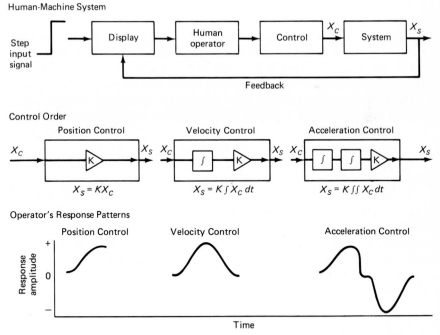

Figure 12-3 At the top is a block diagram of a tracking system receiving a step input signal. The three basic control orders are shown in the center, and the movement patterns of the operator in response to the input are shown at the bottom. X_c is the response of the operator, X_s is the response of the system, and K is the constant of proportionality, or gain. [Adapted from Adams (1967), Fig. 9-6. From *Contemporary Approaches to Psychology,* edited by Harry Helson and William Bevan © by D. Van Nostrand Company, Inc. Reprinted by permission of Wadsworth, Inc.]

so on. The constant of proportionality *(K)* could be different, perhaps defining a more sensitive control where ½ inch movement of the control would move the system 1 foot.

First-order or *velocity (rate) control,* shown in Figure 12-3, has more complex dynamics. In addition to the constant of proportionality, there is one integration between input and output, which means that a control movement defines a velocity output for the system. Figure 12-3 shows that the pattern of control movement is more complex for the velocity control.

Second-order or *acceleration control,* also shown in Figure 12-3, is more complicated still. Here there is double integration between control action and system output. The control movement defines an acceleration of the system. The pattern of control movement, illustrated in Figure 12-3 for the step input, is the most elaborate of all.

Which control order produces the best tracking performance? Figure 12-3 shows that the position control has the simplest response pattern, so proficiency with it should be the best. Position controlling indeed can be as good or better as other control system dynamics (Lincoln & Smith, 1952; Obermayer et al., 1961). Poulton, in his book on tracking behavior, concludes: "A man tracks most accurately with a position control system" (Poulton, 1974, p. 325). There is truth in the statement, but it needs elaboration. When the input is complex, with continually changing positions, velocities, and accelerations, the position control requires that the operator match these complexities, and it may not be easy, although it is probably easier than with a velocity or an acceleration control. When, however, the input has constant velocities that predominate, or constant accelerations, higher-order dynamics might be easier because the control system is accommodating some of the complexities of responding and relieving the operator of them. Suppose, for example, that an operator had to track a constantly accelerating missile with an optical sight. A position control for the sight would require the operator to match the missile's acceleration with a continually accelerating response. With an acceleration control, however, the operator need only insert a simple step value into the control system that would establish an acceleration output of the sight equal to the missile's acceleration.

Lag

A system seldom responds instantaneously to the action of a tracking control. Usually, there is a delay, or *lag* as it is called, between the control movement, the system response, and the corresponding indication on the tracking display. The mathematical function that relates control movement to delayed system output can, in principle, be almost anything, but the lag function that has received research attention because it is characteristic of many vehicles is the exponential lag. The exponential lag is illustrated in Figure 12-4. The exponential function is asymptotic at infinity, which means that there is no finite point at which it ceases and which could be used to define the value of the lag. Some point along the function must be specified, and the convention that is used is the point at which the output travels 63 percent of the asymptotic output level when a step response is inserted in the control system. This value is called the *time constant*.

A pioneering investigator of exponential lag was Levine (1953). He used a one-dimensional tracking task as his research device, with lags ranging from .015 to 2.70 seconds. The display had two small vertical lines as the reference, and the time that the error indicator, a small dot, was kept between the lines was the measurement of performance (time on target). Levine's findings are shown in Figure 12-5. Tracking

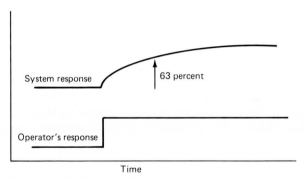

Figure 12-4   Illustration of exponential lag in system response to a step response by the operator. The 63 percent point, discussed in the text, is the time constant of the lag and defines its duration.

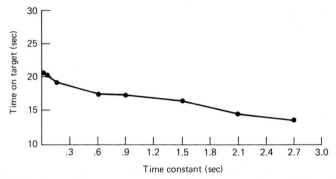

Figure 12-5   Effect of lag on tracking performance. [Source: Levine (1953).]

performance steadily declined as a function of lag. Even small lags of a fraction of a second degrade performance. In a subsequent experiment, Levine et al. (1964) demonstrated that the effects of lag were not overcome with practice. Levine's findings have been replicated in other tracking studies of exponential lag, as well as in tracking studies of other lag functions (Muckler & Obermayer, 1964). There is no question that lag seriously degrades tracking performance.

Quickening

The example of an optical sight tracking a missile that was given above shows how an operator's responding can be simplified with a proper choice of control system dynamics. Another approach to response simplification by design of the control system is *quickening*. Taylor and Garvey (1959), in discussing quickening, contrast a design approach to a training approach. They called the training approach the "Procrustean approach." (It means force fit by extreme methods. Procrustes was a mythical robber who stretched or amputated the limbs of travelers to make them fit his bed.) Taylor and Garvey are saying that training, which adjusts the human to the machine, is the hard way; designing the machine to conform to the human is the easy way. There is no doubt that the design approach is fruitful. An enthusiasm for the design approach can be justified, but we should not forget that even the best of designs require personnel training to optimize system performance (Chapter 18). Ordinarily, it is not a choice between design and training but how best to complement them.

Lags in the system can cause the operator to overshoot and undershoot in her efforts to find the correct control position, which, in turn, can cause system instability. Jagacinski (1977) illustrates lags, and overshooting and undershooting, with the example of adjusting a shower to the desired water temperature. A person will often adjust the shower spigot too rapidly for the delay in the plumbing, and then will have to adjust it back again, and then back again once more. The oscillations that result produce an unstable system. The spigot could be changed very gradually, but that would be a poor solution because the shower system would not change fast enough to achieve the goal of a speedy shower. A solution to this kind of problem is quickening.

Figure 12-6 illustrates quickening for a velocity control system, although of course it could be used for higher-order dynamics as well. Quickening taps the signal of control movement prior to integration, and then summates it on the display with the system output signal. Although the system dynamics remain the same, the operator receives immediate information about the position of the control and does not have to wait for the lag in system response. The overall tracking

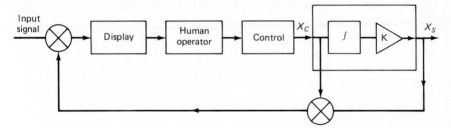

Figure 12-6 Quickening in a rate tracking task. [Adapted from Adams (1967), Fig. 9-7. From *Contemporary Approaches to Psychology,* edited by Harry Helson and William Bevan © 1967 by D. Van Nostrand Company. Reprinted by permission of Wadsworth, Inc.]

response pattern that the operator generates to control the system remains the same but the task at any moment is simple position controlling. Quickening can lead to dramatic benefits in tracking performance (Bailey, 1958; Birmingham et al., 1954; Holland & Hensen, 1956; Taylor, 1960).

A quickened display has disadvantages that may offset its advantages. For all its merits, a quickened tracking display is no longer an accurate status display that shows correct error for the system because the display confounds information directly from the control system with system output. An operator who corrects an error will get immediate elimination of it on the display because of the direct pickup from his control, and it would be easy to assume that the error for the system has been corrected at the same time, but it may not be because of lag. The system error will eventually be corrected as the operator continues the error nulling, but the discrepancy between what the display shows and what the system is actually doing could create a false sense of security in the operator and be potentially dangerous. Other sources of information on the display panel can be required to keep the operator in touch with the system's actual state.

Display Variables

Pursuit versus Compensatory Tracking

There are two basic ways of displaying the error quantity to which the operator responds in a tracking task. One is called *pursuit tracking* and the other *compensatory tracking*. Examples of these two kinds of tracking displays are shown in Figure 12-7. In pursuit tracking the input and the output are displayed separately. The difference between the two is error, which an operator tries to null. With compensatory tracking the error difference is abstracted and used to drive a moving error indicator with respect to a fixed reference. Error nulling is the requirement in both instances, but pursuit tracking has the three quantities of input, output, and error displayed, whereas compensatory tracking displays only error.

Driving your car is a pursuit tracking task. You see the road (input), and you see the position of your car with respect to the road that results from your control actions (output). The inappropriate position of the car on the road (error) is sensed, and the steering wheel is manipulated to correct it. It is hard to visualize driving a car as a compensatory tracking task, but imagine an automobile display where you saw neither the road nor the position of the car on it but only driving error. A compensatory tracking display is unrealistic for an automobile, but it is not unrealistic for many systems. Indeed, a designer has a choice between compensatory tracking and pursuit tracking with many systems. A space vehicle can have a pursuit tracking display where the satellite for rendezvous is represented by one symbol (input), the space vehicle represented by another symbol (output), and the astronaut strives to close the error difference between them. Or the error difference between the input and the output could be abstracted and used to drive the error indicator of a compensatory tracking display. Which is best?

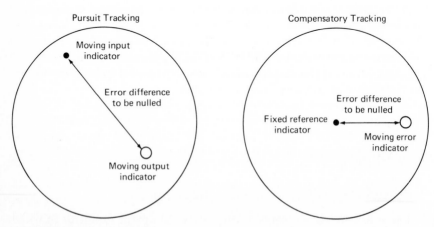

Figure 12-7 Illustration of the differences between pursuit and compensatory tracking displays.

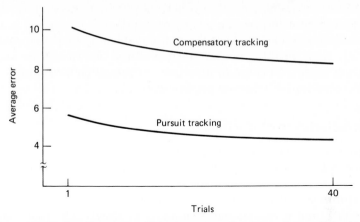

Figure 12-8 Tracking data which show the advantage of pursuit tracking over compensatory tracking. [Adapted from Briggs, G. E. and Rockway, M. R., *Journal of Experimental Psychology* 71, 165–169, Fig. 2. Copyright 1966 by the American Psychological Association. Adapted by permission of the authors.]

A pursuit tracking display almost always produces better performance than a compensatory tracking display (e.g., Briggs, 1962; Briggs & Rockway, 1966; Pew, 1974; Poulton, 1952). Figure 12-8 presents the findings of a study by Briggs & Rockway (1966) that compared compensatory and pursuit tracking. The advantage for pursuit tracking is unmistakable.

Why does a pursuit tracking display produce better performance than a compensatory tracking display? There are two hypotheses to consider:

1. Knowledge of results is better with the pursuit tracking display than with the compensatory tracking display. Knowledge of results is information received about the appropriateness of responding, and it is a strong determinant of human learning (Chapter 5). A pursuit tracking display, which has operator response actions separately displayed in the output indicator, gives clear knowledge of results about the appropriateness of corrective movements. A compensatory tracking display shows neither the input nor the output, only the error difference between them. As a difference score, error is jointly determined by input and output, and knowledge of results can be clouded because change in error at any movement could be the function of either input or output or both.

2. A pursuit tracking display, which shows the input directly, allows the operator to perceive whatever regularities there might be in the input and learn about them. Compensatory tracking, showing only error, has the input pattern obscured. As we will see later in the chapter, learning to anticipate regularities in the input is a determinant of tracking performance.

Characteristics of the Input

The nature of the input to a tracking task, and how it is represented on the display, is a fundamental determiner of tracking proficiency.

Frequency of the Input. An input to a tracking task can be simple or complex, and it can change frequently or slowly. The road you are driving can be straight and uncomplicated, or it can be a twisting mountain road. A pilot can have a comparatively easy tracking task of aligning his aircraft to a runway for landing, or he can have a difficult task of positioning to fire at a maneuvering enemy aircraft whose pilot is determined to escape. Correspondingly, laboratory investigators of tracking have studied simple and complex inputs. Figure 12-9 illustrates the kinds of inputs that have been used in laboratory research. The inputs range from oscillating sine waves of different frequencies to more complex wave forms composed of several sine waves. Sine waves have been most commonly used in tracking research, and manipulation of their characteristics is a good example of how the input can affect tracking performance. The general finding is that the faster the input changes, the poorer the tracking performance (Fitts et al., 1953; Noble et al., 1955; Hartman & Fitts, 1955; Pew, 1974). The study by Noble et al. (1955) is a good example of this line of research. A one-dimensional pursuit tracking task was used, with sine-wave inputs that ranged from .17 to 4 hertz. A sketch of the research task is shown in Figure 12-10. The input cursor, which was propelled by the sine wave, was a short line on a cathode ray tube display, and the output cursor, positioned by the subject's control movements, was another short line. As a pursuit tracking task, the subject attempted to keep the two cursors in alignment as much as possible. The subject's control was a lever with a knob that the subject moved back and forth in the horizontal plane. Three subjects, highly experienced in tracking, were used. After familiarization sessions, 12 one-hour sessions were given where the subjects performed with the various input frequencies. The results are shown in Figure 12-11. First notice the horizontal line across the center of the figure. It is the error when the subject takes his hand off the control and does nothing. Error below this level is improvement, and error above it means that the subject is detrimental

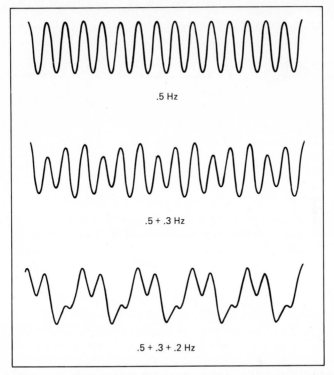

Figure 12-9 Illustration of the input signals used in laboratory research on tracking: A sine wave with frequency of .5 hertz, a combination of .5 and .3 hertz, and a combination of .5, .3, and .2 hertz. [Adapted from Hartman and Fitts (1955), Fig. 3.]

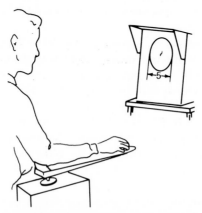

Figure 12-10 One-dimensional tracking task used in the research by Noble et al. (1955). [Adapted from Hartman and Fitts (1955), Fig. 1.]

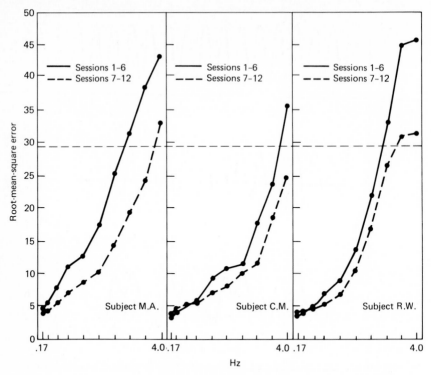

Figure 12-11 For three subjects, tracking error as a function of frequency of the input. [Adapted from Noble et al. (1955), Fig. 1.]

to the system and worsens performance with his response. Figure 12-11 shows that the greater the frequency of the input, the poorer the tracking performance. Performance is so poor at the higher frequencies that the subjects would have been ahead to do nothing. The performance curve for sessions 7 through 12 has a lower error and shows that practice helps but not much, probably because these are highly practiced subjects who cannot improve much more. Experience does little to overcome the demands of difficult input.

The Display Scale Factor. The scale on the display can be fixed when it is from the external world, like the road when you are driving your car, but in many applications the scale on the display is adjustable. What should the amplitude of the input cursor in pursuit tracking or the error cursor in compensatory tracking on the display be? Figure 12-12 illustrates small and large amplitude of the input as it might appear in a pursuit-tracking task. The evidence (Fitts et al., 1953; Hartman & Fitts, 1955) is that tracking performance improves

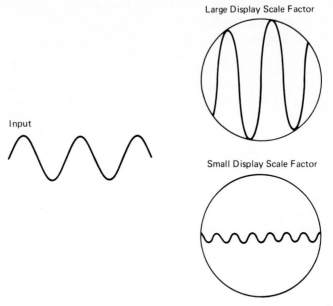

Figure 12-12 How an input can be scaled large or small on a tracking display.

as amplitude on the display increases, and it probably should be as large as the display reasonably allows. Hartman and Fitts (1955) hypothesize that larger amplitude on the display permits better discrimination of amplitudes and rates.

Anticipation of Input Characteristics. When an input is random, the operator has no choice but to wait for each change in it to occur and then respond as rapidly as possible. Being driven by the stimulus, the operator is reactive, and the proficiency that develops is on the motor side. The learning will involve control-display relations (the directional changes in display cursor as a function of direction of control movement), speed, accuracy in positioning the control, and rates and accelerations of movement.

The situation is different, and more perceptual, when the operator has a preview and sees what is coming next in the input signal, or when the input is nonrandom with regularities that the operator can learn. Not only is the input signal itself a determinant of the tracking behavior, but what the operator knows about it in advance is a determinant as well. This topic in tracking is called *anticipation,* and Poulton was a pioneer of it. Surveys of research on anticipation and general discussions of it can be found in Adams (1961, 1967), Bartlett (1951), and Poulton (1957).

A distinction is made between receptor anticipation and perceptual anticipation. *Receptor anticipation* is a preview of the input signal. Seeing the road ahead while you are driving is an example. *Perceptual anticipation* is learning about regularities in the input and preparing yourself for response to them with knowledge acquired in past experience. You can anticipate turns when you always drive the same road home. Both of these kinds of anticipation result in better tracking performance.

Poulton (1964) studied receptor anticipation in pursuit tracking. Previews of both short and long duration were used with both simple and complex inputs. The input, as a wavy line, passed in front of the subject on a moving paper tape, and he attempted to keep a ballpoint pen on it, which was position tracking. A screen could be adjusted for no preview of the tape, or preview of the tape for .5 second ahead or 7.0 seconds ahead. The simple input was a .54-hertz sine wave. The complex input was an irregular waveform that was a composite of four sine waves: .54, .42, .37, and .23 hertz. The complex input had regularities but they would not be as apparent as with the simple sine wave. Both preview intervals were found to benefit the tracking of both inputs.

Poulton also studied perceptual anticipation (Poulton, 1952). The subject viewed a small window with one or two indicators showing in it, and responded with a hand crank. A two-indicator configuration was pursuit tracking, one indicator being the input and the other the output. A one-indicator configuration was compensatory tracking where the subject tried to keep the indicator on a reference line. The same two inputs as described above for Poulton's study of receptor anticipation (Poulton, 1964) were used. Poulton's point of departure was the superiority of pursuit tracking over compensatory tracking, and he hypothesized that having the input separately displayed in pursuit tracking made the learning of input regularities comparatively easy and was a factor in the superiority of pursuit tracking. Without anticipation, a subject will delay response to directional change in the input by at least one reaction time value because the stimulus must change before a response can be initiated to it. With anticipation, the subject can anticipate directional change in the input and get the response under way before it actually occurs. Ideally, with perfect anticipation, the time delay between input change and response to it is zero. Poulton recorded the performances of his subjects and measured the response time between directional change of the input and the response to it. Figure 12-13 shows the distributions of response times for pursuit tracking of the simple sine wave input, and for both early and late in practice. The important point of reference in Figure 12-13 is the arrow,

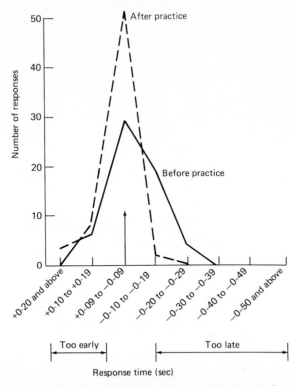

Figure 12-13 Distribution of response times to directional changes of a sine-wave input in pursuit tracking, before and after practice. The arrow is the point of perfect anticipation of the change. [Adapted from Poulton, E. C., *British Journal of Psychology* 43, 222–229 (1952) Copyright 1952 by the British Psychological Society. Reprinted by permission of the British Psychological Society and the author.]

which is the point of perfect anticipation where the subject responds at the instant the input changes. Some anticipation was found early in learning, after relatively little practice, but considerably more is found after practice. Anticipation of input regularities is one of the things that is learned in pursuit tracking. Figure 12-14 has the same measure for compensatory tracking. There is no anticipation early in practice, although some develops with practice but not to the degree found for pursuit tracking. With compensatory tracking the subject sees neither the input nor the output directly, only the error difference between them, and inferring input regularities is difficult. The findings for the complex input were somewhat similar, although there was no effect of practice.

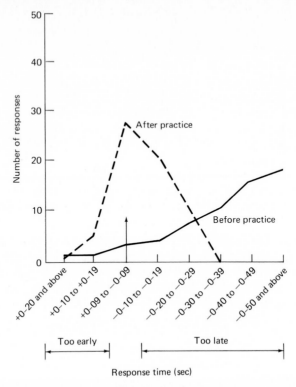

Figure 12-14 Distribution of response times to directional changes of a sine-wave input in compensatory tracking, before and after practice. The arrow is the point of perfect anticipation of the change. Comparison should be made with Figure 12-13, which has corresponding data for pursuit tracking. [Adapted from Poulton, E. C., *British Journal of Psychology* 43, 222–229 (1952) Copyright 1952 by the British Psychological Society. Reprinted by permission of the British Psychological Society and the author.]

Predictor Displays

A predictor display is similar to anticipation in that it helps the operator be ahead of the system and get the response under way early. The heart of a predictor display is a mathematical model of the system that is being controlled, and the model is operated in fast time. The operator's present position of the control is input to the model, the model is used to calculate the position of the system t seconds in the future, say 10 seconds, and this future position is displayed. The display tells the operator what can be expected to happen on the assumption that the control position remains unchanged for the duration of the prediction period. If the operator likes the outcome of his current responding, he leaves the controls as they are and continues on the present course, but if error is predicted, a corrective response is required.

An example of a predictor display might be a landing display for an aircraft, where the position of the aircraft t seconds in the future would be calculated in the approach phase. As the aircraft moved closer and closer to the runway, the pilot would increasingly be able to estimate his touchdown point. Pilots are mentally projecting the future position of their aircraft all the time, but a predictor display could upgrade accuracy of the projection.

Predictor displays have not become as common in systems as one might have guessed from the enthusiasm that was shown for them in the 1950s and 1960s. The concept seems inherently sound. Kelley (1968, Chap. 10) reports research which shows an advantage for tracking with a predictor display. Perhaps the difficulty is accurate specification of the model and its instrumentation. If so, these are temporary difficulties. Perhaps predictor displays will move into prominence as systems such as aircraft become much faster than they are today. An operator may need assistance if he is to stay ahead of a very fast moving system and anticipate response requirements.

Control Variables

There will always be decisions about a tracking control that are particular for a system, but are there general considerations that transcend the particular configuration of a control that is chosen? Two of them are extent of control movement or response scale factors, and tension on the control.

Response Scale Factors

The pioneering work on tracking error as a function of extent of control movement was done by Helson (1949), and research since then has confirmed his findings and conclusions. The origin of Helson's research was in World War II with the need to optimize the controls of antiaircraft weapons. The laboratory analog of the field task was a compensatory tracking task. A simple display had a moving pointer that had to be kept in alignment with a fixed pointer. The control was a handwheel crank. The amount of handwheel motion for full displacement of the pointer on the display could be varied from 5 degrees, which was a sensitive control, to 500 degrees—about a turn and a half—which was an insensitive control. Helson found that a relatively insensitive control gave the least tracking error. There was a tendency for tracking error to increase, however, if the insensitivity became too great, which is expected because too many time-consuming turns of the handwheel would be required. Correspondingly, error increased when sensitivity became too great. Increasing error when either sensitivity and insensitivity became too great led Helson to the *U-hypothesis*. The U-hypothesis says that there is a favorable midrange for control sensitivity, and Helson's research says that the midrange is a relatively insensitive region. Subsequent research by others, using tracking tasks with other kinds of displays and controls, confirmed Helson by finding that tracking was best with comparatively large amplitudes of movement (Fitts et al., 1953; Hartman & Fitts, 1955; Briggs et al., 1957; Andreas et al., 1957).

One hypothesis attributes the benefits of larger amplitudes of movement for tracking to proprioception (Chapter 5). Scientists vary the characteristics of light in studying receptors of the eye, and similarly they vary extents of movement and tensions involved in movement to infer how proprioceptors work. This line of reasoning says that larger extents of movements should be better than small extents because more proprioceptors fire and the brain has better information about positions of the responding limb. A similar contention is made for tension on the control. Proprioception is a speedy source of internal information about the action of the limb and a valuable supplement to information on the display.

Tension on the Control

There can be different kinds of loading on a control, such as a weight, viscosity, or friction, but Poulton (1974, pp. 306–307) regards spring tension on the control as the best, and he is correct. With spring tension the pressure is proportional to displacement, and it is another way to increase information about limb position through proprioceptive feedback. In addition, a spring-loaded control can be self-centering, which means that the operator will always have the same starting position whenever he uses the control.

Using a one-dimensional pursuit tracking task like that in Figure 12-10, Howland and Noble (1953) found that spring tension gave the best tracking performance of several ways to load the control. Briggs et al. (1957) used a compensatory tracking task and had two amounts of spring tension and two extents of movement as experimental variables. The largest amounts of spring tension and extent of movement gave the best tracking performance.

There are lines of basic research on motor behavior that underscore performance advantages for spring tension. A linear positioning task is a laboratory research device that requires a subject to learn to move a slide a specified distance along a track. The slide can have a spring on it. Performance is better with spring tension (Adams et al., 1972; Adams et al., 1977). Adams and Goetz (1973) used a linear positioning task to study variables associated with error detection and correction, the behavior that is at the heart of tracking. Error detection and correction were better with spring tension.

Multidimensional Tracking

A *multidimensional tracking task* is defined as two or more input sources and a dimension of the control system for each. Control systems could have all dimensions on one control, or they could have them on separate controls. Similarly, inputs could be on different displays or the same display.

There is nothing in the research literature to indicate that the variables that have been described above for one-dimensional tracking, which have been featured in previous sections of this chapter, do not apply to multidimensional tracking also. What is different about multidimensional visual tracking is that it will often have two or more stimulus sources, spatially spread out, and this is an added tracking variable.

Spatial Separation of Input Sources

It was said in Chapter 9 that a good workplace design will minimize visual scanning. The same principle applies when the display sources for multidimensional tracking are considered. All research on the topic indicates that proficiency in multidimensional tracking deteriorates as the spatial separation of its inputs increases (Fitts & Simon, 1952; Briggs & Howell, 1959; Adams & Xhignesse, 1960; King, 1961). A qualification of this generalization is that the degrading effects of input separation can be canceled if the stimuli are correlated, and one can be predicted from the other. If, with training, the operator can come to predict what will happen in one source while looking at another, he need not look at it as much to make his response and can avoid the delay of time-consuming eye movements. Adams and Xhignesse (1960) used a two-dimensional pursuit tracking task in an experiment which had the stimuli in the two sources either predictable from one another or not, and the stimulus sources separated by either 5 or 30 degrees.

Tracking performance was negatively related to separation of stimulus sources when inputs were uncorrelated, but separation had no effect on tracking performance when inputs were predictable from one another. King (1961) had a similar finding. Without input correlation the less spatial input separation the better, but with input correlation some spatial separation is allowable.

Summary

A familiar example of tracking is continuous control of a vehicle such as an automobile. Tracking is defined as an externally programmed input signal that defines a system requirement for the operator, and the requirement is met by manipulating a control that generates an output. The input minus output is error, and the operator's task is to null the error. The relationship between system output and control action of the operator is not always proportional but can undergo various transformations. The transformations are called control dynamics. Tracking is usually best when transformations are the least.

Display variables are prominent determiners of tracking performance. Pursuit tracking and compensatory tracking are the two basic ways of displaying the error quantity that the operator must null. Pursuit tracking displays both the input and the output, and the operator responds to eliminate the error difference between them. Compensatory tracking presents only the error difference between input and output. Performance is better with pursuit tracking than with compensatory tracking. Other display variables that were examined were frequency of the input, scale factors, predictability of the input, and predictor displays.

Variables of the control system that are major determiners of tracking performance are extent of control movement and tension on the control. Within reasonable limits, larger extents of control movements and greater tension produce better tracking performance.

Multidimensional tracking has two or more inputs and a dimension of the control system for each. The principles for one-dimensional tracking would appear to apply for two-dimensional tracking also, but two-dimensional tracking has spatial separation of visual input sources as a variable not found for one-dimensional tracking. The less spatial separation the better, although the correlation of inputs, allowing prediction of one input from another and the anticipation of response requirements, overcomes spatial separation of input sources somewhat.

References

Adams, J. A. (1961). Human tracking behavior. *Psychological Bulletin, 58,* 55–79.

Adams, J. A. (1967). Engineering psychology. In H. Helson & W. Bevan (Eds.), *Contemporary approaches to psychology.* Princeton, NJ: D. Van Nostrand, Pp. 345–383.

Adams, J. A., & Goetz, E. T. (1973). Feedback and practice as variables in error detection and correction. *Journal of Motor Behavior, 5,* 217–224.

Adams, J. A., & Xhignesse, L. V. (1960). Some determinants of two-dimensional visual tracking behavior. *Journal of Experimental Psychology, 60,* 391–403.

Adams, J. A., Goetz, E. T., & Marshall, P. H. (1972). Response feedback and motor learning. *Journal of Experimental Psychology, 92,* 391–397.

Adams, J. A., Gopher, D., & Lintern, G. (1977). Effects of visual and proprioceptive feedback on motor learning. *Journal of Motor Behavior, 9,* 11–22.

Andreas, B. G., Gerall, A. A., Green, R. F., & Murphy, D. P. (1957). Performance in following tracking as a function of the sensitivity of the airplane-type control stick. *Journal of Psychology, 43,* 169–179.

Bailey, A. W. (1958). Simplifying the operator's task as a controller. *Ergonomics, 1,* 177–181.

Bartlett, F. C. (1951). Anticipation in human performance. In G. Ekman, T. Husén, G. Johansson, & C. I. Sandström (Eds.), *Essays in psychology.* Stockholm: Almqvist & Wiksells, Pp. 1–17.

Birmingham, H. P., Kahn, A., & Taylor, F. V. (1954). A demonstration of the effects of quickening in multiple-coordinate control tasks. Washington, DC: Naval Research Laboratory, NRL Report 4380, June 23.

Briggs, G. E. (1962). Pursuit and compensatory modes of information display: A review. Wright-Patterson Air Force Base, OH: 6570th Aerospace Medical Research Laboratories, Technical Documentary Report AMRL-TDR-62-93, August.

Briggs, G. E., & Howell, W. C. (1959). On the relative importance of time sharing at central and peripheral levels. Port Washington, NY: U.S. Naval Training Device Center, Technical Report NAVTRADEVCEN 508-2, October.

Briggs, G. E., & Rockway, M. R. (1966). Learning and performance as a function of the percentage of pursuit component in a tracking display. *Journal of Experimental Psychology, 71,* 165–169.

Briggs, G. E., Fitts, P. M., & Bahrick, H. P. (1957). Effects of force and amplitude cues on learning and performance in a complex tracking task. *Journal of Experimental Psychology, 54,* 262–268.

Fitts, P. M., & Simon, C. W. (1952). Some relations between stimulus patterns and performance in a continuous dual-pursuit task. *Journal of Experimental Psychology, 43,* 428–436.

Fitts, P. M., Marlowe, E., & Noble, M. E. (1953). The interrelations of task variables in continuous pursuit tasks: I. Visual-display scale, arm-control scale, and target frequency in pursuit tracking. San Antonio, TX: USAF Air Research and Development Command, Human Resources Research Center Research Bulletin 53-34, September.

Hartman, B. O., & Fitts, P. M. (1955). Relation of stimulus and response amplitude to tracking performance. *Journal of Experimental Psychology, 49,* 82–92.

Helson, H. (1949). Design of equipment and optimal human operation. *American Journal of Psychology, 62,* 473–497.

Holland, J. G., & Hensen, J. B. (1956). Transfer of training between quickened and unquickened tracking systems. *Journal of Applied Psychology, 40,* 362–366.

Howland, D., & Noble, M. E. (1953). The effect of physical constants of a control on tracking performance. *Journal of Experimental Psychology, 46,* 353–360.

Jagacinski, R. J. (1977). A qualitative look at feedback control theory as a style of describing behavior. *Human Factors, 19,* 331–347.

Kelley, C. R. (1968). *Manual and automatic control.* New York: Wiley.

King, W. J. (1961). Coordination in a complex tracking task as a joint function of spatial separation and predictability of stimuli. Unpublished master's thesis, University of Illinois at Urbana-Champaign.

Klass, P. J. (1986). Maneuvering attack system prototype undergoing testing in F-16. *Aviation Week & Space Technology,* June 23.

Levine, M. (1953). Tracking performance as a function of exponential delay between control and display. Wright-Patterson Air Force Base, OH: Wright Air Development Center, WADC Technical Report 53-236, October.

Levine, M., Senders, J. W., Morgan, R. L., Doxtater, L. (1964). Tracking performance as a function of exponential delay and learning. Wright-Patterson Air Force Base, OH: Aerospace Medical Research Laboratories, Technical Report AMRL-TR-64-104, November.

Lincoln, R. S., & Smith, K. U. (1952). Systematic analysis of factors determining accuracy in visual tracking. *Science, 116,* 183–187.

Muckler, F. A., & Obermayer, R. W. (1964). Control system lags and man-machine system performance. Washington, DC: National Aeronautics and Space Administration, NASA Contractor Report CR-83, July.

Noble, M., Fitts, P. M., & Warren, C. E. (1955). The frequency response of skilled subjects in a pursuit tracking task. *Journal of Experimental Psychology, 49,* 249–256.

Obermayer, R. W., Swartz, W. F., & Muckler, F. A. (1961). The interaction of information displays with control system dynamics in continuous tracking. *Journal of Applied Psychology, 45,* 369–375.

Pew, R. W. (1974). Human perceptual-motor performance. In B. H. Kantowitz (Ed.), *Human information processing: Tutorials in performance and cognition.* Hillsdale, NJ: Lawrence Erlbaum. Pp. 1–39.

Poulton, E. C. (1952). Perceptual anticipation in tracking with two-pointer and one-pointer displays. *British Journal of Psychology, 43,* 222–229.

Poulton, E. C. (1957). On prediction in skilled movements. *Psychological Bulletin, 54,* 467–478.

Poulton, E. C. (1964). Postview and preview in tracking with complex and simple inputs. *Ergonomics, 7,* Pp. 257–266.

Poulton, E. C. (1974). *Tracking and manual control.* New York: Academic Press.

Taylor, F. V. (1960). Four basic ideas in engineering psychology. *American Psychologist, 15,* 643–649.

Taylor, F. V., & Garvey, W. D. (1959). The limitations of a "Procrustean" approach to the optimization of man–machine systems. *Ergonomics, 2,* 187–194.

CHAPTER 13

Discrete Controls

Tracking, or continuous manual control, is central to many system operations. Although we covered this in Chapter 12, there are other controls in the workplace also. They are used for such functions as regulation of subsystems like engines, the resolution of emergencies, and starting and stopping operations. The controls for these various functions are often discrete controls, although it would be more exact to call some of them relatively discrete controls. *Discrete controls* assume a small number of states, as in a two-position switch, or a small range of states, as in a knob that is turned to a setting. A tracking control, by contrast, is continuous over a range, sometimes in more than one dimension. Discrete controls are the topic of this chapter.

Figure 13-1 indicates the kinds of controls that will be discussed. Human factors research has not yet included all of the discrete controls in Figure 13-1, so secure recommendations cannot be made about all of them. Nevertheless, the research knowledge that is available provides some general guidance.

Human factors research has documented the difficulties that poorly designed discrete controls can create. Fitts and Jones (1947/1961) collected information on errors that pilots made in operating aircraft controls, and they found that most of the mistakes involved discrete controls. Controls were confused with one another, moved in the wrong direction, adjusted wrongly, or unintentionally activated. Errors like these impaired mission success, put the aircraft in a dangerous attitude, or were deadly (sometimes all three).

The day will come when the system will recognize an operator's voice and all she need do is request a function and it will be done automatically (see Chapter 16 for a discussion of automatic voice recognition).

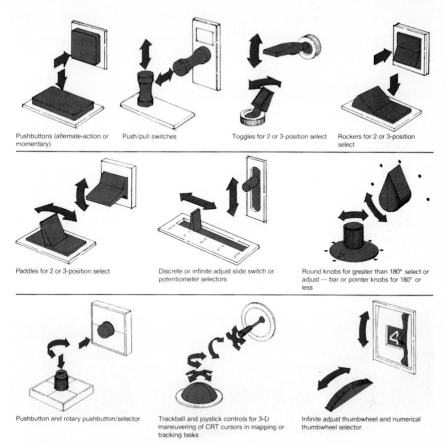

Pushbuttons (alternate-action or momentary)   Push/pull switches   Toggles for 2 or 3-position select   Rockers for 2 or 3-position select

Paddles for 2 or 3-position select   Discrete or infinite adjust slide switch or potentiometer selectors   Round knobs for greater than 180° select or adjust — bar or pointer knobs for 180° or less

Pushbutton and rotary pushbutton/selector   Trackball and joystick controls for 3-D maneuvering of CRT cursors in mapping or tracking tasks   Infinite adjust thumbwheel and numerical thumbwheel selector

Figure 13-1 Representative discrete controls of the kind that are discussed in this chapter. [Adapted from Odom (1984), Fig. 2. Used by permission of Micro Switch, A Honeywell Division.]

Switches and knobs will no longer be needed when this day comes. A design plan for the cockpit of future helicopters has no knobs or switches (Pruyn & Domenic, 1982). Pilot input is planned to be either by keyboard or voice recognition. Until these developments become more of a reality than they are, knobs and switches will dominate the human–machine scene. The inexpensiveness and reliability of knobs and switches will keep them in use even when alternatives become commonplace.

Sometimes speed can be important for response with a discrete control, like a rapid braking response when a car must suddenly be stopped, but fast reaction time is not always required. Reactions with discrete controls can be based on deliberate planning. The starting of your car and the turning on of its lights, or the adjustment of a radio, are unhurried decisions made with goals in mind.

Desirable Characteristics of Knobs and Their Layout

The Discriminability of Knobs

Knobs with similar shapes cannot be discriminated very well in the dark by touch, obviously. Even with vision available, highly similar knobs can be confused. Fitts and Jones (1947/1961), in their study of errors in the operation of aircraft controls, found that failure to discriminate similarly shaped controls was a cause of errors. A study by Jenkins (1947) on the discriminability of knob shapes was an early experiment with an enduring message. Jenkins had a subject feel the

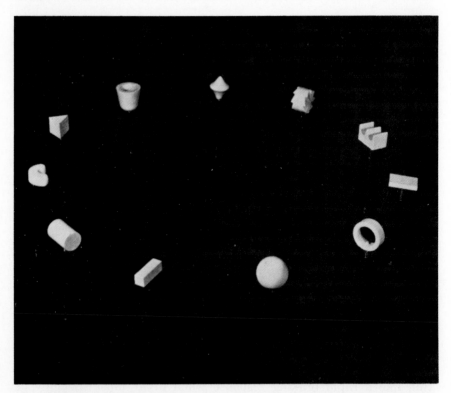

Figure 13-2 Set of perfectly discriminable knobs. [*Source:* Jenkins (1947), Fig. 14.6.]

shape of a test knob for 1 second and then go from knob to knob on a turntable until he found one that he thought matched it. The set of knobs shown in Figure 13-2 had no errors.

Jenkins used a *confusion matrix* to analyze his data. A confusion matrix with hypothetical data is shown in Figure 13-3. The set of four knobs heavily outlined is a perfectly discriminable set of knobs. None of the four knobs was confused with any of the others. The other four knobs were confused with each other some of the time.

The Jenkins study is bedrock information on the discriminability of knobs, but there is more to it than that. An operator of a system does not feel two controls, as Jenkins had his subjects do, and then decide which is the power control. Instead, he thinks to himself that an adjustment of the power control is required and then reaches for it. The operator not only knows what the control feels like, he knows its name, what it does, and where it is. Slocum et al. (1971) present evidence that this more elaborate behavior in realistic situations is facilitated if the knob meaningfully encodes its function. A pilot would benefit if the lever to lower the landing gear had a knob shaped like a wheel, for example. Knobs can be made meaningful in other ways. Labeling

First knob

Comparison knob	A	B	C	D	E	F	G	H
A	100							
B		100						
C			100					
D				100				
E	2	4			90	2		2
F	4		2	2		88		4
G			4		2		92	2
H	1	2		1		2		94

Figure 13-3 Confusion matrix of the kind that is used in research on the discriminability of knobs. The heavily outlined set of four knobs are perfectly discriminable because they are never confused with other knobs. The entry is percent of time that two knobs are judged the same in two successive responses.

makes a knob meaningful for the times that a knob is visible and the operator looks at it when selecting it. The labels should have the same position with respect to all knobs, such as all of them beneath the knobs, so that the operator always knows where to look for them.

The Size of Knobs and Their Spacing

Bradley (1969) reported an experiment that compared the size of circular knobs and their spacing. He concluded that the greatest economy of panel space and the least inadvertent operation of adjacent knobs was with ½-inch knobs that had 1 inch between their edges.

Population Stereotypes

Which way should a knob be turned to move an instrument pointer to the right? To the left? Up? Down? The same questions can be asked of a lever. There are distinct preferences that have been documented for some cases, and four advantages are presumed to follow from using a preferred control–display relationship (Van Cott & Kinkade, 1972, p. 349):

1. Decreased reaction time or decision time.
2. Fewer errors.
3. Better speed and adjustment of knobs and levers.
4. Faster learning

These preferred control–display relationships are called *population stereotypes.*

Population stereotype is an idea that goes back to an early day of human factors engineering. The attitude indicator, which is a moving horizontal bar, is the cockpit instrument that informs a pilot of the orientation of his aircraft. In reporting his research on attitude indicators during World War II, Loucks (1947), had a clear appreciation of population stereotypes, although he did not call them that. When the pilot banks his aircraft to the left, should the horizontal bar tilt to the left as the wings of the aircraft do, or tilt to the right as the horizon line outside the cockpit does? Which of these prominent references for aircraft orientation should determine the movement of the horizontal bar? What is the preferred control–display relationship, or stereotype? Making the bar consistent with the stereotype would mean that learning would be fast for a trainee and errors minimal because the response would be consistent with established response tendencies. When the control–display relationship does not conform to the stereotype, learning would be slow and error-ridden because the trainee is forced to substitute an "unnatural" response for a "natural" one. Even when a pilot is fully trained in an unnatural control–display relationship, there is the possibility that he would regress to the natural one under the pressure of stress or fatigue, which would be an error. Loucks concluded that unnatural control–display relationships are inefficient and a potential source of aircraft accidents.

Population Stereotypes and Stress. The psychological literature has several definitions of stress, but a defensible one, which has status in human factors engineering, is work load stress. Give an operator more events than she can handle per unit of time and she is said to be stressed. Performance will degrade. This work load definition of stress is attractive to human factors engineers because operators of some systems have a great deal to do. It is a design challenge to minimize work load and stress (for additional discussion of work load, see Chapter 8).

Mitchell and Vince (1951) appear to have been the first to define stress in terms of work load and use it to test the hypothesis that there is regression from nonpreferred to preferred modes of responding when stress is present. Stress was imposed by the rate of stimulus events in a discrete tracking task. The subject viewed a moving drum through an opening in a panel. Short stimulus lines on the drum appeared in the opening, above or below a centerline, and the subject had to turn the control knob to move an indicator from the centerline to the stimulus line that had just appeared. The knob could be set to turn in the preferred direction, or in the opposite, nonpreferred, direction. Preferences for direction of control movement had been determined in previous research. The experimental variables were preferred direction of control movement, which is the population stereotype, nonpreferred direction, and stimulus rate. The results are shown in Figure 13-4. When stimulus events were comparatively slow, and subjects were not stressed, there was no difference between preferred and nonpreferred directions of movement, but when events were fast, and movements were fast paced, more errors were made with the nonpreferred direction of control movement. Regression to the population stereotype occurred when work load was heavy.

Simon (1954) tested the regression hypothesis with a work load definition of stress also. In defining stress, Simon assumed that interference with an individual's efforts in goal attainment is stressful, so giving a subject too much to do is interfering and stressful. The primary task was pursuit tracking. Arrays of red and green lights and a knob were a secondary task that combined with the pursuit tracking to increase work load stress. Occasionally, a green light would come on while the subject was tracking and he had to reach and turn the knob to move a red light in alignment with it. Before the experiment a sub-

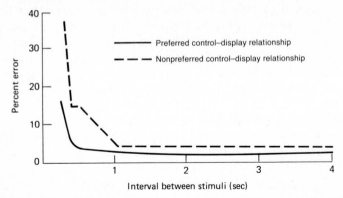

Figure 13-4 Heavy work load is a definition of stress, and Mitchell and Vince (1951) manipulated it by the speed of input events in a discrete tracking task. Performance as a function of speed was compared for preferred and nonpreferred control–display relationships. [Adapted from Mitchell and Vince (1951), Fig. 1. Used by permission of the Applied Psychology Unit, Medical Research Council, Cambridge, England.]

ject's preference for direction of turning the knob was established. Training was then given in the preferred way of knob turning or the nonpreferred way. Under stress subjects who were trained in the nonpreferred way made the most reversal errors in the knob-turning task. As the regression hypothesis says, performance reverts to the population stereotype under stress.

Examples of Control–Display Relationships That Are Preferred. What are examples of control–display relationships that are preferred and can be called population stereotypes, and what is the nature of the research that documents them? A pioneering study was done by Warrick (1947), and it demonstrates how preferences can be clear cut in some cases and indeterminate in others. His research apparatus is shown in Figure 13-5. A light came on in a position other than C, the center, and the subject was required to turn the knob and move the light to C. There was no correct response because only his preference was in question—whatever response the subject made moved the light to center. The direction of the response was recorded. The panel on the left in Figure 13-5 is an example of control–display relationships where the population stereotypes are clear. When a light came on in position A or B, the response was clockwise 94 percent of the time. A light in positions D and E elicited a counterclockwise response 81 percent of the time. The panel on the right is an uncertain control–display relationship. Lights A and B produced a clockwise response only 56 percent of the time. Lights D and E induced clockwise responses 58 percent of the time.

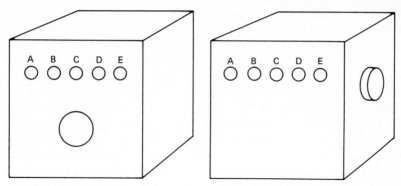

Figure 13-5 Apparatus used in the experiment by Warrick (1947) on control–display relationships. See text for explanation. [*Source:* Warrick (1947), Fig. 9.1.]

Figure 13-6 Apparatus used in the experiment by Fitts and Deininger (1954) on control–display relationships. The subject is responding to a symbolic stimulus, where the number represents a position on a clockface and defines the direction of stylus movement. [Source: Fitts and Deininger (1954), Fig. 1.]

Fitts and Deininger (1954) did a study on preferred control–display relationships for discrete movements. Reaction time was the principal measure. Their experimental apparatus is shown in Figure 13-6. At the start of a trial the subject held a stylus at the intersection of pathways radiating outward. When a signal appeared in the display, the subject moved a stylus along the pathway of choice as rapidly as possible. The choice reaction time of moving the stylus off center was recorded, as was the error of moving on a wrong pathway. The control–display relationships that were investigated are shown in Figure 13-7: spatial and two-dimensional, symbolic (numerical) and two-dimensional, or spatial and one-dimensional. The response directions, shown by the arrows in Figure 13-7, were maximum or in a one-to-one relationship with the display, mirrored or opposite to the one-to-one relationship, or random. To illustrate with the spatial two-dimensional display and the maximum response relationship, a subject would see the circle of eight lights. The upper light would come on and the subject would be required to move up. Or the bottom light would appear and the subject would be required to move down. Sixty-four trials a day were given for two days.

The pertinent results of the Fitts and Deininger experiment on population stereotypes are given in Table 13-1. Table 13-1 has reaction times and errors for day 1 because the preferences that are population stereotypes should be evident at the outset without much learning. A

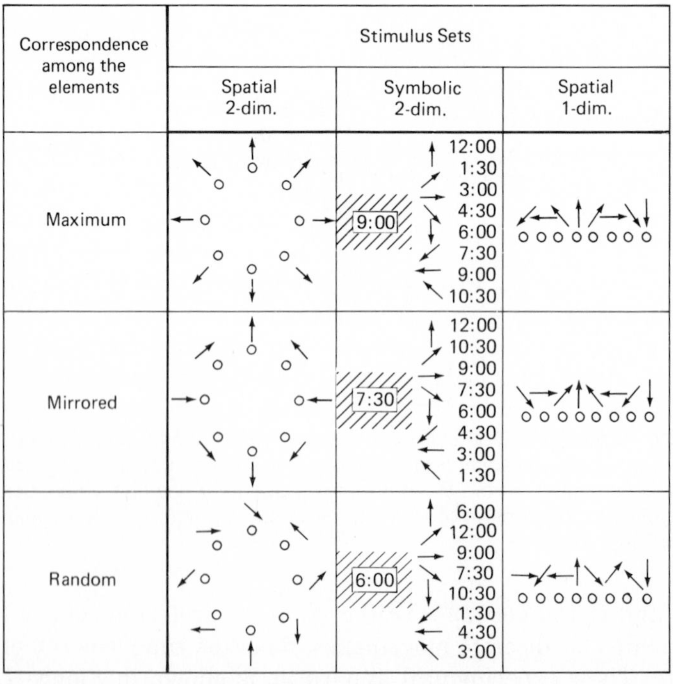

Figure 13-7 Various control–display relationships used in the experiment by Fitts and Deininger (1954). See text for explanation. [Adapted from Fitts and Deininger (1954), Fig. 2.]

TABLE 13.1 Results of Day 1 from the Experiment by Fitts and Deininger*

	Condition		

Spatial Two-Dimensional

	Maximum	Mirrored	Random
Correspondence:			
Reaction time (seconds)	.42	.57	1.26
Errors (percent)	2.2	5.5	20.5

Symbolic Two-Dimensional

	Maximum	Mirrored	Random
Correspondence:			
Reaction time (seconds)	.71	.84	1.00
Errors (percent)	5.9	10.0	12.1

Spatial One-Dimensional

	Maximum	Mirrored	Random
Correspondence:			
Reaction time (seconds)	.85	.93	1.34
Errors (percent)	15.0	19.5	18.9

*Each entry is an average for 10 subjects who each made 64 responses.

Source: Fitts and Deininger (1954), Table 1.

population stereotype is most apparent with the two-dimensional spatial array and maximum response correspondence, where the responses on day 1 were fast and almost error free. It should not be surprising that the two-dimensional spatial array with maximum response correspondence was the strongest population stereotype because the world is built that way. We move the hand up to the top of the desk for a sheet of paper, and then we move it to the right for the pencil on the right-hand side of the desk, and so on. We have a vast experience with direct spatial relationships, so we benefit from it and find these relationships compatible when they are designed in a task that confronts us. In other contexts these benefits of past experience for present performance are called "positive transfer" (Chapter 18, on training, includes a discussion of transfer of training). Experience with other control–display relationships is less, so preferences are weaker, with reaction times slower and errors more frequent.

A clear message from experiments like those of Warrick, and Fitts and Deininger, is that correct design of the control and its display is not enough. The control–display relationship between them must be correctly defined as well.

A Qualification. If experience is a heavy factor in population stereotypes, we would expect population stereotypes to differ from culture to culture because experience will differ among cultures. Courtney

(1986) compared U.S. and Chinese adults for their stereotypic associations of colors with concepts. The dominant associations for U.S. subjects were: cold–blue (96 percent), go–green (99 percent), stop–red (100 percent). The dominant associations of Chinese subjects for the same concepts were: cold–white (72 percent), go–green (45 percent), stop–red (49 percent). Users of data on population stereotypes must be culture conscious.

Conclusions. This survey on control–display relations is illustrative of issues and findings. The topic is not well developed, and this is not surprising considering the many different kinds of controls and displays and relationships between them, but some guidance from research is available, as we have seen. Loveless (1962) has a critical review of research on control–display relationships.

The Layout of Pushbuttons and Switches _____

Pushbuttons and switches are common discrete controls. Pushbuttons are ordinarily on or off, but switches can have more than two positions. Only two-position switches will be considered here, not only because of their widespread use, but because our research knowledge lies there. The research consists of two experiments by Bradley and Wallis (1959, 1960).

The Spacing of Pushbuttons

Bradley and Wallis (1959) had three pushbuttons on a panel. The requirement was for the subject to move his hand from its resting place on the table and press the center button when a light came on. The experimental variables were spatial orientation of the three pushbuttons (vertical or horizontal), button diameter ($\frac{1}{2}$, $\frac{3}{4}$, or 1 inch), and spacing between edges of the button ($\frac{1}{8}$, $\frac{2}{8}$, $\frac{3}{8}$, $\frac{4}{8}$, $\frac{5}{8}$, and $\frac{6}{8}$ inch). The main measures of performance were inadvertently touching or operating an adjacent button, and time to move the hand from the table to the center button. Their results were straightforward: The larger the pushbutton and the more spacing between buttons, the fewer the errors and the faster the response time; aiming the finger at a large target, clearly set off from others, is fast and accurate. Wrong buttons were pressed 10 percent of the time when small buttons of $\frac{1}{2}$-inch diameter were closely spaced $\frac{1}{8}$ inch apart, but no errors were made when 1-inch buttons were spaced $\frac{6}{8}$ inch apart. Horizontally arrayed buttons were more efficient than vertically arrayed buttons.

The Spacing of Switches

Bradley and Wallis (1960) reported an experiment on switches with a design similar to their 1959 experiment on pushbuttons except that it included the direction of throw for the switch as an additional variable. The results were about the same. The greater the spacing between switches, the fewer the errors. A horizontal array of switches was better than a vertical array. The direction of throw for a switch did not matter.

Pushbuttons or Switches?

The comparability of their 1959 and 1960 experiments allowed Bradley and Wallis to ask whether switches or pushbuttons are better. Bradley and Wallis decided in favor of switches. Linearly arrayed switches, with less than an inch between them, averaged fewer errors than corresponding arrays of pushbuttons. Thus, switches are a better choice when panel space is limited. When panel space is generous, and the spacing can be an inch or more, switches and pushbuttons are probably equivalent.

Accidental Activation of Switches and Pushbuttons

The Bradley and Wallis research shows that proper spacing is one way of preventing accidental activation of switches and pushbuttons. Special system circumstances, which task analysis would reveal, might require other steps to prevent accidental activation, however. There are various ways to do it (Seminara & Parsons, 1964), but a common way is to have a hinged guard that must be lifted before the switch or pushbutton can be activated. For maximum security, a switch or pushbutton can be designed with a key to activate it. Response speed is compromised when safeguards like these are used.

Summary _____

Discrete controls such as knobs, switches, and pushbuttons have a small number of states, in contrast to tracking controls, which are continuous over a range, sometimes in more than one dimension. Research was reviewed on the discriminability of knob shapes; the size and spacing of knobs, switches, and pushbuttons; and control–display relationships.

Population stereotypes are preferred control–display relationships, and using them can solve such problems as the direction that a pointer should move when a knob is turned; population stereotypes are more "natural." Using a nonpreferred relationship for a control and its display risks regression to the preferred relationship when the operator is stressed, which is error. Use of the preferred control–display relationship would eliminate the risk. Research on population stereotypes and the regression hypothesis was reviewed.

References

Bradley, J. V. (1969). Optimum knob crowding. *Human Factors, 11,* 227–238.

Bradley, J. V., & Wallis, R. A. (1959). Spacing of pushbutton on-off controls. *Journal of Engineering and Industrial Psychology, 1,* 107–119.

Bradley, J. V., & Wallis, R. A. (1960). Spacing of toggle switch on-off controls. *Journal of Engineering and Industrial Psychology, 2,* 8–19.

Courtney, A. J. (1986). Chinese population stereotypes: Color associations. *Human Factors, 28,* 97–99.

Fitts, P. M., & Deininger, R. L. (1954). S-R compatibility: Correspondence among paired elements within stimulus and response codes. *Journal of Experimental Psychology, 48,* 483–492.

Fitts, P. M., & Jones, R. E. (1947/1961). Analysis of factors contributing to 460 "pilot-error" experiences in operating aircraft controls. In H. W. Sinaiko (Ed.), *Selected papers on human factors in the design and use of control systems.* New York: Dover. Pp. 332–358.

Jenkins, W. O. (1947). The tactual discrimination of shapes for coding aircraft-type controls. In P. M. Fitts (Ed.), *Psychological research on equipment design.* Washington, DC: U.S. Government Printing Office, Army Air Forces Aviation Psychology Program Research Reports, Report 19. Pp. 199–205.

Loucks, R. B. (1947). An experimental evaluation of the interpretability of various types of aircraft attitude indicators. In P. M. Fitts (Ed.), *Psychological research on equipment design.* Washington, DC: U.S. Government Printing Office, Army Air Forces Aviation Psychology Program Research Reports, Report 19. Pp. 111–135.

Loveless, N. V. (1962). Direction-of-motion stereotypes: A review. *Ergonomics, 2,* 357–383.

Mitchell, M. J. H., & Vince, M. A. (1951). The direction of movement of machine controls. Cambridge, England: Applied Psychology Unit, Report A.P.U. 137/50, February.

Odom, J. A. (1984). *Applying manual controls and displays: A practical guide to panel design.* Freeport, IL: Micro Switch: A Honeywell Division.

Pruyn, R. R., & Domenic, R. E. (1982). Conceptual design of the LHX integrated cockpit. Paper presented at the 38th Annual Forum of the American Helicopter Society, Anaheim, CA, May.

Seminara, J. L., & Parsons, S. O. (1964). 17 ways to stop control accidents. *Control Engineering, 11,* 84–90.

Simon, C. W. (1954). Effects of stress on performance in a dominant and a non-dominant task. Wright-Patterson Air Force Base, OH: Wright Air Development Center, WADC Technical Report 54-285, June.

Slocum, G. K., Williges, B. H., & Roscoe, S. N . (1971). Meaningful shape coding for aircraft switch knobs. Savoy, IL: Aviation Research Laboratory, Institute of Aviation, University of Illinois at Urbana–Champaign, Aviation Research Monographs, Volume 1, Number 3. Pp. 27–40.

Van Cott, H. P., & Kinkade, R. G. (1972). *Human engineering guide to equipment design.* Washington, DC: U.S. Government Printing Office.

Warrick, M. J. (1947). Direction of movement in the use of control knobs to position visual indicators. In P. M. Fitts (Ed.), *Psychological research on equipment design.* Washington, DC: U.S. Government Printing Office, Army Air Forces Aviation Psychology Program Research Reports, Report 19. Pp. 137–146.

Keyboard Controls

The keys of a keyboard are discrete controls such as switches or push-buttons, but they are not comparable because they must be activated in long language or numerical sequences, which makes them a very different kind of discrete control. The two main kinds of keyboards are the *alphanumeric keyboard,* with the common typewriter keyboard as the easiest example, and the *numeric keyboard,* such as on a calculator or a pushbutton telephone. A digital computer can have both kinds of keyboards at its workplace. There is, in addition, the *chord keyboard,* as a court stenographer will use, which will be discussed also.

We are on the threshold of a day when a machine will accurately recognize the human voice and do what it says (Chapter 16). Typing, and input to a digital computer, will be vocal, and it will be easy. The adult as a skillful speaker will have no problem in telling the machine what to do, unlike manual typing, which is difficult to learn. Even when automatic voice recognition becomes practical and widespread, we will probably have keyboards because keyboards are low in cost and reliable, and because millions of people are trained to use them.

History of the Typewriter

The typewriter has a long history, beginning with its invention in the eighteenth century, its commercial adoption in the nineteenth century, and its proliferation in the twentieth century. Human factors ideas and research accompanied the twentieth century proliferation. Like so many topics, the past is a route to understanding.

Invention and Early Development

Beeching (1974), a historian of the typewriter, has the origin of the typewriter in the early eighteenth century. Beeching, and others (Boehm, 1984), date the start of the typewriter from a patent that Queen Anne of England granted to Henry Mill in 1714. Little is known about Mill's invention. No drawing or model of it survives.

The first U.S. patent was granted to William A. Burt in 1829. The original drawings and specifications exist, and replicas of his device are in the Smithsonian Museum in Washington, D.C., and in the Science Museum in London. The Burt machine, called the "Typographer," is shown in Figure 14-1. The type for the letters were mounted on a semicircular frame and were positioned by turning a crank on the front. After a letter was positioned, a lever was depressed to bring it in contact with the paper. The Burt machine, and the machines that followed it in the nineteenth century, including the first commercial one by Christopher Sholes and his colleagues Carlos Glidden and Samuel Soulé, were intended to be a legible substitute for handwriting. Indeed, writing with type rather than with the pen is suggested by the

Figure 14-1 Replica of the first typewriter patented in the United States, in 1829. The inventor was William A. Burt. [Adapted from Beeching (1974), Fig. 1.6. Used by permission of W. A. Beeching.]

Figure 14-2 First commercial typewriter, developed in the United States. The production contract for this machine was signed with the Remington Company in 1873. [Adapted from Beeching (1974), Fig. 1.42. Used by permission of W. A. Beeching.]

term "typewriter." There was no conception of fast two-hand touch typing. Hunt and peck was the accepted method.

Sholes and his associates received a U.S. patent in 1868 for the machine that was to be the first production typewriter. In 1873, Sholes and his group signed a contract for the manufacture of 1000 typewriters with the Remington Company, a manufacturer of guns, sewing machines, and farm machinery. This first commercial typewriter is shown in Figure 14-2. The essentials of the modern typewriter are evident—the familiar bank of keys, the visible sheet of paper on a roller.

The world devours an invention when it appears at the right historical moment. Sixty thousand women were working in offices as typists by the 1890s (Beeching, 1974). By 1900 the U.S. government was buying 10,000 typewriters a year (Yamada, 1980). Among the reasons that Yamada (1980) cites for the typewriter's success was that its arrival coincided with industrial expansion in the United States, and that the keyboard was suitable for eventual touch typing.

How the Keyboard for the Typewriter Got That Way

How did the common typewriter for the English language come to have the keyboard that it does? The keyboard that we have has been with us since the days of the Sholes machine, but the reasons for its design are not clear. The keyboard goes by various names. It is called both the *Sholes keyboard* and the QWERTY *keyboard,* after one of its sequences of keys. Here it will be called the *standard keyboard.*

One frequently told story about the standard keyboard is that Sholes had trouble with jamming typebars (the mechanical arm with the letter set in it) when an alphabetic ABC arrangement of the keys was used, so he placed the typebars for letters that regularly go together, such as Q and U, in different quadrants of the typebar circle. The result was the standard keyboard, a major departure from an alphabetic arrangement. There are doubts about this account, however (Noyes, 1983, pp. 266–267). In retrospect, one can wonder, without evidence, why design of the keyboard mattered much to Sholes, whose machine was intended to replace handwriting. The hunt-and-peck method would be used, so he might reason that any keyboard layout would do. (There is latter-day evidence for Sholes' choice. See the discussion below of the study by Norman and Fisher.) Whatever the reason, an international standardization committee met in 1905 and endorsed the standard keyboard. The standard keyboard has had no threatening competition since. Not that anyone needs reminding, but for ease of comparison with other keyboard layouts that will be discussed, here is the familiar standard keyboard:

```
2 3 4 5 6 7 8 9 0
q w e r t y u i o p
a s d f g h j k l ;
z x c v b n m , . /
```

Keyboard Redesign

Why Redesign the Standard Keyboard?

A professional typist, like a good secretary, will do 60 words a minute comfortably. The official world's record for 1 hour of typing, with a 10-word penalty for an error, is 149 words per minute (Boehm, 1984). Why the interest in redesign? There are two reasons. First, it is said that the standard keyboard is not self-evident to users with no training, of which there are many. Naive users use the hunt-and-peck method and

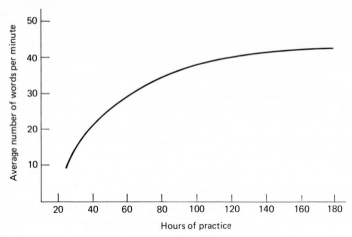

Figure 14-3 Learning curve taken from a high school typing class. [Adapted from Chapman (1919), Diagram 10.]

could benefit from a keyboard that conforms to population stereotypes about letter sequences. This is the argument for an alphabetic keyboard, which conforms to the logical sequence that is the alphabet (see section that follows on the Alphabetic Keyboard).

Second, and most important, the standard keyboard is difficult to learn. A usable speed might evolve with the hunt-and-peck method, but fast, coordinated two-hand touch typing needs formal instruction and long practice. Figure 14-3 shows the learning curve of a high school typing class for 180 hours of training (Chapman, 1919). About 100 hours of training is required to reach 40 words per minute, which is a minimal speed for a professional typist. The rate of increase beyond 100 hours is slow. An extrapolation of the curve in Figure 14-3 indicates that hundreds of hours of practice are necessary to attain the 60 to 80 words/minute expected of a professional typist. The difficulty of learning skillful touch typing on the standard keyboard places it in the league with complex athletic skills. Hundreds of hours of practice are required to be an acceptable tennis player, and the same is true of touch typing. August Dvorak, of whom more will be said in the next section, wrote a book about typewriting behavior and said that typing was the most time-consuming subject in any educational program at any level (Dvorak et al., 1936, p. v).

Dvorak's Simplified Keyboard

A serious challenge to the standard keyboard was launched by August Dvorak. Dvorak was a professor at the University of Washington in Seattle and director of a project on advancement of the teaching of typing. He cared about training methods, but he was also interested in keyboard design as a route to better performance. He was aware of the methods of motion and time study that Taylor and Gilbreth had developed (Chapters 1 and 8). Using these methods, Dvorak found that the standard keyboard required inefficient, time-consuming movements, just as Taylor and Gilbreth had found for industrial jobs. Dvorak (Dvorak et al., 1936, Chap. 9; Dvorak, 1943) established a number of shortcomings of the standard keyboard:

1. Most of the population is right-handed. The right hand is the strongest hand, yet the left hand does 57 percent of the work on the standard keyboard.
2. Some fingers are stronger and more skillful than others. The assignment of work load is not proportionate to finger strength and skill.
3. Only 32 percent of the typing is done on the home row. Fifty-two percent of the typing is done on the upper row and 16 percent on the lower row—unnecessary movements are required.
4. There are times when an entire hand is idle, leaving the other hand to do the work. Too often the hand doing the work is the comparatively weak left hand.

Dvorak's answer was the *simplified keyboard,* which he patented in 1932 with W. L. Dealey. Fifty-six percent of the words were now typed with the stronger and more skillful right hand. Assignment of work to fingers was proportionate to strength and skill. Seventy percent of the typing was done on the home row. Finger motions were reduced by 90 percent. Here is the simplified keyboard:

```
7 5 3 1 9 0 2 4 6 8
? , . p y f g c r l /
a o e u i d h t n s -
' q j k x b m w v z
```

Is Typing Performance Better on the Simplified Keyboard?
Dvorak's use of motion and time analysis was good, but he knew that human performance data were required also. The performance data that Dvorak presented (Dvorak et al., 1936, Chap. 9; Dvorak, 1943) were not especially convincing. He needed learning curves over hundreds of hours for the standard keyboard and his simplified keyboard so that learning rates could be compared, and he needed data on transfer of training from the standard keyboard to the simplified keyboard. How much difficulty will typists, experienced in the standard keyboard, have in transferring to the simplified keyboard? The learning data that he relied on were fragmentary, as were his transfer data.

Notwithstanding, the simplified keyboard was convincing enough for the U.S. government to sponsor a study by Strong in 1956 (for summaries, see Noyes, 1983; Alden et al., 1972; Yamada, 1980). Strong used two groups of subjects, all skilled in the standard keyboard. Following pretraining on the simplified keyboard, one group was given 100 additional hours on the simplified keyboard. The second group continued on the standard keyboard. The findings were not decisive enough for Strong to recommend the simplified keyboard.

Adoption of the simplified keyboard was slowed by the Strong study, but it refuses to die. Lingering interest in it continues in modern times (Alden et al., 1972, p. 283; Cooper, 1983, Chap. 1; Yamada, 1980). The *Guinness Book of Records* (Boehm, 1984) reports that Mrs. Barbara Blackburn typed 150 words per minute on the standard keyboard over 50 minutes, but typed 170 words per minute on the simplified keyboard.

Reasons Why the Simplified Keyboard Has Not Been Adopted. Dvorak's use of motion and time analysis has merit, and his simplified keyboard probably has the virtues that he ascribes to it. There are good reasons, however, why the simplified keyboard has not been adopted.

1. The research that compared performances on the simplified keyboard and the standard keyboard has been unsatisfactory. Dvorak relied on minor studies of performance and on anecdotes. The experiment by Strong is another minor study, insufficient for the conclusion that was drawn from it.

   A conceptually simple but time-consuming experiment is required. The experiment should have two control groups and several experimental groups. One control group, composed of inexperienced typists who have had no prior typing experience of any kind, would receive only extensive practice on the simplified keyboard, until its performance curve leveled off and approached its limit. A second control group of inexperienced typists would receive the same training on the standard keyboard. Several experimental groups, each with subjects having different amounts of past experience on the standard keyboard, would be trained on the simplified keyboard until their performances have leveled off and approach a limit. How do the learning curves of the two control groups compare with each other? Which curve reaches the higher limit and at what rate? How does past experience on the standard keyboard impair performance of the experimental groups? What does it take in training on the simplified keyboard to overcome the impairment? A hint of the retraining that would be required is a study by B. J. Lessley (reported by Yamada, 1980, pp. 188–189) which found that experienced typists took about 100 hours of practice on the simplified keyboard to reach the level of proficiency which they held on the standard keyboard. The trend of the retraining curve suggested that eventual level of performance on the simplified keyboard would be higher than on the standard keyboard.

2. The standard keyboard has all the faults that Dvorak found for it, but it has an unwitting virtue that can partially offset the faults. Dvorak found that successive movements by opposite hands were faster than successive movements by the same hand, and he criticized the standard keyboard for not having enough of it (Dvorak, et al., 1936, p. 218). Nevertheless, Dvorak found that 48 percent of successive movements on the standard keyboard were between hands, and this percentage is probably high enough to be a positive contributor to speed. Dvorak, seeking optimization with his simplified keyboard, increased the percent of successive movements between hands to 67 percent. That successive movements between hands are faster than successive movements of the same hand has become an established finding (Fox & Stansfield, 1964; Sternberg et al., 1978; Gentner, 1983; Larochelle, 1983; Ostry, 1983).

3. The standard keyboard is not optimum, and it is difficult to learn, but impressive skill levels can be reached on it nevertheless. Typists appear willing to pay the high cost of high skill.

The Alphabetic Keyboard

The *alphabetic keyboard* is believed to be desirable for naive typists. Everyone is highly familiar with the orderly string that is the alphabet, so it is reasoned that untrained typists would find a keyboard with an alphabetic layout the easiest of all. Butterbaugh and Rockwell (1982), for example, compared three versions of the alphabetic keyboard for an aircraft cockpit. Performances on all of them were high, in seeming confirmation of the hypothesis that the alphabetic keyboard is best for naive typists. Norman and Fisher (1982) extended the Butterbaugh experiment and included other keyboard layouts as well.

The Norman and Fisher Experiment on Keyboard Layout with Naive Subjects. Using inexperienced typists, Norman and Fisher (1982) compared two alphabetic keyboards with the standard keyboard and a keyboard with randomly arranged keys. The subjects were pretested on the standard keyboard to be sure that they had no typing skill, and then were tested on the four keyboards. The keyboards, in addition to the standard keyboard, were:

Alphabetic Keyboard

    a b c d e f g h i j
    k l m n o p q r s
    t u v w x y z

Diagonal Alphabetic Keyboard

    a d g j m p s v y
    b e h k n q t w z
    c f i l o r u x

The Random Keyboard

    c y i f m g z d n j
    q o x h b t r w l
    v a u p k e s

The test for a keyboard was 10 minutes of typing. The score was number of letters typed. The standard keyboard was the best of all. The two alphabetic keyboards had a 10 percent advantage over the random keyboard, but it is a small advantage and was considered of no practical significance. Nor did the subjects have a dominant preference for the alphabetic keyboard.

Why is the standard keyboard best for naive typists? A possibility is that naive typists have been exposed to the standard keyboard even though they never learned to use it, and they might have benefited from the preacquaintance. Another possibility is that the standard keyboard may not be good but that the alphabetic keyboard is poor by comparison. Norman and Fisher give the example of typing the word *flower* on the alphabetic keyboard. That the alphabet is broken down into three sequences for keyboard arrangement destroys the logical order of the alphabet. The letter *l* is not to the right of *f* as it would be in the alphabet but to the left. The *w* is not to the right of the *o* but to the left. The alphabet has no more "logic" than the standard keyboard when it is structured as a keyboard and used in the typing of text. The same is true of its variant, the diagonal alphabetic keyboard.

The Chord Keyboard

The standard keyboard, and variations of it, have a one-to-one visual–motor code, where each character has a distinctive keypress response. For whatever problems the standard keyboard might have, mystery about the code is not one of them. More complex codes are possible, of course, and implications for keyboard design go along with them. A *chord keyboard* is an example. A chord keyboard may use the one-to-one visual–motor code for some of the characters, but the name derives from characters, or groups of characters, that are represented by two or more keys being depressed, such as a chord on a piano.

There are two kinds of chord typewriters in common use (Arnott et al., 1979). The U.S. public is familiar with the stenograph machine, which has a chord keyboard and is used for verbatim recording in courtrooms; it is a kind of shorthand machine. The stenograph machine has 23 keys. The corresponding British device, which is the same in principle, is called the palantype machine. The palantype machine has 29 keys. Machines like these with their complex codes are not easy to learn, but the codes can be efficient and high typing rates are possible with enough practice. Seibel (1972, p. 328) estimates that typing rate on a chord keyboard can be 150 percent of the standard keyboard.

Recent Developments in Chord Keyboards

One-Hand and Two-Hand Five-Key Chord Keyboards. The stenograph machine and the palantype machine are for two-handed use, but one-handed chord keyboards are possible also. Gopher and Raij (1986) compared one- and two-handed chord keyboards, which they developed for research purposes, with the standard keyboard for the typing of Hebrew. The one-handed keyboard had five keys, where a letter was entered as a chord combination of one to five fingers pressed simultaneously. With five keys, there is a total of 31 chord combinations. The two-handed keyboard had the five keys for each hand, so the user could enter a letter with either hand, or enter one letter with one hand and another letter simultaneously with the other hand.

Three groups of subjects, one for each kind of keyboard, had 34 one-hour sessions of typing from a popular novel, and their learning curves are shown in Figure 14-4. The performance on the standard keyboard is about the same as has been found by others (e.g., Figure 14-3), but performances on the one-handed and the two-handed chord keyboards were better than on the standard keyboard. Error rates were low for all groups, 1 to 2 percent.

The superiority of chord keyboards over the standard keyboard is an expected finding (Seibel, 1972, p. 328), but unexpected was the lack of difference between one-handed and two-handed chord keyboards. The capability of typing two letters simultaneously should have given better performance on the two-handed keyboard, it would seem. One possible reason for lack of difference is insufficient practice; the superiority of the two-handed keyboard might eventually emerge. Very

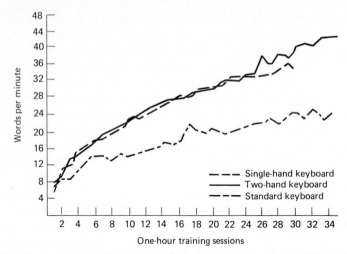

Figure 14-4 Learning curves for typing on one-hand and two-hand chord keyboards and the standard keyboard. The superiority of chord keyboards is clearly demonstrated. [Adapted from Gopher and Raij (1986), Fig. 4.2. Used by permission of D. Gopher.]

extensive practice, however, may not work because of the inherent difficulty of typing two letters simultaneously. One letter can easily be typed with one hand, and two letters can be successively typed with two hands, but can two letters be typed simultaneously with two hands? Whether we can learn to maintain two codes in mind simultaneously and use them to activate two simultaneous movements is not known.

The Letter-Shape Keyboard. Chord keyboards have had specialized uses, as with the stenograph and palantype machines, and they have not been easy to learn, but R. C. Sidorsky (1974) had the idea of an easy chord keyboard that anyone could use with minimal training. He was not thinking of replacing the standard keyboard but of a computerized battlefield. A vision for the military future is the computerized battlefield, where commanders will have computers as aids for their decision making. Input to the computers could be by soldiers (e.g., frontline observers) untrained in typing, and their slow use of the standard keyboard would be a bottleneck. Consideration has been given to the use of predefined messages that could be inserted in the computer with the flick of a switch, but it is a rigid approach. A keyboard, easy to learn, which could be used for any kind message, would promote the processing of battlefield information.

After an uneven history in psychology, mental imagery as a mechanism of learning and memory is on sound scientific footing today

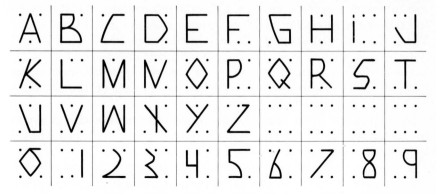

Figure 14-5 The letter-shape keyboard devised by Sidorsky uses visual imagery to specify the form of the keypress for an alphanumeric character. There are three keys, and one to three of them must be pressed twice. The small dots indicate the keys that are *not* pressed. The keys that are successively pressed have a movement pattern that approximates the visual pattern of the letter. [*Source:* Sidorsky (1974), Fig. 1.]

(Chapter 6). What is interesting about Sidorsky's thinking is that he turned mental imagery toward conceptualization of a chord keyboard. Sidorsky no longer defined the keyboard with a one-to-one visual–motor code like the standard keyboard, or with the more complex visual–motor codes of the stenograph and palantype machines, but by a code mediated by mental imagery. Mental imagery has most frequently been studied as an agent of verbal memory, but Sidorsky turned it toward the control of skilled movements. Sidorsky called the new keyboard "alpha-dot," but it is more descriptively called the *letter-shape keyboard* (Gopher, 1984).

Sidorsky's letter-shape keyboard had five keys and was operated by one hand. The alphanumeric characters were formed by the three middle keys, called the character keys. Nonalphanumeric characters such as punctuation and mathematical symbols were defined by the other two keys, operated by the thumb and little finger. Each alphanumeric character was formed by two strokes of one to three character keys, where the keypresses were determined by the shape of the character; the code was governed by visualization of the character's form. Figure 14-5 shows the three keys and how two presses were defined by the shape of the character. The relationship between shape of the letter and the pattern of the two keypresses might be crude, but a relationship exists, nevertheless.

Sidorsky conducted a feasibility study of his new keyboard. The code was not difficult to learn. Subjects were able to learn it within an hour and a half. After only 5 hours of practice the subjects were typing about one character a second, or 11 to 13 words per minute, which is twice the minimum rate that the U.S. Army requires for field messages. These subjects had some proficiency on the standard keyboard, and their performance on it was compared with performance on the letter-shape keyboard. Depending on type of message, performance on the letter-shape keyboard was 60 to 110 percent of that on the standard keyboard.

Interest in the letter-shape keyboard shifted from the U.S. Army to the Israeli Air Force. The Israeli Air Force was interested in a keyboard for aircraft cockpits that allowed touch typing for input to a computer with one hand while flying the aircraft with the other. Ease of learning was required. D. Gopher, who carried out this research for the Israeli Air Force, used two presses of three keys for the characters, as Sidorsky had done. A fourth key for computer-editing functions such as cursor control was operated by the thumb. Gopher reasoned that a letter-shape keyboard might work better for Hebrew than English because the "squareness" of Hebrew letters made the translation of a visual pattern to a movement pattern easy. The correspondence of visual pattern and movement pattern is less direct for the English alphabet because the forms of the characters are more complex. Figure 14-6 illustrates the ease with which the Hebrew alphabet can be translated to the letter-shape motor code. Sidorsky had some success in translating the English alphabet to the letter-shape motor code (Figure 14-5), but the tailoring seems better with Hebrew.

Gopher (1984) reported an experiment conducted by A. Schelach where a group that learned the letter-shape code was compared with a group that learned an arbitrary code that would not benefit from imagery. The purpose of the experiment was to show the potency of imagery visualization of letters. The letter-shape code was learned in 7 minutes, but the arbitrary code took 50 minutes. Twenty hours of practice were given in typing from a novel, and the learning curves for the two groups are shown in Figure 14-7. The letter-shape code was superior from the beginning of practice, with no indication that the curve for the arbitrary code would ever catch up. The terminal rate for the letter-shape group was 25 words per minute, and 19 words per minute for the group with the arbitrary code. A comparison of the Israeli subjects of the letter-shape group with U.S. high school students on the standard keyboard is inexact, but nevertheless, a comparison of Figure 14-3 with Figure 14-7 shows that 25 words per minute is attained in less than half the time on the letter-shape keyboard. The outcome of Gopher's research was installation of the letter-shape keyboard in the cockpits of some Israeli aircraft.

The motivation for the letter-shape keyboard was specialized military uses, but spin-off for the civilian sector can be a possibility. What is the possibility that the letter-shape keyboard might challenge the

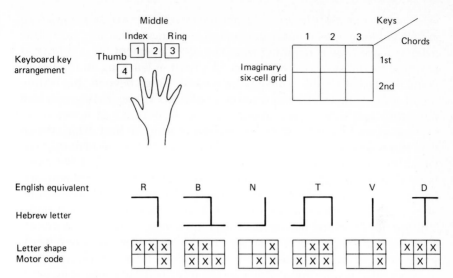

Figure 14-6 Gopher's adaptation of Sidorsky's letter-shape keyboard for use with Hebrew. See text for explanation. [Adapted from Gopher (1984), Fig. 2. Used by permission of Springer-Verlag New York, Inc.]

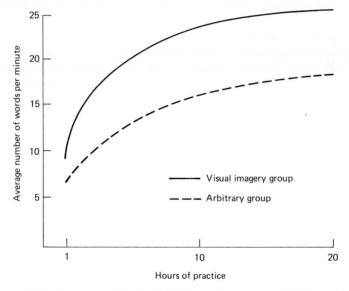

Figure 14-7 Experiment by A. Schelach on the typing of Hebrew with a chord keyboard. The visual imagery group used the letter-shape keyboard. The other group had an arbitrary code for the same chord keyboard. Performance on the letter-shape keyboard was superior throughout the 20 hours of training. [Adapted from Fig. 4 in Gopher, D., pp. 195–208 in Prinz, W. and Sanders, A. F., eds., *Cognition and Motor Processes,* Springer-Verlag, New York, 1984. Reprinted by permission of Springer-Verlag, Heidelberg.]

standard keyboard? Complete replacement of the standard keyboard seems unlikely. One mental image and two keystrokes for each character is more time consuming than the one movement for a character required by the standard keyboard. The limit of learning on the letter-shape keyboard would appear to be modest. To go beyond this ceiling might require more efficient coding, analogous to the stenograph and palantype machines, where certain groups of letters and words could be represented by images and keystrokes. Another limitation, which was discussed above, is that successive movements alternating between two hands are faster than successive movements of the same hand. This means that the one-hand operation of the letter-shape keyboard is inherently slower than a two-handed keyboard, everything else equal. Notwithstanding, these limitations on speed may not make a difference for many people; not everyone needs the speed of a professional typist. Many people must shy away from a typewriter or a word-processing system because they use the laborious hunt-and-peck method. Such people might welcome the intermediate speed of an easy-to-learn letter-shape keyboard.

Conclusion on Chord Keyboards. Replacement of the standard keyboard, if it ever happens, is likely to come from a chord keyboard, whether it is based on mental imagery or some other kind of coding scheme, because the chances of negative transfer from past experience on the standard keyboard is likely to be small. The simplified keyboard has inescapable negative transfer from experience on the standard keyboard because the same old movements must be reprogrammed, and relearning can be a slow process. A chord keyboard, on the other hand, requires entirely new movement patterns, not the redirection of old ones, so the likelihood of interference is small.

Numeric Keyboards

The kind of 10-digit keyboard for numerical entry that is found on the pushbutton telephone is

$$
\begin{array}{c}
123 \\
456 \\
789 \\
0
\end{array}
$$

The pushbutton telephone will also have letters of the alphabet on these keys, so alphanumeric entry is possible also, but the primary use is numerical entry.

Calculators usually have the 10 digits arranged like this:

789
456
123
0

Membrane numeric keyboards, where a conductive membrane is separated from a printed circuit board by less than 1 millimeter and where the operator presses a flat surface and has no travel of the finger, are occasionally found. Keys, requiring force and travel for activation, predominate, however.

Research on the Layout of Numeric Keyboards

Several studies on the layout of the numeric keyboard have been done, and they provide some guidance, although justification for a particular layout is not strong. One reason is that findings are mixed, and another is that investigators have used different measures of behavior which do not always agree. Two kinds of measures have been used: preference for keyboard layouts and performance measures of the speed and accuracy of keying. Preference measures do not always agree with performance measures (Deininger, 1960). Moreover, preferences can change with experience.

A study by Lutz and Chapanis (1955) used the preference measure. Variously arranged blank 10-key layouts were presented to subjects who wrote in the digits as they would like to see them arranged. There was preference for numbers in horizontal rows, with an advantage for the telephone layout (on p. 302), but only a slight one. Other studies have used performance measures, and they do not point a clear way. Deininger (1960) evaluated a number of numeric keyboard layouts with performance measures and found that wide latitude in design is allowable. Conrad and Hull (1968) found better performance for the telephone layout, but Nakatani and O'Connor (1980) could not confirm it.

Sensory Feedback and Keying Efficiency

Can keying efficiency be improved by increasing the sensory feedback associated with responses? Auditory feedback accompanying each button press has been tried with no success. The same tone, a distinctive tone, or a click for each button press has no appreciable effect on keying behavior (Deininger, 1960; Pollard & Cooper, 1979). Another research route has been manipulation of the force and the travel distance of buttons in an effort to influence proprioceptive feedback. No effect on keying efficiency has been found (Deininger, 1960).

Conclusion on Numeric Keyboards

At present the best recommendation is to leave the numeric keyboards in common use as they are; there is no basis for change. The daily use of pushbutton telephones and calculators will entrench these keyboards in the habits of people, and if they are not yet population stereotypes, they are likely to become so.

Summary ─────────────────────────────────

The key of a keyboard is a discrete control, but keys differ from discrete controls such as pushbuttons and switches because they must be activated in long language or numerical sequences. The two main kinds of keyboards are the alphanumeric keyboard as on a typewriter, and the numeric keyboard as on a calculator or a pushbutton telephone. Another kind of keyboard is the chord keyboard.

Redesign of the typewriter keyboard has been a research issue for about 50 years. The keyboard that we now have on the typewriter for the English language is called the standard keyboard. The standard keyboard is difficult to learn. The most serious challenge to the standard keyboard has been the simplified keyboard, which was designed by Dvorak. Dvorak conducted a motion and time analysis of the standard keyboard and found it wanting. The simplified keyboard may be better than the standard keyboard, but research has not made the point strongly enough to warrant adoption of the simplified keyboard.

The alphabetic keyboard, where keys are laid out in ABC fashion, is an alternative keyboard assumed to have special merit for naive typists, but research has not verified the assumption.

A keyboard like the standard keyboard or the alphabetic keyboard has a one-to-one visual–motor code, but a chord keyboard derives its name from characters or groups of characters that are represented by two or more key presses, as a chord on a piano. Machines used by court stenographers for verbatim recording have chord keyboards. The most interesting development in chord keyboards is the letter-shape keyboard, where the pattern of keypresses is based on visualization of the form of the alphanumeric character, not a one-to-one visual–motor code. Research has found learning of the letter-shape keyboard to be rapid. So far, chord keyboards have found only specialized uses.

Research on numeric keyboards has not been decisive. There is no compelling reason to change the keyboard layouts of pushbutton telephones and calculators that are in common use.

References

Alden, D. G., Daniels, R. W., & Kanarick, A. F. (1972). Keyboard design and operation: A review of the major issues. *Human Factors, 14,* 275–293.

Arnott, J. L., Newell, A. F., & Downton, A. C. (1979). A comparison of Palantype and Stenograph for use in a speech transcription aid for the deaf. *Journal of Biomedical Engineering, 1,* 201–210.

Beeching, W. A. (1974). *Century of the typewriter.* New York: St. Martin's Press.

Boehm, D. A. (Ed.). (1984). *1984 Guinness Book of Records.* New York: Sterling.

Butterbaugh, L. C., & Rockwell, T. H. (1982). Evaluation of alternative alphanumeric keying logics. *Human Factors, 24,* 521–523.

Chapman, J. C. (1919). The learning curve in typewriting. *Journal of Applied Psychology, 3,* 252–268.

Conrad, R., & Hull, A. F. (1968). The preferred layout for numeral data-entry keysets. *Ergonomics, 11,* 165–173.

Cooper, W. E. (1983). *Cognitive aspects of skilled typewriting.* New York: Springer-Verlag.

Deininger, R. L. (1960). Human factors engineering studies of the design and use of pushbutton telephone sets. *Bell System Technical Journal, 39,* 995–1012.

Dvorak, A. (1943). There is a better typewriter keyboard. *National Business Education Quarterly, 12,* 51–66.

Dvorak, A., Merrick, N. L., Dealey, W. L., & Ford, G. C. (1936). *Typewriting behavior.* New York: American Book Company.

Fox, J. G., & Stansfield, R. G. (1964). Digram keying times for typists. *Ergonomics, 7,* 317–320.

Gentner, D. R. (1983). Keystroke timing in transcription typing. In W. E. Cooper (Ed.), *Cognitive aspects of skilled typewriting.* New York: Springer-Verlag. Pp. 95–120.

Gopher, D. (1984). The contribution of vision-based imagery to the acquisition and operation of a transcription skill. In W. Prinz & A. F. Sanders (Eds.), *Cognition and motor processes.* New York: Springer-Verlag. Pp. 195–208.

Gopher, D., Raij, D. (1986). Typing with a two hand chord keyboard—will the QWERTY Keyboard become obsolete? In D. Gopher (Ed.), Experiments with a two hand chord keyboard—the structure and acquisition process of a complex transcription skill. Haifa, Israel: Faculty of Industrial Engineering and Management, Technion—Israel Institute of Technology, Technical Report HEIS-86-5, August. Pp. 46–63.

Larochelle, S. (1983). A comparison of skilled and novice performance in discontinuous typing. In W. E. Cooper (Ed.), *Cognitive aspects of skilled typewriting.* New York: Springer-Verlag. Pp. 67–94.

Lutz, M. C., & Chapanis, A. (1955). Expected locations of digits and letters on ten-button keysets. *Journal of Applied Psychology, 39,* 314–317.

Nakatani, L. H., & O'Connor, K. D. (1980). Speech feedback for touch-keying. *Ergonomics, 23,* 643–654.

Norman, D. A., & Fisher, D. (1982). Why alphabetic keyboards are not easy to use: Keyboard layout doesn't matter much. *Human Factors, 24,* 509–519.

Noyes, J. (1983). The QWERTY keyboard: A review. *International Journal of Man–Machine Studies, 18,* 265–281.

Ostry, D. J. (1983). Determinants of interkey times in typing. In W. E. Cooper (Ed.), *Cognitive aspects of skilled typewriting.* New York: Springer-Verlag. Pp. 225–246.

Pollard, D., & Cooper, M. B. (1979). The effect of feedback on keying performance. *Applied Ergonomics, 10,* 194–200.

Seibel, R. (1972). Data entry devices and procedures. In H. P. Van Cott & R. G. Kinkade (Eds.), *Human engineering guide to equipment design.* Washington, DC: U.S. Government Printing Office. Pp. 311–344.

Sidorsky, R. C. (1974). Alpha-dot: A new approach to direct computer entry of battlefield data. Arlington, VA: U.S. Army Research Institute for the Behavioral and Social Sciences, Technical Paper 249, January.

Sternberg, S., Monsell, S., Knoll, R. L., & Wright, C. E. (1978). The latency and duration of rapid movement sequences: Comparisons of speech and typewriting. In G. E. Stelmach (Ed.), *Information processing in motor control and learning.* New York: Academic Press. Pp. 117–152.

Yamada, H. (1980). A historical study of typewriters and typing methods: From the position of planning Japanese parallels. *Journal of Information Processing, 2,* 175–202.

PART **VI**

The Computer in the Workplace

Design of the Human–Computer Workplace

There is no need to view the human–computer workplace as qualitatively different from older workplaces, requiring human factors engineering to forget about the principles of the past and begin anew. As before, the computer is embedded in a system with one or more goals. The computer presents information on a display for an operator's response. There are controls that the operator uses for response to the displayed information. Viewed in this way little has changed, and much of what we know about design of the workplace applies. What is different is that the computer has memory for storing information and programs for operating on it. The operator must communicate with the computer and request information, or request that operations be performed on information. The operator now has a machine to supplement his own capabilities for storing, processing, and using information.

In this chapter we emphasize use of a digital computer with a word processing program, or, simply, *text editor*. Digital computers were originally conceived to aid mathematical analysis, but now they have roles of processing different kinds of information, and text material is one of them. The text editor and its user are a system that is easy to understand because the goal is writing a document, or retrieving information about documents, and we are all familiar with it. Moreover, text editors are widely used today, so many readers will appreciate the human factors issues. These human factors issues are not unique to text editing but can be general ones because text editors are interactive systems, like many kinds of human–computer systems, where the user and the computer have a dialogue. Text editing is a paradigm for many computer tasks.

The Task of Writing a Document _____

We have all written a document such as a letter or a report many times, so we all are familiar with the steps involved even though we seldom notice them. Foregoing a formal task analysis, it is useful to itemize the major steps in the preparation of a document and to compare how they are done manually and with a text editor.

Steps in the Manual Writing of a Document
Get paper from storage.
Put paper on the desk.
Get a pen, pencil, or typewriter.
Plan the format of the document—size of margins, single or double spacing, page numbering, sections, headings, and so on.
Write a draft version, using the format.
Check draft version for typographical errors, spelling errors, grammatical errors, omissions, poor organization, stylistic inelegancies.
Revise draft version. Correct errors, add material, delete material, rearrange paragraphs and sections to improve organization by cutting and pasting, make style changes.
Retype draft version.
Perhaps one or more additional draft versions, continuing corrections and revisions. Retyping of them.
Assemble final version.
Type the final version.
Make a file copy of the final version.
File a copy of the final version.
Throw draft version(s) away.

Steps in the Computer Writing of a Document
Turn on computer
Insert word-processing program.
Insert commands to open a document window on the screen (a blank "page").
Specify format of the document.
Write a draft version, storing it in computer memory as it is being written.
Convert the draft version into the final version. Identify and correct errors, make additions and deletions, improve organization of document with rearrangement of sentences, paragraphs, and sections.

Store final version in computer memory.
Print out final version.

Manual and Computer Writing Compared. Once you have used a text editor, the differences between manual writing and computer writing will seem larger than this simplistic task analysis indicates. More than an elegant typing machine, a text editor is an editorial machine. Editorial work includes typographical and spelling errors, but it is more a matter of structure, organization, format, and style of the document, and a text editor can do an editorial operation at the press of a key or two. Repeated typing of draft versions is a thing of the past. Retrieval of information and filing are swift and accurate. Instead of pages, perhaps hundreds of them, littering the desktop and the file cabinet, the pages are filed with invisible neatness in computer memory. Instead of slowly finding a page amid a pile of pages, retrieval from computer memory is easy and nearly instantaneous.

Is computer writing superior to manual writing? The answer is usually "yes," but it should not be taken as a self-evident answer. We have been writing manually for a long time with success, and it pays to ask if the gains with a text editor are sufficient to warrant the expense of automation. One study, for example, found no difference between computerized and noncomputerized writing. Gould (1981) found that the composition and writing of letters by touch typists with text editors took about 50 percent longer than manually. The writers took longer in almost all phases of letter writing, primarily because they made more modifications and because of the commands that had to be given to the computer. The quality and style of the letters were judged about the same for both methods of writing.

Granting this finding for letters, there nevertheless is little doubt that computerized writing has the edge when a wider range of documents is considered. The text editor emerges with unmistakable advantages when the ease of storing a long document or many documents, and of retrieving information from them, are considered. Manual writing bogs down when a long document must be corrected and rewritten, or when many documents and forms must be accessed.

The Languages of Communication
with a Computer

Operators have always "communicated" with their systems by moving a control or throwing a switch, and so have "told" their systems what to do. The user must tell the system what action must be performed or the object of the action. The same is true of computers. The difference is that there must be a language of communication with the computer. A computer programmer has used the code that is a programming language to write a program for the computer, and a user's code must now be specified as a way of communicating uses of the program. There would be no problem if the computer programmers were also the users because they would be no strangers to the code world of the computer, but the problem comes when ordinary individuals use the system. Few of us want to be bothered with the details of automotive engineering that are implicit in the car that we drive. We want a car that is easy to use and which accomplishes our transportations goals efficiently. The same is true of a computer-based system. Ordinary individuals are not interested in computer science or in learning about it. We do not want to be bothered with abstract codes such as those used by the computer programmer. We want dialogue with the computer to be easy and natural. Unfortunately, easy communication with a computer is a goal that has not yet been reached. Progress has been made, but much remains to be done before computer-based systems such as text editors are self-evident to the beginner. How to ease communication with computers has attracted research.

The most common way of communicating with text editors is with command languages. Another way is through selection from a menu on the display. First, command languages.

Command Languages

A *command language* is a set of instructions for the computer which has a specified vocabulary and grammar. The language defines the legal commands for the computer; all statements outside the language are illegal commands. A command language can vary in complexity from abstract codes to the natural language of everyday use. Abbreviations, as intermediate between an abstract command language and natural language, are also used. The input of a command language is with a standard keyboard (Chapter 14) that has a few keys added for special functions.

As for any language of the earth, the distinction can be made between *syntax* and *semantics*. The rules for the grammatical structure of a sentence of a natural language can be described (syntax) independently of the meaning of the sentence (semantics). The same is true of an artificial command language for a computer. The input actions at the keyboard by the user (syntax) are independent of what it means for the computer functions that follow (semantics). Many syntactical arrangements of keyboard inputs can be associated with a computer function, just as many sentences in a natural language can have the same meaning. The issue is identification of a command language that is easy for the user to learn and remember.

In extreme form, here is the problem with command languages. Shneiderman (1987, p. 138) discusses a system that requires the user to type this command to delete blank lines:

GREP -V $ FILEA > FILEB

The command to get a printout of a particular kind was

CP TAG DEV E VTSO LOCAL 2 OPTCD-3871 X-GB12

The problem is illustrated in less extreme form by an electronic mail system where messages are sent from one person's computer to another's (Shneiderman, 1987, p. 48). Pressing the key marked RETURN was required to terminate a paragraph. The command CTRL-D signified termination of a letter. Neither of these commands has any meaningful bearing on the functions that they perform.

The computer knows only what it is programmed to know, so one language can be as good as another from the computer's point of view, but the burden of abstract command languages is heavy for the user. The issue is like that of keyboards (Chapter 14). The commonplace standard keyboard has shortcomings and is difficult to learn, requiring long hours of boring practice and many frustrating mistakes. The world needs an easy keyboard, and it also needs an easy command language.

The Experiment by Ledgard et al. A main research direction for command languages has been comparison of abstract and natural command languages. Ledgard et al. (1980) tested the hypothesis that "an interactive system should be based on familiar, descriptive, everyday words and legitimate English phrases." Putting their hypothesis another way, they wanted to capitalize on the established language capabilities of the human and not require her to learn an alien symbolic system.

Two text editors of equal power were used. One required a relatively abstract language and was called the *notational editor*. The other, using English phrases, was called the *English editor*. To illustrate the difference between the two, the notational editor used the command LIST;* to display all lines from the current line to the last line of text. The English editor required the command LIST ALL LINES. To correct the typing error of KO for OK, the notational editor had RS:/KO/,/OK/. The English editor had CHANGE ALL 'KO' TO 'OK'. All subjects had familiarity with typing, and their experience with computers ranged from little or no experience to over 100 hours. After familiarization with a text editor, the commands that would be required, and the editing task of correcting text, the subjects were given 20 minutes to correct text material.

Averaged over all subjects, the results of the study were for the English editor and the notational editor, respectively, 63 percent versus 48 percent for percentage of editing tasks completed, and 7.8 percent versus 16 percent for wrong commands. More experienced users performed better than inexperienced users on both editors, which is to be expected, but perhaps not expected is that experienced users showed a performance advantage for the English editor just as inexperienced users did. The investigators concluded (p. 563): "We feel that the experienced user when faced with a new system or with a system that he or she uses only intermittently is in much the same position as the novice user, and so needs good human engineering just as much."

The Experiment by Grudin and Barnard. Grudin and Barnard (1984) compared five kinds of commands for the editing of errors in proverbs. A subject might be given *hange is as a good as a pjqf rest* and be required to correct it so that it reads *a change is as good as a rest.* The commands are shown in Table 15-1, ranging from natural language to meaningless consonants. The subjects had three daily sessions, with eight proverbs presented at each session. A subject had the same kind of command throughout. Instructions in the use of the commands were given on the first session. The text editor had a HELP key (HELP commands are discussed in a later section), where, at any time, the subject could have all the command names, or all the command names and their functions, displayed as a reminder. At each session a subject was given a memory test of the command names and their functions.

Figure 15-1 shows the average time to complete a proverb as a function of type of command and number of sessions. As Ledgard et al. (1980) had found, editing performance with the natural language commands was best. Learning occurred for all types of commands, so maybe performance with the other commands would catch up to natural language commands eventually, but the learning demands are obviously greater.

Other measures of performance showed the same trends as time to complete a proverb. Some types of commands were inefficient because they increased the number of extra transactions. For example, an incorrect command could be inserted and then canceled, or greater use of the HELP facility could be made. The natural language commands had the fewest extra transactions and the consonant commands the most, with the other types of commands falling in between. The same trends were found for errors. The results of the memory tests favored natural language commands also, which is a finding that deserves

TABLE 15.1   Five Sets of Commands Names for Text Editing That Were Compared Experimentally

Specific Names	Abbreviations	Unrelated Names	Pseudowords	Consonant Words
front	frn	stack	blark	ksd
back	bck	turn	lins	rip
insert	nsr	afford	aspalm	trp
delete	dlt	parole	ragole	fnm
prefix	prf	chisel	clamen	drs
append	pnd	uncurl	extich	gpf
merge	mrg	rinse	dapse	lts
split	spl	throw	thrag	ctf
fetch	ftc	light	manst	nrb
store	str	brake	fluve	rcn

Source: Grudin & Barnard (1984), Table 1. Reprinted with permission of The Human Factors Society.

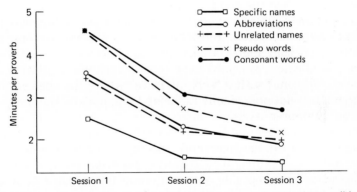

Figure 15-1   Average time to complete each proverb for the three editing sessions. [Adapted from Grudin and Barnard (1984), Fig. 2. Copyright 1984 by the Human Factors Society, Inc. and reproduced by permission.]

highlighting because the forgetting of commands is common. Almost all the natural language commands could be recalled each day, as well as the functions they defined. Consonants and their functions were most poorly recalled, with the others falling between them.

Abbreviations. The experiments by Ledgard et al. (1980) and Grudin and Barnard (1984) demonstrated that natural language commands are best, but the issue is not so simple. Natural language commands are wordy and time consuming to enter. Novices would welcome the ease of natural language commands, but experienced users are impatient and like abbreviated codes that can be entered rapidly. Single-letter commands are the shortest of all, and they can be made memorable, like *P* for PRINT and *C* for COPY, but there are not enough letters to do the job. One might think that longer abbreviations which readily imply the full command are the answer, but they did not fare as well as natural language commands in the Grudin and Barnard experiment, even though an effort was made to design them so that the full command name and its meaning could be recaptured (Table 15-1). Perhaps a better design of abbreviations in the Grudin and Barnard experiment would have made them equivalent to natural language commands, but the route to better design is not self-evident because people do not agree on what abbreviations should be. Hodge and Pennington (1973) found a great deal of variability in the abbreviations which their subjects made from meaningful words, and the longer the words, the less the agreement. About two-thirds of the subjects agreed on an abbreviation when the words had four or five letters, and only about one-third agreed when the words had eight or nine letters. Streeter et al. (1983) had their subjects abbreviate command terms that were used in a large computer system and found only 37 percent agreement. It is not surprising that abbreviations have an intermediate level of success in the Grudin and Barnard experiment. Perhaps a solution to this problem will come with a "soft interface" where the user can specify his own abbreviations, ideally tailoring commands to the individual.

Some computer systems have *automatic command completion*, where the user types the first few letters of a command, perhaps only enough to be a unique command, and the computer infers the whole command. This is efficient by requiring fewer keystrokes, but the full command or an acceptable abbreviation of it still must be known so that the user can type enough of it for the computer to recognize it.

Selection from a Menu

A *menu* is a list of commands on the display from which the one that is needed is chosen, like choosing the answer to a question on a multiple-choice examination. The items of the menu are usually in natural language, which is an advantage. Figures 15-2 and 15-3 show the menu for formatting paragraphs that is used on the word processing program Microsoft® Word for the Apple® Macintosh™ computer, which is a menu-based system. All the information about formatting paragraphs cannot be conveniently displayed at the same time, so the menu is hierarchically arranged. Selecting the FORMATS or TABS command in

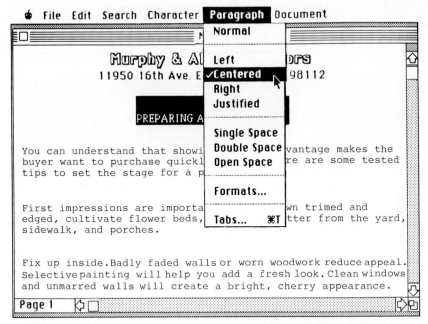

Figure 15-2 Example of a menu used for text editing with the word processing program Microsoft Word for use with the Apple Macintosh computer. The menu shown is for the formatting of paragraphs. [Adapted from the Microsoft Word user's manual. Used by permission of the Microsoft Corporation.]

Figure 15-2 causes another menu with more choices to appear. Figure 15-3 shows the menu that appears when the FORMATS command is chosen.

There is no memory requirement for menu selection other than to know the relationship between the term on the display and the computer function that it implies. Menu selection is easy and accurate, and it is considered effective for beginners or occasional users of computers. Experienced users can be impatient with menus because the request for a menu and selection from it can be more time consuming than typing a command from a command language. Some believe that computer programs should be designed so that the user can change from menu selection to a command language as experience is acquired. There are programs that allow this, but whether they optimize the performance of users is not known.

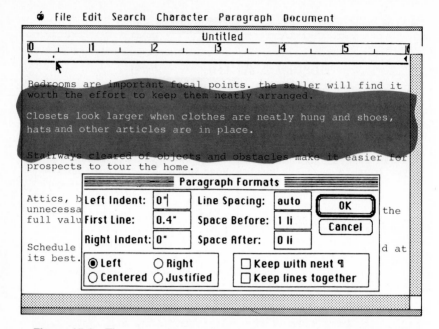

Figure 15-3 The menu shown in Fig. 15-2 is arranged in a hierarchy with these additional commands for the formatting of paragraphs. Choosing the FORMATS command in Fig. 15-2 produces the submenu shown here. A hierarchical arrangement allows the efficient display of many commands that would crowd a single menu. [Adapted from the Microsoft Word user's manual. Used by permission of the Microsoft Corporation.]

The Depth–Breadth Trade-off. Design of the menus in Figures 15-2 and 15-3 is not difficult because the amount of information is small. When the amount of information is large, however, perhaps involving hundreds of items, there is an issue of how to design the menu. The issue is illustrated in Tables 15-2 through 15-5. Suppose that you had many recipes in a data base and that you decided to find a seafood recipe for dinner. How should these data be organized so that you can use a menu efficiently to search and find the kind of fish that you want and how you want it cooked?

All the recipes are shown in Table 15-2, and they could be presented, all at once, just as shown, in a noncategorized, nonhierarchical, one-level presentation. Choose one and the menu is replaced with the recipe. Table 15-3 has the same items structured and presented simultaneously, in a categorized, one-level presentation. The choices could appear two at a time, as in Table 15-4, in a two-choice/five-level hierarchy. Or the choices could appear two or four at a time, in a three-level hierarchy, as in Table 15-5.

TABLE 15.2   Example of a Noncategorized One-Level Menu

Broiled haddock fillets	Poached oysters
Broiled shrimp	Baked shrimp
Baked cod fillets	Steamed cod fillets
Broiled cod fillets	Steamed oysters
Poached whole catfish	Baked haddock fillets
Steamed haddock fillets	Broiled crabs
Poached crabs	Poached whole trout
Steamed crabs	Baked mussels
Baked crabs	Steamed shrimp
Broiled mussels	Poached shrimp
Broiled whole catfish	Broiled oysters
Baked whole catfish	Steamed whole trout
Poached mussels	Poached haddock fillets
Steamed mussels	Baked oysters
Broiled whole trout	Steamed whole catfish
Poached cod fillets	Baked whole trout

TABLE 15.3   Example of a Categorized One-Level Menu

Fillets	Whole Fish	Soft Shellfish	Hard Shellfish
Broiled cod	Broiled catfish	Broiled shrimp	Broiled oysters
Broiled haddock	Broiled trout	Broiled crabs	Broiled mussels
Baked cod	Baked catfish	Baked shrimp	Baked oysters
Baked haddock	Baked trout	Baked crabs	Baked mussels
Poached cod	Poached catfish	Poached shrimp	Poached oysters
Poached haddock	Poached trout	Poached crabs	Poached mussels
Steamed cod	Steamed catfish	Steamed shrimp	Steamed oysters
Steamed haddock	Steamed trout	Steamed crabs	Steamed mussels

TABLE 15.4   Example of How a Two-Choice/Five-Level Hierarchical Menu Might Be Organized

Choice 1 appears:	Fish	Shellfish
You choose Fish.		
Choice 2 appears:	Fillets	Whole fish
You choose Fillets.		
Choice 3 appears:	Cod	Haddock
You choose Cod.		
Choice 4 appears:	Oven	Stove top
You choose Oven.		
Choice 5 appears:	Broiled	Baked
You choose Baked.		

TABLE 15.5   Example of How a Three-Level Hierarchical Menu With Two and Four Choices Might Be Organized

Choice 1 appears: You choose Fillets.	Fillets	Whole fish	Soft shellfish	Hard shellfish
Choice 2 appears: You choose Cod.	Cod	Haddock		
Choice 3 appears: You choose Baked.	Broiled	Baked	Poached	Steamed

Should the menu have depth or breadth? Time is required to step through many levels, so too many levels would be undesirable. On the other hand, presenting too many items at once could be a cluttered display, costly in visual search time. Some trade-off would seem to be called for, and this has been called the *depth–breadth trade-off*.

Snowberry et al. (1983) conducted an experiment of the depth–breadth trade-off. She presented a target word on a computer display and had the subjects search a data base of 64 words for it. Three groups of subjects had hierarchical presentation—a two-choice/six-level hierarchy, a four-choice/three-level hierarchy, and an eight-choice/two-level hierarchy. A fourth group had a categorized one-level presentation, (as illustrated in Table 15-3), and a fifth group had noncategorized one-level presentation. Her measures of performance were search time and percent correct. Snowberry et al. found the best performance of all for the categorized one-level condition. The noncategorized one-level condition had slow search time. Among the groups that had hierarchical search, the group with the eight-choice/two-level hierarchy was the best. A small number of levels, with a few options at each level, is better than many levels with binary choice at each level.

The superiority of the categorized one-level presentation has been supported in other experiments (Seppälä & Salvendy, 1985) and analyses (Papp & Roske-Hofstrand, 1986), and a moment's reflection shows why. With a noncategorized one-level presentation, where the viewer stops searching when the item has been found, the search will cover half the items on average (Lee & MacGregor, 1985). When, however, a categorized one-level presentation is used, as in Table 15-3, search time is minimized because the data are organized perceptually and the user can direct his search immediately to the region of the display that has the wanted item. With learning, searching efficiency can be even higher. We would expect learning to benefit the searching of a noncategorized one-level display also, perhaps eventually reaching the level found for the categorized one-level display. Whether a one-level presentation is categorized or not, there is no stepping through levels, which costs in additional time.

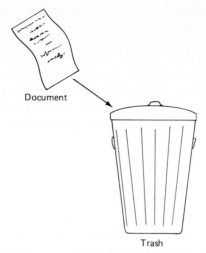

Document

Trash

Figure 15-4   Example of icons used with the Apple® Macintosh computer. Moving the document icon to the trash icon throws the document away. [Adapted from Macintosh™ icons © Apple® Computer, Inc. Used by permission.]

Icons

Icons are stylized drawings of the things that they represent, and are another way of communicating with the computer. An example of icons used in a text editor is shown in Figure 15-4. Figure 15-4 illustrates one of the ways that a document is thrown away with the word-processing program Microsoft® Word for the Apple® Macintosh™ computer. There is a document icon and a trash icon. Moving the document icon to the trash icon removes it from computer memory.

Icons are an easy way to communicate with a computer. Icons are pictorial, and we saw in Chapter 10 that pictorial displays can be effective for information processing. Icons that are designed well can be a useful option, but they cannot do the whole job. Rubinstein and Hersh (1984, pp. 77–80) recommended that icons be used for concrete functions. The concrete function of throwing a document away can be effectively handled with an icon, but an icon would not lend itself very well to the formatting of a table.

Assistance to the User

The learning of a computer-based system can be a frustrating experience for a novice. Confronted with a densely coded command language having a substantial vocabulary, the beginner is faced with a learning requirement before a document as simple as a letter can be written. Many of the things that come without conscious thinking in manual writing, like the width of margins and the spacing of lines, must be deliberately commanded with a text editor. Even menu selection, which is easy, does not result in instantaneous learning of a text editor. The day undoubtedly will come when a text editor will be as simple as manual writing, but meanwhile the user must look for help to ease the frustrations of learning. Obviously, a reference manual that is well written with clear examples is a learner's good friend. Other aids are HELP information, display feedback, and reminders.

HELP **Information**

A commercially available computer program will come with a reference manual, but the manual can be summarized in the form of HELP information, which is an abbreviated reference manual that the user can request from computer memory; the commands and brief statements of their functions are on call. The HELP information will replace the text on the display when it is requested, and when the user is finished with it he commands its removal and returns to the text. HELP information seems like an instructor at your side, ready to refresh your memory when you forget what to do.

As good an idea as HELP information appears to be, there is no evidence that it serves its purpose and that users rely on it very much. Rosson (1984) questioned 121 users of a text editor and found that only 3 percent of them found the HELP facility to be the most important aid during learning. Only 6 percent said that HELP information was the preferred aid. The reference manual was the most important and the most preferred learning aid. Rosson's questionnaire findings are supported by laboratory research. Cohill and Williges (1985) compared several ways of providing assistance to the learner of a text editor and found that help was better than no help. On the other hand, a comparison of the reference manual with HELP information showed an advantage for the reference manual.

As instant assistance at the fingertip, why is not HELP information more effective? One possibility that has been discussed (Cohill and Williges, 1985, p. 342; Rosson, 1984, p. 473) is that the HELP information replaces the text material on the screen and denies a comparison of the problem with its solution. The user must memorize the HELP information and remember it until he can return to the display and try it. A reference manual, however, allows side-by-side comparison. Cohill and Williges believe that the solution lies in a split screen, where the user sees text in one half of the display and HELP information in the other. Finally, a reference manual has more information and can communicate it better. A good reference manual is readable, with plenty of examples to clarify the commands and their uses.

Display Feedback

Nothing is more frustrating than to enter a command and have nothing happen. Was the command illegal? Was the command received? Has the computer failed? A worthy principle is that the system should indicate the actions that have been taken and the correctness of them, and should indicate the occurrence of necessary delays.

If an illegal command has been typed, the computer should not be indifferent to a command out of bounds but should instruct the display to say something like ILLEGAL COMMAND. If a legal command does not produce the desired action almost immediately, and there is to be a delay, the display should indicate it by displaying something as simple as WAIT, PLEASE. A small clockface has been used to indicate delay.

The same rules apply to menu selection. When an item has been selected, the item should brighten, or dim, or have a check mark appear beside it so that the entry is acknowledged and the user sees the correctness of her response.

Reminders

A human–computer system such as a text editor is an interactive system that can appraise some of the user's actions and engage in a dialogue about them. A *reminder* is a query about a user's action which may be legitimate but which deserves a comment nevertheless. A reminder may be about an error, but not necessarily.

A computer might display a reminder such as DON'T FORGET TO SAVE YOUR DOCUMENT if you start to shut the computer down without having committed the material to permanent storage and are in danger of losing it. You may have fully intended to lose the material, but on the other hand, you may have wanted to save the document and forgot to do it. The reminder gives time to reevaluate. Norman (1981), who writes of the shortcomings of communicating with the UNIX™ system, discusses how the pressing of *q* for quitting work erases all the work that had been done in the session because the program assumes that the user would have saved the material if he had wanted it saved. A reminder could easily prevent a disaster.

There are limits to reminders. Too many reminders can be bothersome. Reminders about trivial matters are not worth it. Notwithstanding, appropriately conceived reminders can be a valuable part of human–computer interaction.

Error Detection and Correction in Text Editing

There is nothing a text editor can do about it if you are not thinking clearly and so are not typing good content. The text editor does not know your thoughts or what they should be. Errors in spelling, grammar, and style are less profound than errors of content, but they need correcting nevertheless, and there are programs for text editors that help in their detection and correction.

Speller Checkers. Programs that check errors in spelling are common, and they are called *speller checkers*. A speller checker is a computerized dictionary, and each word of the document is verified against a dictionary entry. A word is drawn to a user's attention when the word is not found in the dictionary, which means that the word is misspelled and should be corrected or that the word is outside the bounds of the dictionary, like specialized technical terms or proper names. Speller checkers often allow the main dictionary to be augmented with an adjunct dictionary of technical terms and proper names so that they can be checked automatically.

There are two basic kinds of speller checkers. One kind runs through the document and checks it for errors after the document is finished. The other kind checks each word as the typing is in progress and signals the user with a tone that an error has just been made. Some typists may be grateful for information on errors as they are made, but others may find it of dubious merit for two reasons. One reason is that an individual well trained in a skill can detect and correct many of his own errors. Skilled typists can detect and correct about 60 to 90 percent of their own errors (West, 1967; Rabbitt, 1978; Neal & Emmons, 1984), and they seldom need a machine to tell them about it. Second, a beeping tone every time an error occurs can be annoying.

Readability and Style Checking. The detection and correction of spelling errors are necessary but they are the least of editorial work. A document can have punctuation errors, have instances of poor grammar or style, or be wordy, too abstract, or badly organized. Bell Laboratories has developed a set of computer programs called Writer's Workbench Software that will check a document for these deficiencies, as well as spell checking (Frase, 1983; Macdonald, 1983; Gingrich, 1983). The need for these programs is greater than might be apparent. Technical writing is a major chore for modern industry. An instruction sheet or a manual on how to use a simple item is a minor chore, but technical documentation in support of a major human–machine system can be several hundred thousand pages. The writing and editorial work is a big, time-consuming, expensive job. Computers cannot write the document, but Writer's Workbench Software is helpful in the editorial work. Technical manuals have not been distinguished by their readability. Poor technical manuals will slow the use and repair of human–machine systems.

After analyzing the characteristics of good writing, Writer's Workbench Software emerged with two main sets of programs: proofreading and style.

Proofreading Programs

1. *Speller checker.*
2. *Punctuation.* Detect such errors as failure to capitalize the first letter of sentences.
3. *Consecutive occurrences of the same word.*
4. *Wordy phrasing.* Detect clumsy phrasing such as "bring to a conclusion" and indicate that "end" is a more readable choice.
5. *Split infinitives.* Infinitives split by adverbs, such as "to quickly decide," is poor form.

Style Programs

1. *Stylistic information.* This program contains indices of style, such as the average length of words and sentences, and readability.
2. *Stylistic analysis.* Style statistics are compared to a standard of excellence in writing. Changes are recommended when the document deviates too far from the standard.
3. *Abstractness.* The words of the document are tested against a list of words rated as abstract. Suggestions are made for changes when too many abstract words are found.
4. *Organization.* The headings, paragraph boundaries, and first and last sentences of each paragraph are printed as an abstract of the paper so that the writer can better see the paper's organization.

Gingrich (1983) conducted an evaluation of Writer's Workbench Software and found that writers willingly used it, liked it, and found it helpful. Writer's Workbench Software is not the only development of its kind (Macdonald, 1983, p. 1897). Programs of this kind are becoming available commercially.

Controls

The keyboard is, of course, the main input device for text editors, but there is more to data entry than alphanumeric characters; the user often must indicate where action is required on the display. The user may wish to indicate where a word should be positioned, or indicate a letter that should be deleted. The control requirements are different from those discussed in Chapters 12 and 13, so different kinds of controls have appeared. Some systems use the keys of an augmented keyboard for positioning an indicator on the display. Others use special devices such as a joystick or "mouse" for indicating. Touching the display is used. Which control is the best?

Experimental Findings

The Experiment by Card et al. Four common control devices that are used with text editors were compared in an experiment by Card et al. (1978; 1983, Chap. 7). The controls are illustrated in Figure 15-5. The *mouse* had two wheels on the bottom, at right angles to each other. One wheel coded movement in the x direction and the other in the y direction, and the x–y coordinates of a cursor on the display were coded correspondingly as the user moved the mouse on the tabletop alongside the keyboard. There was a linear relationship between mouse movement and cursor movement, making it a position control. The *joystick* was a two-dimensional isometric control that did not move when force was applied (in contrast to an isotonic control, which does), but the force was registered by a strain gauge which determined cursor position. The greater the force exerted, the faster the cursor moved, making it a rate control (see Chapter 12 for a reminder about the distinction between rate and position controls). The *step keys* were a five-key cluster that was an addition to the keyboard. The central home key returned the cursor to the upper left corner of the text, and the other step keys were direction keys to move the cursor in x–y directions. Pressing a direction key caused it to move one space, but holding it down caused it to enter a repeating mode and move through a number of spaces at a constant speed (a position control). The *text keys* were labeled Paragraph, Line, Word, Character, and Reverse. The paragraph key caused the cursor to move to the beginning of the next paragraph. The line key moved the cursor downward in the same position in the next line. The word key moved the cursor forward one word. The character key moved the cursor forward one character. Holding the reverse key down while pressing another text key caused the cursor to move opposite to the usual direction. Like step keys, holding a text key down caused it to enter a repeating mode and move the cursor at a constant speed (a position control).

The task was to move the cursor to a target, starting with the hand on the space bar of the keyboard. A page of text was displayed on the screen and a word was highlighted as the target. The words varied from 1 to 10 characters in length. The starting position of the cursor

Figure 15-5 Pointing devices that were used in the experiment by Card et al. (1978; 1983, Chap. 7). See text for explanation. [*Source:* Card et al. (1978), Fig. 1. Reprinted by permission of Taylor & Francis Ltd.]

from the target varied from 1 to 16 cm. The subject pressed a button when the cursor was in position. The button was on top of the mouse, but with the other controls a special button was pressed with the left hand.

Measures of performance were errors and speed. An error was failure to hit a target. One measure of speed, called homing time, was the time from when the hand left the space bar until the cursor began to move. The other speed measure, called positioning time, was time from when the cursor began to move until the selection was made. Total time was the sum of these two speed measures.

The main results are shown in Table 15-6. The mouse shows a deficit in homing time—getting the hand from the keyboard to the mouse—but the deficit is overcome in the fast positioning time, making total time for the mouse the fastest of all. Also, errors for the mouse were the fewest. The joystick is ranked second, with the two key configura-

TABLE 15-6 Results of the Experiment by Card et al. That Compared Four Selection Devices for Text Editors*

Device	Homing Time (seconds)	Positioning Time (seconds)	Total Time (seconds)	Error Rate (percent)
Mouse	0.36	1.29	1.66	5
Joystick	0.26	1.57	1.83	11
Step keys	0.21	2.31	2.51	13
Text keys	0.32	1.95	2.26	9

*Entries are averages for four subjects.
Source: Adapted from Card et al. (1978). Reprinted by permission of Taylor & Francis Ltd.

tions poorest of all. English et al. (1967), in a pioneering study of controls for text editing, also found the mouse superior.

What can be said of this experiment? It is a good human factors experiment whose findings can be useful to someone making a decision about commercially available controls for text editors. At the conceptual level the study highlights variables that future investigators of text editing controls might consider. In Chapter 12 we discussed variables that can be important for the controls of text editors also. One is control–display ratio, or control sensitivity—amount of cursor movement on the display relative to control movement. Control–display ratio was uncontrolled in the study by Card et al. and deserves control or systematic manipulation. Another variable from the tracking domain that deserves consideration is control order. Control order was unsystematic in the Card et al. study, with the mouse and the keyboards being position controls and the joystick a rate control.

Variables such as these will be considered by investigators in time, just as they have been for other systems. In the early days of a system, which is where human–computer systems today are historically, the features of workplaces may be designed arbitrarily and quickly, perhaps under the competitive pressures of commerce. Eventually, there is time for scientific reflection, and when that time comes, basic human factors variables receive research attention.

The Touch Panel. A menu requires selection from a display, and another way to do it is the touch panel (Pickering, 1986; Whitefield, 1986). There are several engineering techniques for making a display touch sensitive (Pickering, 1986). The user merely has to point at the display and touch an item to select it. The response is easy and natural, without an intervening mechanism.

Karat et al. (1986) experimentally compared the mouse and the keyboard, as Card et al. had done, with a touch panel for menu selection. In two experiments they found that performance was best with the touch panel.

The Human–Computer Workplace _____

The visual display unit (VDU) is the heart of a human–computer system because actions by the user and information from the computer are displayed on it (the VDU is usually a cathode ray tube, but see the Appendix of Chapter 10 for a review of alternative displays). A typical text editing workplace has the VDU in the center on a table, a keyboard, a chair, and a copy holder for holding printed text in a readable position. Depending on the system, there can be a selection device such as a mouse alongside the keyboard.

A review of human factors research on VDUs (Helander et al., 1984) is not enthusiastic about the investigations so far, perhaps because the rapid increase in number of human–computer workplaces has caused some investigators, in their zeal to help, to be hurried. Generalizations are tentative, but a few observations are appropriate. Main areas of research have been workplace layout, physical discomfort, and health hazards.

Workplace Layout

There do not appear to be problems in the layout of human–computer workplaces that have not been encountered, and solved, before in the field of human factors (Chapter 9). Human factors engineering has sought general principles of workplace layout over the decades, and insofar as the search has been successful, the principles should apply to human–computer workplaces like any other workplace.

A recommendation for human–computer workplaces is that the main elements of the workplace should be adjustable so that an operator can optimize them for herself. Recent articles (Grandjean, 1984; Marek et al., 1984; Wyatt, 1985) say that the main dimensions of the workplace, such as height of the table, height of the chair, height of the keyboard, position of the keyboard, position of the VDU, and angle of the VDU, should all be adjustable. This is good advice as long as it is kept in mind that it costs more to make things adjustable than not.

Physical Discomfort

A questionnaire to workers who use VDUs will elicit complaints of muscular discomfort, visual discomfort, and so on (Helander et al., 1984). There is no doubt that some workers experience physical problems in their daily use of VDUs, but to know this is only a beginning. Any kind of sustained experience may produce discomfort, and there may be nothing exceptionally discomforting about VDUs.

Gould and Grischkowsky (1984) compared the proofreading of text in hard copy and on a VDU. They had their subjects rate their feelings on a questionnaire. There was a significant increase in discomfort over the day, but there was no difference between hard copy and a VDU. Other indices of discomfort were sought in measures of vision and videotaping of subjects while they worked to see if body movements might reveal evidences of discomfort. None showed a difference between hard copy and the VDU. There was a performance difference, however. The proofreading of hard copy was faster by 20 to 30 percent. Faster proofreading with hard copy is a general finding for which the variables have not yet been isolated (Gould et al., 1987).

Howarth and Istance (1985) obtained the same finding as Gould and Grischkowsky. VDU users reported an increase in visual discomfort over a workday, but so did non-VDU users. The investigators say (p. 148): "We do not conclude that VDU users suffer no discomfort, but rather that the VDU is not a causal factor to any greater extent than is the paper in a typewriter."

Smith et al. (1981), in their study of health complaints by VDU users, were unable to find the VDU a causal agent. Human factors principles were often ignored in the design of modern VDU-centered workplaces, they observed. Some may find it tempting to place the blame for complaints on the VDU, but there are other possibilities. The issue is whether VDUs are causal in the discomfort or whether other, covarying characteristics of the workplace, are the cause. There can be variables correlated with the VDU that are producing the effect and making the VDU appear to be the cause. A keyboard that is placed too high will cause muscular discomfort, and it is easy to blame it on the nearby VDU.

Health Hazards

There is a fear that VDUs with cathode ray tubes are a radiation hazard. As fears so often are, the radiation fear appears to be unrealistic; there is no evidence for it (Murray et al., 1981). Radiation levels are low relative to current occupational exposure standards.

Summary ————————————————————————

Access to a digital computer in the workplace gives the operator an impressive machine to supplement his own capabilities for storing, processing, and using information. Some of the issues surrounding a digital computer in the workplace are traditional, but communication with the computer is a new topic for human factors engineering, and it has attracted research. The issue is ease of communication for a user who is not a computer scientist. The two principal methods of communication are a command language and a menu.

A command language is a set of instructions for the computer which has a specified vocabulary and grammar, and its input is with an augmented standard keyboard (Chapter 14). Computer programs will often prescribe commands unrelated to function. Research has shown that abstract commands, such as LIST;* to display all lines from the current line to the last line of text in text editing, are more difficult than natural language commands such as LIST ALL LINES.

A menu is a list of commands on the display from which the operator chooses the one needed, like choosing an answer on a multiple-choice examination. The items of the menu are usually in the natural language. In contrast to command languages, there are few learning or memory requirements, which is an advantage for the nonspecialist user of computers. The research issue for menu is how best to organize the material on the display for selection.

Other aids to the nonspecialist user of computers are HELP information, display feedback, and reminders. HELP information is an abbreviated reference manual that the user can request from computer memory. HELP information seems like a useful idea, but it is not used very much. Display feedback avoids ambiguity by telling the user what is happening. A reminder is a query about user action.

The mouse, the joystick, and two arrangements of five keys on the keyboard have been compared experimentally as input devices for text editing. The mouse was the fastest, with the fewest errors. The touch panel, where the operator touches the item on the display that she wants selected, also has merit.

The visual display unit is central to human–computer systems because actions by the user and information from the computer are displayed on it. Workers who regularly use computers complain of visual discomfort, but research has found that the visual discomfort is the same as working with hard copy. The discomfort is real but the visual display unit is not causal. Nor is there convincing evidence that visual display units with cathode ray tubes are a radiation hazard.

References

Card, S. K., English, W. K., & Burr, B. J. (1978). Evaluation of mouse, rate-controlled isometric joystick, step keys, and text keys for text selection on a CRT. *Ergonomics, 21,* 601–613.

Card, S. K., Moran, T. P., & Newell, A. (1983). *The psychology of human–computer interaction.* Hillsdale, NJ: Lawrence Erlbaum.

Cohill, A. M., & Williges, R. C. (1985). Retrieval of HELP information for novice users of interactive computer systems. *Human Factors, 27,* 335–343.

English, W. K., Engelbart, D. C., & Berman, M. L. (1967). Display-selection techniques for text manipulation. *IEEE Transactions on Human Factors in Electronics, HFE-8,* 5–15.

Frase, L. T. (1983). The UNIX™ Writer's Workbench Software: Philosophy. *Bell System Technical Journal, 62,* 1883–1890.

Gingrich, P. S. (1983). The UNIX™ Writer's Workbench Software: Results of a field study. *Bell System Technical Journal, 62,* 1909–1921.

Gould, J. D. (1981). Composing letters with computer-based text editors. *Human Factors, 23,* 593–606.

Gould, J. D., & Grischkowsky, N. (1984). Doing the same work with hard copy and with cathode-ray tube (CRT) computer terminals. *Human Factors, 26,* 323–337.

Gould, J. D., Alfaro, L., Barnes, V., Finn, R., Grischkowsky, N., & Minuto, A. (1987). Reading is slower from CRT displays than from paper: Attempts to isolate a single-variable explanation. *Human Factors, 29,* 269–299.

Grandjean, E. (1984). Postures and the design of VDT workstations. *Behaviour and Information Technology, 3,* 301–311.

Grudin, J., & Barnard, P. (1984). The cognitive demands of learning and representing command names for text editing. *Human Factors, 26,* 407–422.

Helander, M. G., Billingsley, P. A., & Schurick, J. M. (1984). An evaluation of human factors research on visual display terminals in the workplace. In F. A. Muckler (Ed.), *Human factors review: 1984.* Santa Monica, CA: Human Factors Society. Pp. 55–129.

Hodge, M. H., & Pennington, F. M. (1973). Some studies of word abbreviation behavior. *Journal of Experimental Psychology, 98,* 350–361.

Howarth, P. A., & Istance, H. O. (1985). The association between visual discomfort and the use of visual display units. *Behaviour and Information Technology, 4,* 131–149.

Karat, J., McDonald, J. E., & Anderson, M. (1986). A comparison of menu selection techniques: Touch panel, mouse and keyboard. *International Journal of Man–Machine Studies, 25,* 73–88.

Ledgard, H., Whiteside, J. A., Singer, A., & Seymour, W. (1980). The natural language of interactive systems. *Communications of the ACM, 23,* 556–563.

Lee, E., & MacGregor, J. (1985). Minimizing user search time in menu retrieval systems. *Human Factors, 27,* 157–162.

Macdonald, N. H. (1983). The UNIX™ Writer's Workbench Software: Rationale and design. *Bell System Technical Journal, 62,* 1891–1908.

Marek, T., Noworol, C., Gedliczka, A., & Matuszek, L. (1984). A study of a modified VDT-stand arrangement. *Behaviour and Information Technology, 3,* 405–409.

Murray, W. E., Moss, C. E., Parr, W. H., & Cox, C. (1981). A radiation and industrial hygiene survey of video display terminal operations. *Human Factors, 23,* 413–420.

Neal, A. S., & Emmons, W. H. (1984). Error correction during text entry with word-processing systems. *Human Factors, 26,* 443–447.

Norman, D. A. (1981). The trouble with UNIX™. *Datamation, 27,* November, 139–150.

Papp, K. R., & Roske-Hofstrand, R. J. (1986). The optimal number of menu options per panel. *Human Factors, 28,* 377–385.

Pickering, J. A. (1986). Touch-sensitive screens: The technologies and their application. *International Journal of Man–Machine Studies, 25,* 249–269.

Rabbitt, P. (1978). Detection of errors by skilled typists. *Ergonomics, 21,* 945–958.

Rosson, M. B. (1984). Effects of experience on learning, using, and evaluating a text editor. *Human Factors, 26,* 463–475.

Rubinstein, R., & Hersh, H. (1984). *The human factor.* Bedford, MA: Digital Press.

Seppälä, P., & Salvendy, G. (1985). Impact of depth of menu hierarchy on performance effectiveness in a supervisory task: Computerized flexible manufacturing system. *Human Factors, 27,* 713–722.

Shneiderman, G. (1987). *Designing the user interface: Strategies for effective human–computer interaction.* Reading, MA: Addison-Wesley.

Smith, M. J., Cohen, B. G. F., Stammerjohn, L. W., Jr., & Happ, A. (1981). An investigation of health complaints and job stress in video display operations. *Human Factors, 23,* 387–400.

Snowberry, K., Parkinson, S. R., & Sisson, N. (1983). Computer display menus. *Ergonomics, 26,* 699–712.

Streeter, L. A., Ackroff, J. M., & Taylor, G. A. (1983). On abbreviating command names. *Bell System Technical Journal, 62,* 1807–1826.

West, L. J. (1967). Vision and kinesthesis in the acquisition of typewriting skill. *Journal of Applied Psychology, 51,* 161–166.

Whitefield, A. (1986). Human factors aspects of pointing as an input technique in interactive computer systems. *Applied Ergonomics, 17,* 97–104.

Wyatt, D. G. (1985). Analysis of the VDT–worker interface. *Behaviour and Information Technology, 4,* 215–230.

CHAPTER 16

Automation

In common usage and in the literature of human factors engineering (e.g., Parsons, 1985, p. 99) the term *automation* means the substitution of a machine function for a human function. Substitutions of this kind are common, but the definition is descriptive and fails to capture the trade-offs that are involved when system designers weigh the capabilities of humans and machines and decide whether a function should be performed solely by a machine, solely by a human being, or by a mix of the two. Sometimes the decision is easy because there are functions that machines can perform best and functions that humans can perform best, as we will see in this chapter. Other times the decision will be subtle. Are there circumstances of system operation where the human operator will have so much to do under time pressure that she will perform badly and should have part of her job automated? Even though the operator can perform each of the functions well, she does poorly because of work load. Viewed in these ways, automation is part of a larger decision-making process in system design.

The topic of the relative capabilities of the human and the machine has been in the human factors literature for almost 40 years (for a review, see Price, 1985), but the topic was an idle one in the early years because machines were poor competitors of human functions. That is not the case today. The digital computer, and the fields of artificial intelligence and robotics nurtured by it, have developed machines that are competitors with some human functions; the topic is no longer idle. A considerable portion of what follows comes from the field of artificial intelligence, so it deserves definition at the outset. *Artificial intelligence* is a discipline that develops machines which simulate functions that we associate with intelligent human behavior (e.g., Barr & Feigenbaum, 1981, 1982; Cohen & Feigenbaum, 1982; Winston, 1984). The functions are usually cognitive or perceptual, and the machines are usually digital computers, although there is no strict necessity for either. The interest is in simulating the complex cognitive functions that would qualify as thinking and the complex perceptual functions of interpreting our visual and auditory worlds. Artificial intelligence is an enterprise that system designers are cultivating, and as it matures, system designers will increasingly have choice between humans and machines to perform system functions. This chapter is about those choices.

A field that is allied with artificial intelligence is *robotics*. Edwards (1984, p. 45) defines an industrial robot as "reprogrammable, multifunctional manipulators designed to move parts, tools or specialized devices through variable programmed motions for the performance of a variety of tasks." In other words, robots are programmable devices that execute acts in the spatial world. The challenge for robotics is not mindless movement but intelligent action, of the kind that a human being would perform. Increasingly in robotics there will be a union of machines that move with machines from artificial intelligence that process knowledge.

This chapter is an examination of the relative capabilities of the human and the machine, and so is a commentary on the systems decision-making process. Rather than long lists of what humans do best and what machines do best, getting into gray areas, it is best to stick to the point and concentrate on shorter lists of unmistakable differences. Discussions of topics within lists will review modern technologies as a way of addressing machines that are competitive with human functions.

Ways in Which Humans are Superior to Machines _____

Thinking and Problem Solving

There is no consensus on a definition of human thinking and reasoning. Trying for a definition, nevertheless, thinking and reasoning is mental problem solving that uses knowledge to assign degrees of importance to the alternative paths to a goal. The thinking person is capable of mentally reviewing the alternative implications of a large amount of information and making decisions about them. An industrial executive is thinking when she weighs the fiscal, personnel, production, and marketing alternatives for the maximizing of profit. A football captain is thinking when he evaluates the alternatives and chooses the next play in his effort to score a goal. General experience and specific training provide the knowledge base for thinking.

Given this problem-solving definition of thinking, are there machines that think? Some of the research done by the field of artificial intelligence has been in the realm of thinking. Computers have been programmed to play chess, checkers, and poker, not because anyone cares about machines that play games but because games are problem-solving activities and a route to the scientific understanding of automated thinking. The best example of machines that think is expert systems. There is no question that an expert thinks when he solves a problem, and a computer that can imitate him is undeniably thinking.

Expert Systems. Can a computer be programmed to imitate a human expert of a topic? If a machine is to imitate the behavior of experts, we should first ask what experts know and do.

Experts usually know a lot about a little; they have a deep understanding of a narrow domain. They know many facts and relations among facts, which is their knowledge base. They keep their knowledge base up to date by learning. The expert reasons about her knowledge, and her powers of reasoning grow with experience. An expert is fortunate if she can start with a lot of information about a problem, but often she must work from piecemeal information, making educated guesses about a solution; experts perceive patterns that escape the novice when data are fragmentary. Experts have a feel for the reliability of their answer. They understand knowledge that is in dispute and know when to qualify their answer. Experts can communicate what they know to others, justifying and qualifying as required. Attempts to build machines, called *expert systems,* to do all of this, have led to a subdivision of artificial intelligence called *knowledge engineering.*

How do you build an expert system? First you pick a topic that can have a high payoff and in which there are experts who know things that novices do not. The topic should be narrow enough so that it can be accomplished in a reasonable period of time. In addition, the task should be cognitive. Topics that experts routinely explain to others work well. Having picked the topic, the knowledge engineer finds one or more experts and has them verbalize the facts about it and how the facts are used. This sounds easy, but it is not. First, the amount of knowledge that an expert has on a topic can be large, even a narrow topic. Second, experts may use intuitive hunches, not always being able to verbalize all the facts that he uses or all the uses that he makes of them. A knowledge engineer will have many sessions with an expert, helping him to express himself, having him describe the facts and what he does with them. The knowledge engineer then takes this information and puts it into two categories: (1) the knowledge base, which has the facts; and (2) the inference engine, as it is called, which is what the expert does with the facts to solve problems. The separation of facts from the use of facts is important because the same methods of inference might be used on an entirely different knowledge area, so the same inference engine can be used, producing a new expert system with relative ease (Van Horn, 1986, Chap. 7). Another reason for keeping the knowledge base and the inference engine separate is that it is easy to add to the knowledge base when it is distinct from the inference engine (Davis, 1984, pp. 23–24). All of this information is then programmed for a computer.

MYCIN was an expert system that was built in the 1970s to help in the diagnosis of infectious diseases (Barr & Feigenbaum, 1982). MYCIN would be most useful to a physician who is not an expert on infectious diseases and who needs immediate help for a stricken patient, although it could also have its uses for an expert on infectious diseases, to confirm his decisions or to help him when he forgets. The format of MYCIN is a dialogue between computer and physician. After collecting routine information about the patient, such as name and age, the computer and the physician might have an exchange which, in part, would look like this (Barr & Feigenbaum, 1982, Chap. 8):

COMPUTER. Have you been able to obtain positive cultures from a site at which Fred Smith has an infection?
PHYSICIAN. Yes.

INFECTION-1

COMPUTER. What is the infection?
PHYSICIAN. Primary-bacteremia.
COMPUTER. Please give the date and approximate time when signs or symptoms of the primary-bacteremia first appeared (month/day/year).
PHYSICIAN. May 5, 1988.
COMPUTER. The most recent positive culture associated with the primary-bacteremia (Infection-1) will be referred to as

CULTURE-1

From what site was the specimen of Culture-1 taken?
PHYSICIAN. Blood.
COMPUTER. Please give the date and time when this blood culture (Culture-1) was obtained (month/day/year).
PHYSICIAN. May 9, 1988.

COMPUTER. The first significant organism from this blood culture (Culture-1) will be referred to as:

ORGANISM-1

Enter the identity of Organism-1.
PHYSICIAN. Unknown.

The computer asks the physician many questions like this, including questions on symptoms and laboratory tests, and the physician answers them. When MYCIN has amassed enough information, a recommendation is made about the cause of the infection:

COMPUTER. My therapy recommendation will be based on the following probable infection(s) and potential causative organism(s):

Infection-1: Primary-Bacteremia
Item 1. The identity of Organism-1 may be *Pseudomonas-Aerugniosa*.
Item 2. The identity of Organism-1 may be *Klebsiella-Pneumoniae*.
Item 3. The identity of Organism-1 may be *E. Coli*.
Item 4. The identity of Organism-1 may be *Bacteroides-Fragilis*.
Item 5. The identity of Organism-1 may be *Enterobacter*.
Item 6. The identity of Organism-1 may be *Proteus-Non-Mirabilis*.

After additional questions the MYCIN program offers a recommendation for therapy:

COMPUTER. My preferred therapy recommendation is as follows: In order to cover for Items 1, 2, 3, 5, and 6:

Give: Gentamicin.
Dose: 119 mg (1.7 mg/kg), q8h IV (or IM) for 10 days.
Comments: Modify dose in renal failure.

In order to cover for Item 4:
Give: Clindamycin.
Dose: 595 mg (8.5 mg/kg), q6h IV (or IM) for 14 days.
Comments: If diarrhea or other GI symptoms develop, patient should be evaluated for possible pseudomembranous colitis.

The physician, using this program in an advisory capacity, is then offered next-choice therapy by MYCIN. If the physician is dissatisfied with MYCIN's recommendations, she can interrogate MYCIN's reasoning. MYCIN's reasoning is as a human being would do it, so the physician can readily follow why the system came to the decision that it did. An expert system called ACE (Automated Cable Expertise), developed by AT&T Bell Laboratories for studying the records of past telephone cable troubles to identify faults for preventive maintenance, cannot explain its reasoning to the user as MYCIN does (Miller, 1984). The credibility of ACE lies in the recommendations that it makes, not in justifying its actions.

MYCIN is a good example of a useful expert system. Another medical expert system that has been devised manages patients who require digitalis for heart ailment (Digitalis Therapy Advisor). PUFF is an expert system for the diagnosis of pulmonary disorders, and an expert system called PROSPECTOR has been devised to help geologists prospect for minerals (Brachman et al., 1983).

It pays to list the advantages of expert systems, potential and real, even though some of them are obvious (Basden, 1983):

1. An expert system can have the composite knowledge of the best experts of the world, being better than any one of them.
2. Expertise can be easily duplicated. There may be few human experts on a topic, but expert systems can be duplicated indefinitely.
3. An expert system is accessible. It does not get sick or go on vacation; it is always there when you need it.
4. An expert system is reliable. It does not forget or find itself under stress and making mistakes.
5. An expert system is consistent. For a given input the same factors will be considered each time and the same weight will be given to them.
6. An expert system, being computer based, has the potential of being faster than the human. Speed is not a consideration for MYCIN, but speed could be an important consideration for an expert system to help an aircraft pilot in decision making.

Offsetting the advantages are disadvantages:

1. Expert systems are difficult to build. The statements that become the knowledge base and the inference engine can run into the thousands. Eliciting the statements from experts, and programming them, are large undertakings.
2. The assimilation of new knowledge may not be easy for an expert system; they do not learn but have to be restructured and reprogrammed. The world is dynamic. Research creates new knowledge. Systems, such as automobiles, radar, and airplanes undergo modifications and model changes. The human expert will often

acquire the new information easily and integrate it into the old, but the designers of an expert system may not find it easy.
3. The assumption of expert systems is that the human expert can verbalize everything that he knows on a topic. This may not be so.

This latter point deserves elaboration because it is an unresolved controversy in psychology, and designers of expert systems have yet to engage its implications. It is obvious that a verbal description of how you ride a bicycle is a crude account of the skill, as is an artist's description of how she paints a picture. Equally obvious is that we can verbalize a vast amount about immediate experience and the storehouse of events in memory. Between these anchor points is disputed territory. Nisbett and Wilson (1977) contend that individuals are often unaware of stimuli or the responses that they make to them. When asked to report how a stimulus affected a response, they might provide a plausible account of what the relationship might be, not necessarily what it actually was. Ericsson and Simon (1980) are more optimistic, although they have exceptions about the accessibility of information to verbal reports, such as a highly practiced task that is performed automatically without conscious intervention. Berry and Broadbent (1984) did not find a relationship between performance on a task and questions about it, but Sanderson and Vicente (unpublished manuscript) countered by finding a stronger relationship when the questions were more penetrating and probing. Berry and Broadbent (1984) and Kellogg (1982) argue that a requirement to pay conscious attention to mental processes during performance of a task, such as verbalizing the steps of a process aloud as a teacher might do, increases the chances of a verbal report being accurate. And so it goes. Expert systems will be slowed in reaching maturity until this chasm is crossed with more research. Expert systems will fail to be truly expert if the expert mechanic relies on the sound of the engine and the expert surgeon the color of the tissue, and neither know that they do it.

Validation data can document the discrepancy between human expert behavior and the performance of expert systems, but it is not abundant. To illustrate validation data, the expert system called Digitalis Therapy Advisor was evaluated with the case histories of 50 heart patients, and the results were judged by a panel of five live experts. The panel had a preference for the recommendations of the attending physician, but Digitalis Therapy Advisor was the same or better 60 to 70 percent of the time (Barr & Feigenbaum, 1982, p. 211).

Conclusion. Back to the original question: Can machines think? Writing in 1950, the British mathematician A. M. Turing said that the question was meaningless (Turing, 1950, p. 442). The real issue, he said, was whether a machine could replace a human function with indistinguishable results. Using this criterion, Turing believed that by the end of the century we would be able to speak of machines thinking without fear of contradiction. Turing's prediction was sound. Expert systems are not yet commonplace but soon will be. At present, however, it is fair to conclude that human beings think better than machines.

A qualification to be kept in mind is that the human being is not a particularly good decision maker. After examining the research on human decision making, Wickens (1984, p. 114) said that his review ". . . suggests a fairly pessimistic view of human performance in decision making and diagnosis" and that ". . . performance probably lies somewhere between chance and optimal, shifting further from the optimum as complexity and stress increase." Wickens's pessimism comes from suboptimal human performance in such decision-making domains as integrating information over time, integrating information over several sources, limited ability to entertain more than a few hypotheses, biases in choosing hypotheses, overconfidence in judgment, and erroneously inferring causal relationships. For all the genuine experts that there are in the world, we should be cautious in acclaiming the decision-making powers of the human. That human decision-making powers are often modest suggests that machines might come to rival the human being in some knowledge areas sooner than we might think.

Pattern Recognition

The swift, accurate perception of visual and auditory patterns is so effortless for us that we seldom think of it as the high skill that it is. Unerringly, we recognize complex objects such as houses, trees, or the faces of friends. The ease with which we process the auditory patterns of speech is astonishing. Our vast storage of pattern information, and the speed of access to it, is one of the reasons why we operate so superbly in this world. Machines that could operate with but a fraction of our capabilities in pattern perception would be an enormous achievement.

There is a large interest in automatic pattern perception because practical applications are beyond imagination. Machines that accurately recognize visual patterns could read any printed page regardless of typography, interpret X rays, interpret weather satellite data, find a face among the millions of criminal faces on file, or do quality control in a factory by separating poorly manufactured parts (inappropriate pattern) from properly manufactured ones. The accurate recognition of auditory patterns would mean that we could fly an airplane by talking to it in our natural language, or operate a text editor by talking to it;

the system would do what you asked. How good are machines at automatic pattern perception?

Visual Pattern Recognition. First consider simple visual patterns and the ways that machines recognize them. When the patterns are invariant, like the numerals on bank checks, the method will be *template matching*. The computer has the patterns stored as template standards, and when the input representing a visual pattern matches a template exactly, the pattern is recognized. Figure 16-1 illustrates template matching. Automatic machine recognition, however, is more complicated when the visual pattern is not rigidly standardized. Consider the recognition of triangles, where the triangles can differ in size and orientation, as in Figure 16-2. The character of the pattern is invariant but all sizes and orientations of it must be accommodated. *Feature matching* is used with this kind of problem. The computer has in storage the general properties of triangles, such as "a closed plane figure with three straight-line sides." In addition, the computer must have routines that permit identification of closed figures and straight lines. Instead of a single matching operation as with template matching, the computer program takes the input and processes it through a series of operations—find a straight line, find a corner, count the straight lines, and so on—that lead to a decision about whether something is or is not a triangle. The requirement is that the pattern be a triangle; size and orientation do not matter.

What of complex scenes, of the kind that humans process moment to moment, where there are a number of objects of different sizes, shapes, and orientations, and where there is light and shadow, color and texture? Although the field of artificial intelligence has some capability for the interpretation of naturalistic scenes, most of the research on scene interpretation has been done with simple geometric forms, as shown in Figure 16-3. Edges and lines are identified by template matching or feature matching, and their meeting at corners is specified. Texture as regions of uniformity, if it is part of the scene, can be indicated. How does the computer program know the objects of a scene once having specified them? How does it "understand"? One approach is to give the program expectations for certain kinds of objects and test to see if the objects that have been analyzed meet the expectations, much as a human being has expectations of what the objects of a scene might be. Another approach uses "heuristics," or rules of thumb, where a calculated guess is made about the objects, the successes and failures noted, and then a guess is tried again, and so on (Raphael, 1976, Chap. 7). Successful though methods like these have been, automatic pattern recognition is primitive relative to a human being. To give some idea of the state of the art, the Defense Advanced Research Projects Agency of the U.S. Department of Defense has a truck-size research device

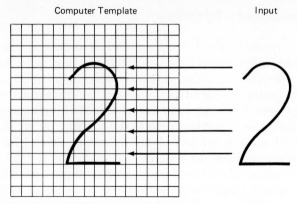

Figure 16-1 Representation of automatic pattern recognition by template matching. Recognition requires that the input pattern correspond to the pattern in the computer.

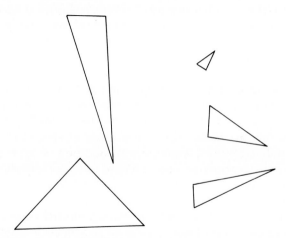

Figure 16-2 Automatic pattern recognition by feature matching could identify any of these triangles, regardless of size or orientation.

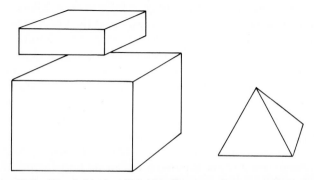

Figure 16-3 The field of artificial intelligence often uses block forms in arrangements like these for research on scene interpretation.

under development called the *autonomous land vehicle* (Stein, 1986). The unmanned autonomous land vehicle drives itself with information derived from a color television camera whose input is interpreted by an automatic pattern recognition machine. Interpreting the edges of the road, the vehicle successfully traveled a 2.2-kilometer road at 10 km/hr, turned around, and returned to the starting point. This driving task would be considered trivial for a human driver, but it is an achievement for artificial intelligence.

Even when object identification and scene interpretation become sophisticated, there remains the forbidding problem of context. The meaning of an object can depend upon its setting. Suppose that an automatic pattern perceiver mounted on an aircraft was programmed to fire a missile when an enemy tank was recognized. The decision to fire would be correct if the context of the tank was an empty football field, but what if the tank was parked beside a children's hospital? This would be no problem if the pilot was doing the perceptual evaluation and making the firing decision, but it would be a forbidding problem for present-day pattern recognition machines.

Auditory Pattern Recognition. The status of automatic auditory pattern recognition parallels that for visual patterns, and understandably, because the problem is conceptually the same. For all the auditory patterns of the world, it is recognition of speech patterns that draws research attention. Automatic recognition of speech would be the core of a *voice interactive system* where the speaker tells a machine what to do and it recognizes the speech and does it.

Commercially available speech recognition machines are almost all based on template matching and recognize a limited vocabulary of isolated words, not connected words as in natural speech. They are called *speaker-dependent systems* because a speaker who is to use the device must first speak a vocabulary for the machine and define the templates in the unique set of auditory patterns that is her own voice. The vocabulary is small, ordinarily 100 to 200 words. When the speaker then uses the machine and utters the words of the vocabulary, they are matched to the templates and are recognized. A user must speak the words distinctly, with pauses between them, because running words together creates compound verbal units that have no template counterpart. Accuracy of recognition can vary. Tests of isolated word recognition, using a 20-word vocabulary in a benign environment, found error rates from .2 to 12 percent for six speaker-dependent systems (Flanagan, 1984, p. 33). Recognition errors occur often enough to justify an error correction mechanism so that the human operator has feedback about recognition mistakes and a way to correct them (McCauley, 1984). Schurick et al. (1985) evaluated an automatic voice recognition system that had several experimental types of visual and auditory feedback with error correction. They found correct performance to average 97.3 percent with feedback and error correction and 70.3 percent without it.

A *speaker-independent system* can be addressed by anyone; it does not need a prior sample of the speaker's voice as a speaker-dependent system does. Speaker-independent systems are a more difficult engineering feat than speaker-dependent systems, so their development is not as far along.

One approach to a speaker-independent system is the use of feature matching. The basic unit of the spoken language is the phoneme (Chapter 11). Phonemes can be described by 13 two-valued features, and these features can be programmed into a voice recognition machine as the basis of feature matching. Any speaker of the language should be able to use the machine. A problem with this approach is that the utterance of phonemes can depend on the context of surrounding phonemes, so a phoneme's features may not always be constant. The same phoneme /t/ has different acoustic characteristics in *tea, eaten, steep, beater,* and *tweed* (Flanagan, 1984, p. 20). This effect is known as *coarticulation,* and it can occur within words or across word boundaries. Laboratory tests of speaker-independent systems with small vocabularies have had error rates of 5 to 10 percent (Flanagan, 1984, p. 33).

Beyond the recognition of isolated words is the recognition of continuous speech as in sentences. There are several approaches that have been taken (Lea, 1980). One is *word spotting,* which ignores some of the speech and detects only key words. Another is *highly constrained sequences of words,* where each word is recognized but the vocabulary is small and the sentence structure is formalized. Pilots and air traffic controllers speak to each other in a highly constrained way, for example. Recognition by word spotting and highly constrained sequences of words is helped if the words are spoken deliberately and articulated clearly. The most sophisticated approach to continuous speech is *speech understanding systems.* A speech understanding system, like a human being, will recognize individual words, but it goes beyond and recognizes the connected words in sentences spoken by sloppy speakers, which is most of us. The acoustic characteristics of everyday connected speech are strongly determined by the surrounding phonetic context and can have slight resemblance to the individual words taken in isolation. Words are run together, and syllables are dropped or have their identity lost at word boundaries, so an approach based on the recognition of individual words will not work. Klatt (1977) uses the sentence *Did you hit it to Tom?* in illustration. Some speakers might speak slowly and articulate each word distinctly so that recognition technology for isolated words might be used, but there are plenty of us who would say *Didja hit itto Tom?*. Some word boundaries have now disappeared, and new "words" have surfaced. The decoding problem is difficult because *didja* is not a word in the English language, and *itto* is not a word either, although it is two legitimate words that are run together. One approach to this problem is to appreciate that a language has grammatical constraints and to use the constraints to help the recognition process. Thus, in the English language, the verb follows the subject of the sentence, and a noun, adverb, or adjective will follow *the,* as in *the house, the likely event,* or *the strong man.* A sequence such as *the are* is not allowed. Knowledge about language constraints can be built into a speech understanding system and used to help the recognition process with ideas about what might be expected next. Voice recognition systems for isolated words operate only on the auditory input and so use only bottom-end processing, but linguistic knowledge adds top-end processing (Chapter 11).

How good are speech understanding systems? The best evidence comes from a competition held by the Advanced Research Projects Agency of the U.S. Department of Defense (Klatt, 1977). Four systems were evaluated. The main criteria for a system were that many speakers could use it, that it would have a vocabulary of 1000 words, and that it would understand continuous speech with error of less than 10 percent. The competition was won by a system called HARPY that was developed by Carnegie–Mellon University. Speaking 184 sentences with a vocabulary of 1011 words, three men and two women had an error rate of 5 percent. HARPY was not completely speaker independent because each speaker was allowed to speak 20 sentences so that some adjustment could be made to the particular characteristics of the voice. Central to HARPY's conception was a network of 15,000 states that was specified by the constraints of the English language for all possible sentences of the vocabulary. A sentence was defined by the most plausible paths through the network for phonetic segments of the utterance.

Conclusion. Impressive though developments in automatic pattern recognition are, and as useful as these machines are becoming, there is no question that their capabilities are far below corresponding human capabilities.

Perceiving, Understanding Natural Language, and the Execution of Intelligent Action

Suppose I said "Go downstairs to the kitchen. Pick up the apple and the orange on the table. Bring me the heavier one." The request would be so elementary that you would think nothing of it. A child would think nothing of it. Could a machine—call it a robot—be devised to do it? Not a machine specifically defined to perform this particular task but a machine that could do a range of ordinary tasks of the kind that humans do many times a day. Probably not. This is what Waltz (1982) called "common sense," and the fields of artificial intelligence and robotics are far from it. Consider what is involved.

The robot would have to understand speech. A speech-understanding system might be devised for these simple tasks provided that the commands were kept simple. Beyond speech, the robot would have to understand natural language. Understanding of natural language by a machine is enormously complex because it entails semantics, or meaning, and not merely looking up meaning in a dictionary. Meaning can involve large networks of information. Meaning turns importantly on context because many words and sentences have meaning only in context. The word *book* has one meaning in the sentence *I read the book* and another in *Book me in the hotel*. Weizenbaum (1976) gives the example of the sentence *The house blew it*. The sentence is grammatical but it is nonsensical taken by itself. Only within the context of an entire story might one know that *house* means *gambling house,* and that a gambling house failing to stop a gambler's scheme was the colloquial meaning of the sentence. No dictionary look-up of the words, or a narrow look at context by examining nearby sentences, would yield the sentence's meaning. The field of artificial intelligence has had minor successes in narrow, circumscribed domains of natural language understanding (Barr & Feigenbaum, 1981, Chap. 5), but there is nothing that can qualify as useful natural language understanding like that which a human being has.

The robot would require high powers of pattern recognition. There would be a requirement for the interpretation of three-dimensional scenes such as rooms with objects in them. Identification of stairs, table, apple, and orange would be required. The concept of objects being on a surface, such as the apple and the orange on the table, must be perceptually understood. From the discussion of visual pattern recognition above, it is unlikely that much of this is possible. Moreover, touch perception would be required for the handling of objects.

Locomotion would be required. Wheels are an easy form of locomotion when a robot has to move along a smooth surface, but what of stairs? Legs would probably be required. With all the locomotion of machines over the ground that we have in this world, you would think that we might sometimes see legs as well as wheels on vehicles, particularly for places where wheels cannot go (McGhee, 1976). Not so. The reason is that the coordinated movements of legs is a very complex engineering problem. Investigators at The Ohio State University

(McGhee, 1976; McGhee & Waldron, 1985; Messuri & Klein, 1985) have had an extensive research project on a six-legged walking machine called the *adaptive suspension vehicle*. More a proof-of-concept than development of a prototype, the adaptive suspension vehicle is an electromechanical system that requires a complex of computers and a human operator for its control. A maximum speed of 8 miles per hour is planned. Operator input should be minimal over relatively smooth terrain, but rough terrain requires intense operator involvement as the terrain is evaluated for foot placement and as the legs and body of the vehicle are controlled. The adaptive suspension vehicle is a research project that is a distance from practical applications. The inescapable impression from the project is that walking machines are very complex.

Conclusion. Mankind has always been fascinated with machines that imitate living things, from toys with animal and humanlike properties, to Mary Shelley's Frankenstein, to robots found in our modern factories (Heppenheimer, 1985). For all the talk about robots, there are only about 20,000 of them in the United States (Holusha, 1986). None can perform the intelligent, coordinated activities that humans perform many times every day. It will be a long time before we have a robot that can walk downstairs to the kitchen, pick up the apple and the orange from the table, and return with the heavier one.

Learning

Some machines are totally rigid and can perform only one function, but not all machines. A computer program can execute choices and branch off in one direction or another depending on the outcome of a comparison. But whether the machine is rigid or flexible, all the choices and outcomes must be anticipated by the designer. Machines have no useful power to change as a result of experience, although research is being done on the topic.

A remarkable characteristic of living organisms, even some insects but most notably humans, is the power of learning. Certain kinds of experiences change the characteristics of our responding. Biologically, learning can be viewed as adaptation. All organisms have instinctive patterns built in for handling some contingencies of the world, such as the pupil of the eye which automatically contracts and expands to accommodate changing levels of illumination, but the world is an uncertain place and instincts cannot handle all the events and relationships that occur. Learning is adjustment to new events and relationships that are not anticipated by the organism's instinctive structure.

The field of artificial intelligence is conducting research on computer learning. Examples of their laboratory efforts so far are rote learning, such as learning a list of words; inductive learning, where a rule is inferred from different instances of it; and concept learning, where the computer learns to affirm all instances of a category (e.g., all triangles) and reject all else. A computer has been programmed to learn the playing of checkers (Cohen & Feigenbaum, 1982, Chap. 14). As noteworthy as these accomplishments are, they are trivial alongside human capabilities in learning. Representative ways of human learning are Pavlovian conditioning; reward (reinforcement) learning, where the consequence of a response determines the behavioral change; observational learning or imitation; learning to recognize a pattern; and conceptual learning. Computer learning will remain far below human learning capabilities for a long time.

Memory

Digital computers have memories that are analogous in functioning to human memory, but a comparison of them is qualitative. The capacity of a digital computer's memory can be specified exactly, but not human memory. The capacity of human memory is unknown but is very large. Beginning at birth, our human memory system stores enormous amounts of information. We add new information daily through learning. Pattern recognition is known in the field of artificial intelligence for the large amounts of computer memory that it consumes, but we interpret a large variety of complex patterns in each of our sense modalities every day, so we must have correspondingly large amounts of memory for it. Moreover, human memory is multidimensional, and we store information in each of the dimensions. We have memory for movements. We remember verbal material, concepts, different kinds of emotions, sounds, touch, and odors. The capacity of human memory is vast. Computers can also store vast amounts of information, such as census data or tax records; nevertheless, they seem second to human memory and the large amounts of information that it can store, the different kinds of information that it can retain, and the speeds with which the information can be accessed.

Packaging

Weighing in the range of 50 to 100 kilograms, the human being is a remarkably lightweight subsystem considering all the complex functions that it can perform. Machines that could do all the things that humans do, if they existed, would fill warehouses.

Ways in Which Human Beings Are Inferior to Machines

Physical Strength

Humans are physically weak. An ordinary forklift vehicle can lift and carry more than can a human being.

Reaction Time and Movement Time

Human reaction time is slow (Chapter 5). Under ideal conditions, where there is only one stimulus and the human subject is ready to respond, the reaction time will be about 120 milliseconds. One hundred twenty milliseconds is an eternity in the world of machines.

Reaction time is the speed of getting a response started, and movement time is the speed of the response in action. Machines can move from A to B faster than a human being, whether the A–B distance is 10 centimeters or 10 kilometers.

Limitations of the Senses

The human senses have remarkable capabilities in their sensitivity and in the perception of patterns, color, and depth, but they have limitations also. The eye can see only to the horizon, but radar can "see" hundreds of miles. Sonar can "see" underwater, whereas the eye sees almost nothing. Machines can detect colorless and odorless gases and radiation, which our senses have no power to detect whatsoever.

Fragileness

Human tolerance limits for impact, heat, air, radiation, and acceleration are low. Protecting the fragile human being can be complex and expensive, as in the life support system for a manned space vehicle. It is easy to build a machine that is tougher than a human being.

Mental Operations

Insofar as computers can execute mental operations, they do it swiftly. Most certainly they are remarkable in the speed of arithmetic calculations. A computer takes more time for more complex computations, but the solution time is undeniably fast.

Instability

Humans cannot sustain constant performance. Fatigue limits human endurance for physical work. Powers of sustained attention are limited. Forgetting occurs. Motivation wanes. Emotions intrude and disorganize behavior. There can be no doubt that machines commonly maintain performance longer and better than do their human counterparts. A pilot can fly an airplane all day, but the autopilot is a better choice for the job.

Individual Differences

Quality control by humans is low. Differences among the abilities that humans have are large. A quality control engineer would not tolerate differences among machines that are as large as we have among humans. As Chapter 17 will show, psychological tests are used to select individuals from among a population of personnel so that those selected are more alike than choices from an unselected population would have been, but the variability that remains is substantial, nevertheless. Identical twins provide the best way to minimize human variability, but they are in short supply.

Work Load

The amount of work that machines can do per unit of time can be controlled by design and can be large. The amount of work that a human can do per unit of time is an active research area in psychology, and it is clear that human performance degrades when events from too many sources occur too fast (Chapter 8). Humans overload easily.

Modern Uses That Are Made of our Knowledge About the Capabilities of Humans and Machines _____

Some uses of the comparative capabilities of humans and machines are straightforward and well established. The decision is clear when a system requirement is to see 100 miles or at night—use radar. A system is under development to fly a fighter aircraft automatically during high-*g* combat maneuvers when the pilot is unable to do so (Klass, 1986). Assigning a human being to the task when natural language understanding or learning is required is a clear decision, also. Less obvious, and more challenging, are cases where human capabilities must be combined with the capabilities of machines to optimize system functioning. The most interesting of these cases involves human work load, where the human operator needs automatic help to avoid overload and the degraded performance that results from it. The most active area of development is the cockpit of advanced aircraft. Commercial pilots can be very busy under certain conditions of flight, such as landing, flying in bad weather, or maybe worst of all—landing in bad weather. Military pilots have all the problems of commercial pilots and the stressful burdens of combat as well. The solution is seen in the products of artificial intelligence.

Voice Recognition

The potential advantages of voice recognition systems are several (Lea, 1980). Speech is our most natural way of communicating. No training is required. Speech is a high-capacity output. Speech allows multimodal responding because you can do things with your hands and feet while speaking. Darkness makes no difference, nor do unusual environmental conditions like weightlessness.

The Rockwell/Collins Corporation has a highly automated facility for the manufacturing of military electronic components such as solid-state circuit boards (Klass, 1987). Quality control is semiautomatic, and a speaker-dependent voice recognition is a part of it. Each component is labeled with a bar code, like groceries in a supermarket. An inspector passes a wand over the code and the component's identification is read and stored in a computer. The inspector then uses a microscope to inspect the component's miniaturized elements. When a fault is found, the inspector aligns the cross-hairs of the microscope with the fault and describes it into a microphone (e.g., "bad solder joint"). The voice recognition system recognizes "bad solder joint" and stores it in the computer along with the positions of the cross hairs that identify the fault's location. Later, when the component is repaired, the rework operator waves a wand over the component's bar code to identify the fault, its location appears on her computer display, and the component is repaired with neither the inspector filling out paperwork nor the rework operator processing it.

A major application of voice interactive systems will be in the aircraft cockpit. The French Air Force is planning it for their advanced fighter aircraft and helicopter aircraft (Lenorovitz, 1984). The systems will recognize either isolated words or connected speech, and the pilot will use it to command such aircraft functions as selection of weapons, choice of radio frequencies, and operation of the autopilot and radar.

Using the available vocal response mode unburdens the hands and frees them for their primary duties of controlling the aircraft. Also planned is an unburdening of the pilot's busy eye movements. Instead of looking at aircraft instruments for altitude and fuel supply, he can ask for them, and a synthetic voice will respond with the answer. Boeing Commercial Aircraft Company is exploring similar ideas for transport aircraft of the 1990s (Merrifield, 1986).

One difficulty with voice interactive systems is that they will have to operate in practical environments—unlike the quiet laboratories where they were developed—and how well they will perform is hardly known (Flanagan, 1984). How good will voice recognition be in situations that are noisy, where there is tension in the voice, or where the operator is apprehensive and speaks with a higher pitch? Factors such as these can be expected to degrade system performance, so application may be slowed until more understanding is gained.

Expert Systems

Following closely on applications of voice recognition will be applications of expert systems in the cockpit of advanced aircraft (Stein, 1986). The Advanced Research Projects Agency of the U.S. Department of Defense has a "Pilot's Associate Program" aimed at the development of artificial intelligence technologies for the cockpit of the 1990s (Kandebo, 1986; Stein, 1986). The "pilot's associate" turns out to be expert systems of one kind or another, not only to ease the pilot's work load but to give him capabilities that are more expert than his own (one of the potentials of expert systems). These systems, if they become reality, will make the pilot more of an executive in the cockpit, concentrating on higher-level decisions. Five on-board expert systems are planned:

System status monitoring: diagnosis of actual and anticipated malfunctions; recommendations for corrective procedures

Situation assessment: decision making about the general combat situation

Tactical planning: decision making based on the immediate tactical situation, the pilot's intentions, and mission goals

Mission planning: advice on a mission plan that balances mission goals, threats, risks, target values, and route options

Pilot–vehicle interface: monitoring of performance; managing the content, format, and mode of information that is presented to the pilot

Rules of Thumb for Automation _____

Here are four rules of automation (that would be qualified by the cost factors of any situation):

Rule 1. Use machines for functions that the human cannot perform.
Rule 2. Use machines for functions that the human cannot perform well during periods of excessive workload.
Rule 3. Use machines to assist in work that has so many sequential operations that human productivity is low.
Rule 4. Avoid unnecessary automation. Do not use machines for functions that the human can perform well.

Rule 1 says get radar if you want to see 200 miles. Rule 2 says that if a fighter pilot has to fly at an altitude of 300 feet at 600 miles per hour, navigate, find targets, and arm and discharge weapons, he deserves help from one or more automatic subsystems because he may not be able to do the job without them. Rule 3 says that if a product requires 100 operations in its manufacture, the worker would benefit from automation that would allow ten a day to be made instead of the one that could be made without it. Rule 4 says that the high technology of automation can be a fad. Machines are a distance from the accomplished thinking and pattern recognition which a human being does routinely and swiftly, and it is faddish (and expensive) to substitute lesser machines for such functions, even when progress in the development of these machines and interest in them is considerable.

Summary _____

Automation is the substitution of a machine function for a human function, but this definition fails to capture the trade-offs that can confront system designers when they ask whether a human being or a machine can perform a function the best. The capabilities of humans and modern machines must be appreciated as background for such design decisions. The decisions were simple in the past because machines had little capability for performing human functions, but the discipline of artificial intelligence has changed that. The allied discipline of robotics has the potential for changing it more. Prominent human capabilities were compared with the capabilities of modern machines that challenge them.

Ways in Which Humans Are Superior to Machines

Human superiority in thinking and problem solving is being challenged by the development of expert systems by the field of artificial intelligence. Expert systems meet the criteria of machines that can think and solve problems, and useful expert systems are being developed, but the human being remains superior. The same is true of pattern recognition. Artificial intelligence is making good progress in machines for visual and auditory pattern recognition, but the human being is superior. Learning, memory, and speech understanding are also capabilities that are superior in humans. Most impressive of all, humans have all these complex functions organized in a compact package and can use them in the execution of intelligent action.

Ways in Which Humans Are Inferior to Machines

For all their splendid capabilities, humans are inferior to machines in a number of ways. Humans are physically weak, slow in response time, and fragile. The human senses have their limitations. Mental operations can be slow. Changing states such as motivation and fatigue make humans unstable. Quality control is low because human individual differences are large. Human tolerance of work load is low.

Rules of Thumb for Automation

Four rules of automation were offered:

1. Use machines for functions that humans cannot perform.
2. Use machines for functions that humans cannot perform well during periods of excessive work load.
3. Use machines to assist in work that has so many sequential operations that human productivity is low.
4. Avoid unnecessary automation. Do not use machines for functions that humans can perform well.

References

Barr, A., & Feigenbaum, E. A. (Eds.). (1981). *The handbook of artificial intelligence.* Volume 1. Los Altos, CA: William Kaufmann.

Barr, A., & Feigenbaum, E. A. (Eds.). (1982). *The handbook of artificial intelligence.* Volume 2. Los Altos, CA: William Kaufmann.

Basden, A. (1983). On the application of expert systems. *International Journal of Man–Machine Studies, 19,* 461–477.

Berry, D. C., & Broadbent, D. E. (1984). On the relationship between task performance and associated verbalizable knowledge. *Quarterly Journal of Experimental Psychology, 36A,* 209–231.

Brachman, R. J., Amarel, S., Engleman, C., Engelmore, R. S., Feigenbaum, E. A., & Wilkins, D. E. (1983). What are expert systems? In F. Hayes-Roth, D. A. Waterman, & D. B. Lenat (Eds.), *Building expert systems.* Reading, MA: Addison-Wesley. Pp. 31–57.

Cohen, P. R., & Feigenbaum, E. A. (Eds.). (1982). *The handbook of artificial intelligence.* Volume 3. Los Altos, CA: William Kaufmann.

Davis, R. (1984). Amplifying expertise with expert systems. In P. H. Winston & K. A. Prendergast (Eds.), *The AI business: The commercial use of artificial intelligence.* Cambridge, MA: MIT Press.

Edwards, M. (1984). Robots in industry: An overview. *Applied Ergonomics, 15,* 45–53.

Ericsson, K. A., & Simon, H. A. (1980). Verbal reports as data. *Psychological Review, 87,* 215–251.

Flanagan, J. L. (Ed.), (1984). *Automatic speech recognition in severe environments.* Washington, D.C.: National Academy Press.

Heppenheimer, T. A. (1985). Man makes man. In M. Minsky (Ed.), *Robotics,* Garden City, NY: Omni Publications. Pp. 29–69.

Holusha, J. (1986). Robot sales increased to $442.7 million in 1985. *The New York Times,* April 22.

Kandebo, S. W. (1986). Lockheed stresses crew participation in pilot's associate development. *Aviation Week & Space Technology,* July 7.

Kellogg, R. T. (1982). When can we introspect accurately about mental processes? *Memory & Cognition, 10,* 141–144.

Klass, P. J. (1986). Maneuvering attack system prototype undergoing testing in F-16. *Aviation Week & Space Technology,* June 12.

Klass, P. J. (1987). Automated military avionics factory reducing flaws in workmanship. *Aviation Week & Space Technology,* October 26.

Klatt, D. H. (1977). Review of the ARPA Speech Understanding Project. *Journal of the Acoustical Society of America, 62,* 1345–1366.

Lea, W. A. (1980). The value of speech understanding systems. In W. A. Lea (Ed.), *Trends in speech recognition.* Englewood Cliffs, NJ: Prentice-Hall. Pp. 3–18.

Lenorovitz, J. M. (1984). Voice command unit enters new tests. *Aviation Week & Space Technology,* January 30.

McCauley, M. E. (1984). Human factors in voice technology. In F. A. Muckler (Ed.), *Human factors review: 1984.* Santa Monica, CA: Human Factors Society. Pp. 131–166.

McGhee, R. B. (1976). Robot locomotion. In R. M. Herman, S. Grillner, P. S. G. Stein, & D. G. Stuart (Eds.), *Neural control of locomotion.* Advances in Behavioral Biology, Volume 18. New York: Plenum Press. Pp. 237–264.

McGhee, R. B., & Waldron, K. J. (1985). The Adaptive Suspension Vehicle Project: A case study in the development of an advanced concept for land locomotion. *Unmanned Systems,* Summer.

Merrifield, J. T. (1986). Boeing explores voice recognition for future transport flight deck. *Aviation Week & Space Technology,* April 21.

Messuri, D. A., & Klein, C. A. (1985). Automatic body regulation for maintaining stability of a legged vehicle during rough-terrain locomotion. *IEEE Journal of Robotics and Automation, RA-1,* 1–10.

Miller, F. D. (1984). Introducing an expert system into the workplace. In G. Kohl & S. J. Nassau (Eds.), *Combining human and artificial intelligence: A new frontier in human factors.* Proceedings of a symposium sponsored by the Metropolitan Chapter of the Human Factors Society, New York, November 15, 1984. Pp. 41–51.

Nisbett, R. E., & Wilson, T. D. (1977). Telling more than we can know: Verbal reports on mental processes. *Psychological Review, 84,* 231–259.

Parsons, H. M. (1985). Automation and the individual: Comprehensive and comparative views. *Human Factors, 27,* 99–112.

Price, H. E. (1985). The allocation of functions in systems. *Human Factors, 27,* 33–45.

Raphael, B. (1976). *The thinking computer: Mind inside matter.* San Francisco: W. H. Freeman.

Sanderson, P. M., & Vicente, K. J. (Unpublished manuscript). Verbalisable knowledge and task performance: How can we elicit the expert's knowledge?

Schurick, J. M., Williges, B. H., & Maynard, J. F. (1985). User feedback requirements with automatic speech recognition. *Ergonomics, 28,* 1543–1555.

Stein, K. J. (1986). Technical survey: Artificial intelligence. *Aviation Week & Space Technology,* February 17.

Turing, A. M. (1950). I. Computing machinery and intelligence. *Mind, 59,* 433–460.

Van Horn, M. (1986). *Understanding expert systems.* New York: Bantam Books.

Waltz, D. L. (1982). Artificial intelligence. *Scientific American.* October.

Weizenbaum, J. (1976). *Computer power and human reason.* San Francisco: W. H. Freeman.

Wickens, C. D. (1984). *Engineering psychology and human performance.* Columbus, OH: Charles E. Merrill.

Winston, P. H. (1984). *Artificial intelligence.* Second Edition. Reading, MA: Addison-Wesley.

PART VII

Personnel Selection and Training

CHAPTER 17

Personnel Selection

From roots in industrial engineering that reach back to the nineteenth century, workplace design has been a central element in human factors engineering. The emphasis is justified because, as we have seen in Parts I through VI, the ease with which the operator receives information and responds to it is basic to the efficiency of a human–machine system. There is more to efficiency, however. The efficiency of a system can be helped with good personnel selection. When a system like an automobile is built for the mass market, the main route to system efficiency is design because the users cannot be selected. There are, however, many industrial and military systems where personnel are selected from a pool of applicants, and how well these systems operate is importantly dependent on the people who are chosen to run them. This chapter is about that selection process.

Personnel selection in some form has always taken place; it is nothing new. Interviews have been a common way to gather information about a job candidate. Many find it self-evident that a person such as a supervisor, who is skilled in the job, should be able to gather the appropriate information and make a good decision about a candidate. Research in personnel psychology has shown this assumption to be an oversimplification (Landy, 1985, Chap. 5). The interview is a complex social process that involves the attitudes, motivations, perceptions, expectations, and stereotypes of both interviewer and interviewee. The interview can be a proper source of data to be combined with other data, like that from application forms and psychological tests, but the interview by itself is an unreliable way of making personnel decisions. In the past 100 years the science of psychology has moved personnel selection toward reliance on objective psychological tests and away from subjective approaches such as the interview.

Today we are all familiar with psychological tests at the level of the test taker, but not everyone appreciates the technical side of tests. Ill-considered praise or blame of tests is less likely when there is familiarity with the scientific reasoning that lies behind testing. Some history of testing will precede an account of the essentials of testing technology.

Historical Background

Psychological testing had its origins in civil service examinations, the assessment of academic achievement, and the scientific study of individual differences in human behavior (DuBois, 1970).

Origins of Psychological Tests

Civil Service Examinations. Competitive examinations for government positions have roots as far back as 2200 B.C. in China. At that time the Chinese emperor tested his officials every third year to determine their fitness for office. In 1115 B.C. the tests were music, archery, horsemanship, writing, arithmetic, and the rites and ceremonies of public and private life. By 1370 A.D. the testing of applicants for government positions in China emphasized knowledge of Confucian classics and the ability to write poems and essays. About 3 percent of the applicants were said to have passed the test (DuBois, 1970, pp. 3–5).

British diplomats, visiting China in the nineteenth century, returned with an appreciation of testing for government service and recommended that it be tried in Great Britain. Open competitive examinations were first given to applicants for the Indian Civil Service in 1833. The British experience with testing was noticed in the United States and was seen as a way of avoiding cronyism in appointment to government positions. Testing was authorized by the U.S. Congress when it established the Civil Service Commission in 1883. The Congress urged that tests fairly measure the capacity and fitness of the individuals for the jobs. Tests for a number of government jobs were soon in use.

University Examinations. The academies of ancient Greece and Rome did not appear to have formal examining procedures, but by the Middle Ages, when university degrees came into existence, oral examinations were administered to determine eligibility for a degree.

The Jesuit Order of the Catholic Church pioneered written examinations. In 1599, the Jesuits published rules for the conduct of written examinations in lower schools. European and U.S. universities began to use written examinations in the nineteenth century.

The Scientific Study of Individual Differences in Behavior. DuBois (1970), in his scholarly treatise on the history of mental testing, regards civil service examinations and university examinations as important for the acceptance of formal examining methods, not for the scientific development of psychological tests. Scientific testing as we know it today had its roots in the study of individual differences, and the pioneer was Sir Francis Galton (1822–1911) of Great Britain. One of Galton's interests was the measurement of individual differences in human capabilities. He developed tests to measure such capabilities as the discrimination of weights, strength, and the upper limit of hearing.

The U.S. psychologist James McKeen Cattell (1869–1944) did his graduate work in Europe, and, for a time, was an assistant of Galton's. Cattell came to be an influential experimental psychologist, and the interest in individual differences that he brought with him on return to the United States left its historical mark. Most of the tests that Cattell developed were of sensory and motor capabilities. A number of Cattell's students went on to be prominent contributors to the developing science of individual differences.

The tests that Galton and Cattell developed were in the spirit of experimental psychology and its interest in sensory and motor functions, but Alfred Binet (1857–1911) of France went in a different direction and developed a test of higher mental processes, called the intelligence test. The motivation for the test was "abnormal children," as Binet called them, who could not profit from academic instruction. Items of graded difficulty, such as memory for sentences, remembering sequences of digits, and verbal comprehension, made up the test, and the score was the child's "mental age." Scores on the test were found related to achievement in school, and they became useful for identifying the academic potential of students. Under the inspiration of Binet's work, Lewis M. Terman (1877–1956) of Stanford University in the United States developed a revision of Binet's test, which came to be called the "Stanford–Binet" intelligence test. The Stanford–Binet test was published in 1916 and became a standard instrument for decades, securing intelligence testing in schools and psychological clinics.

Military Uses of Psychological Tests

The American Psychological Association is the main professional organization of psychologists in the United States, and Robert M. Yerkes

(1876–1956) was its president when World War I started. Yerkes formed committees to ask how psychology could contribute to the war effort, and the consensus was that psychological tests could perform a valuable national service. Taking items from various tests in use, the Army Alpha Test was devised as a test of intelligence that could be administered to groups. The test correlated well with scores on the Stanford–Binet test, ratings by superior officers, and amount of schooling. Administered to over 1 million men, the military made major use of the Army Alpha Test for personnel decisions.

The success of the Army Alpha Test did much to secure the use of psychological tests in practical situations, and it was undoubtedly a factor in the subsequent proliferation of tests. In the postwar years, tests were developed to measure interests, dimensions of personality, musical aptitude, aptitude for business occupations, and manual skills, to name a few.

All of the branches of the U.S. armed forces had psychological testing programs in World War II. The principal development was the Army General Classification Test, which was a group test that was an updated version of the Army Alpha Test. The Army General Classification Test was administered to over 9 million men. Tests of motor skills were developed for the classification of aircrews.

Conclusion

Today we have many good psychological tests that are successful in making decisions about the assignment of people to positions. Tests are widely used, as we all know. A selling point of scientifically constructed tests is their objectivity, in contrast to the notorious unreliability of subjective approaches to personnel decisions, but objectivity is a small part of the story. The primary importance of psychological tests is that they can contribute to system efficiency by selecting the most able among candidates and rejecting the others.

Definition and Rationale of Psychological Tests _____

Psychological Tests Defined

A *psychological test* is a sample of a person's behavior taken under standardized conditions. A *printed test* with booklet format is a common kind of psychological test. A test score is almost always a composite of performance opportunities, and a printed test will have a number of items. There is no necessity for it, but printed items are usually

Figure 17-1 Complex Coordination Test that was in the test battery used for aircrew selection in World War II. [*Source:* Melton (1947), Fig. 4.1.]

scored pass–fail, with one point for pass and zero for fail. The number of items passed is the score on the test. A common kind of printed test is a multiple-choice test, where the test taker chooses from several alternative answers to an item. Printed tests will measure mental skills. Examples are a verbal test that will measure the comprehension of verbal materials, a perceptual test to evaluate the interpretation of patterns and forms, or a mathematics test to measure mathematical capabilities. Some printed tests are *speed tests,* where the score is number of items completed successfully in a limited time period. Others are *power tests,* where plenty of time is given to complete the test. A product of our time is *adaptive testing,* where the test is flexibly adapted to the individual by a computer. A procedure that might be used in adaptive testing is for a person to start with an item of intermediate difficulty. If the response is correct, the next item will be more difficult, but if the response is wrong, the next item will be easier. This procedure is repeated through a short list of items. The score is based on the characteristics of the items that are passed, such as their level of difficulty. Computerized tests appear to do as well as conventional printed tests and with a smaller number of items and less testing time (Anastasi, 1982, pp. 301–305).

A *motor test* is for the measurement of movement skills. A test taker might be given a series of simple manual manipulations to perform, such as picking up objects in one box and moving them to another as fast as possible. Threading nuts on bolts as fast as possible could be another kind of simple motor test. The number of items accomplished in the allotted time is the score. Some motor tests are more elaborate and require complex coordinations. Figure 17-1 is an example of a complex motor task that was used for aircrew selection in World War II (Melton, 1947, Chap. 4). It is called the *Complex Coordination Test.* Three lights which define three movements of the control—two dimensions of the stick controlled by the hand and one of the rudder bar controlled by the feet—appear on the display panel. When the movements are made successfully, three different lights appear, which define a new pattern of movements to be made, and so on. The speed with which this is done on a trial of, say, 3 minutes is the trial score. Several trials will be given, and an overall score will be the test score.

The Rationale of Psychological Tests

A job requires a particular configuration of abilities, and some people are more favorably weighted on the abilities than others and so perform better. Psychological tests are designed to measure abilities, and one or more of them are assembled to measure the abilities required of the job. The applicants with highest scores on the tests are most favorably weighted on the abilities and are selected for the job.

An applicant often wonders about the relationship between items of the test and the job. What does interpreting a paragraph about the history of England, or moving objects from one box to another as fast as possible, have to do with the job? The relationship is indirect, through abilities. An *ability* is a behavioral repertoire that is a determiner of performance in the various tasks that require the ability. A person has a number of abilities, and tasks require them in different combinations. Examples of abilities are verbal, mathematical, perceptual, and manual dexterity. The similarities between test items and the job may be superficial or absent, yet they can require the same abilities. Moving objects from one box to another as fast as possible will measure speed of motor movement. The job can have speed of movement as one of its ability dimensions, although the worker would never be required to move objects from one box to another as fast as possible. Interpreting a paragraph on the history of England could measure the verbal ability that would be required to interpret a technical manual on how to operate a machine. Personnel psychologists have techniques for identifying and measuring different abilities.

Reliability

A test is a measuring instrument and, like any measuring instrument, must be reliable. *Reliability* means that essentially the same measures will be obtained from one measurement occasion to the next. A bathroom scale that weighed you as 200 pounds one day and 100 pounds the next would be thrown out. An elastic yardstick that measures 5 inches on one occasion and 4 inches the next would be thrown out. The same is true of a psychological test. The order of individuals ranked by the test should be the same from one testing occasion to another. It is a poor measuring instrument where the person with highest ability gets the top score on one test occasion and an average score on the next. All of this says that the correlation between test scores should be high from one test occasion to the next. A correlation coefficient (defined in Chapter 1), when it is used to specify measurement stability, is called the *reliability coefficient*.

Errors of measurement are a cause of imperfect reliability. The sources of measurement errors can be many. You feel well the first time a test is administered and have a bad cold the second time. There may be an environmental factor such as the testing room being air-conditioned on the first test occasion and not on the second. Variations in the level of distracting noise can be a factor. Changes in motivation can lower reliability. Abilities are assumed to be stable characteristics of behavior, and the test aims to measure them, so these unwanted error influences affect test performance and distort the measurement of abilities. The test is unusable when errors of measurement become too great and the reliability coefficient becomes too small. Doubts about the usefulness of a test begin to arise when its reliability coefficient drops below .70.

Methods of Measuring Test Reliability

One type of reliability measurement is *test–retest reliability*. The same test is given again, after a time period such that the test responses will not easily be remembered, and the two sets of scores are correlated. Test–retest reliability is like using the same yardstick for measurement on two different occasions. Test–retest reliability has problems, however. Practice effects can affect the test scores of some individuals. Remembering some of the test items and the responses that were made to them can spuriously raise the reliability of a test by making test scores more alike from one test occasion to the next than they ordinarily would be. The memory difficulty can be lessened by lengthening the interval between test occasions, but it is not clear what this interval should be.

If practice and memory for items and their answers is cause for worry, two *parallel forms* of the test should be used to determine reliability. Test developers have techniques for developing equivalent forms of a test, where the two forms have corresponding but different items. The assumption is that the items of each test are a fair sample from a population of items with a common content. The use of parallel forms for measuring reliability is equivalent to using one yardstick for measuring a length on one occasion and a different yardstick for measuring the same length on another occasion.

A third approach is *split-half reliability*. The test is divided into halves, such as odd items and even items. Each half is scored, and the two scores are correlated. To continue the yardstick analogy, split-half reliability is like taking a measurement with one end of a yardstick and then repeating it with the other end. With each of the tests being half the size, the reliability coefficient will be lower than on the full test because a few samples of behavior are less reliable than many samples. The reliability coefficient is corrected for length with the Spearman–Brown formula:

$$r_{nn} = \frac{nr_{11}}{1 + (n - 1)r_{11}}$$

where r_{11} is the reliability that was calculated from the halves and r_{nn} is the estimated reliability of a test n times as long. The value of n would be 2 in the conventional case of estimating the reliability of a full-length test from its halves, in which case the formula simplifies to

$$r_{nn} = \frac{2r_{11}}{1 + r_{11}}$$

Validity

What a test measures and how well it measures constitute the validity of a test. Criterion validity, content validity, and construct validity are the three principal types of validity.

Criterion Validity

A measure of performance on the job is called the criterion, and *criterion validity* is defined as the correlation between a test and a criterion. The correlation is called the *validity coefficient*. Does the test predict performance on the job or success in training?

Criterion validity involves two sets of scores. One set is from a psychological test. The other set is the criterion, and typically it is a measure of performance on a job or success in a training program. Measures of job performance might be monthly sales in dollar volume, the number of parts machined per week, or ratings of proficiency by a supervisor. Time to complete a training program, or grades in a training program, can be criteria. Remember that a test does not have a single validity. A test can predict many criteria. Verbal skills are important in many jobs, and a test of verbal ability can be valid for all of them.

Criterion measures must be reliable, just like the scores of a psychological test, so the reliability of the criterion must be assessed just like the reliability of a test. A supervisor's checklist of performances on dimensions of a job as the criterion might have its reliability evaluated by the test–retest method. Research on the reliability of a criterion measure such as number of parts machined might be a determination that number of parts machined in a week is more reliable than number of parts machined in a day. The validity coefficient will be reduced if either the psychological test or the criterion measure, or both, have low reliability.

Content Validity

Tests used for the selection of system personnel usually require a criterion, but not all tests do. An achievement test in arithmetic that is given to all primary school children in a city is not correlated with a criterion. A job knowledge test to appraise how much a worker knows about her job is not correlated with a criterion. *Content validity* is when a test has a fair sample of the content domain. Test developers work with source materials for the topic and with subject matter experts to ensure the accuracy and coverage of the items.

Construct Validity

Construct validity is the least precise of the three types of validity, but is not without importance. *Construct validity* is the extent to which a test measures a theoretical construct or trait. An intelligence test has construct validity. Binet began the intelligence test by validating it against success in school, but the concept of intelligence is considered a general mental capability that goes beyond schools and is evident in a wide range of situations. Anxiety is a dimension of human personality, and a test of anxiety that measured it could have construct validity. The construct validation of tests such as these could be sloppy, where intelligence and anxiety is anything that someone says that it is. Examples of sloppiness can be found in any science, but fortunately there are substantive ways of determining construct validation.

One route to construct validation is via the variety of experiences that converges on the concept defined by the test. An intelligence test consistently predicts performance in situations that analysts judge as intelligent. Success in an academic setting and in scientific research both require intelligence. Those people who have academic and scientific success should both do well on an intelligence test. Similarly, a test of anxiety should consistently predict the anxiety of people in a variety of settings.

Another route to construct validation is that a test should correlate well with some tests and not others if its construct validity is to be demonstrated. A test of mechanical comprehension with construct validity for mechanical capabilities should correlate well with other tests of mechanical ability but not with a test of reading comprehension.

There are experiment approaches to construct validation. A person with disabling anxiety should score higher on a test of anxiety before psychotherapy than after. Or, if psychological theory says that anxiety is a motivational force that energizes behavior, those who score high on a test of anxiety should perform better on a laboratory task than those who score low. Conversely, if theory says that anxiety is a disorganizing agent for behavior, the high scorers on an anxiety test should perform poorly on the laboratory task.

Face Validity

Face validity is not validity in the technical senses that have been described above. Rather, it is the appearance of the test, making the test appear to measure what it actually measures. Face validity can be useful in eliciting the test taker's cooperation. A test of reading comprehension with items on the history of England might encounter resistance from machinists. It would be better for the test material to be about the operation of machines, close to the interests of the machinists. Face validity will not make a test more valid, but the importance of eliciting the interest and cooperation of the test takers cannot be discounted.

Example Data on Reliability and Validity ⎯⎯⎯⎯⎯⎯⎯

During World War II and after, the U.S. Air Force had a large-scale testing program for the selection of aircrew personnel. Both printed and motor tests were used. Air navigation is importantly mathematical, and a valid printed test was a general mathematics test (Guilford, 1947, Chap. 6). An example item was

$$(3x - 1)(2x + 2) =$$
a. $6x^2 - x + 2$
b. $6x^2 + 4x - 2$
c. $6x^2 - 4x - \frac{1}{2}$
d. $3x^2 - x + 3$
e. $x^2 + 2x - 6$

This test had a reliability of .92 and validity in the .40 range for predicting success in the classes of navigator training school.

The Complex Coordination Test shown in Figure 17-1 was a motor test that the Air Force used for the selection of pilots. The reliability was .87, and validity was in the range .35 to .40 for predicting the success of trainees in the classes of flight training school (Melton, 1947, Chap. 4).

A major testing program to select high school graduates for college is the American College Testing Program (ACT), which is second only to the Scholastic Aptitude Test (SAT). The ACT is administered to over

1 million applicants a year. In addition to biographical and interest information, the ACT has four academic tests:

1. The *English Usage Test* for understanding the conventions of written English. Its reliability is .92.
2. The *Mathematics Usage Test* for the measurement of mathematical reasoning ability. The reliability is .91.
3. The *Social Studies Reading Test* for the measurement of reasoning, analysis, and problem solving in the social sciences. The reliability is .88.
4. The *Natural Sciences Reading Test* for the measurement of interpretation, analysis, and reasoning in the natural sciences. The reliability is .85.

A composite score based on these four tests has a validity of . 40 to .50 for the prediction of college freshman grades (Mitchell, 1985, pp. 28–36).

Personnel Decisions _____

The validity coefficient is an important index of test effectiveness, but how does it relate to the practical problem of personnel selection? Ordinarily, there are more applicants than jobs. A psychologist will administer a valid psychological test, and starting from the top test score and working down, the organization will hire until all the jobs are filled. The validity coefficient is always less than 1.0, which means that the test is imperfect. Notwithstanding, tests with validities in the neighborhood of .30 to .40, which is modest, can be successful. But even with modest validity it would seem easier to guarantee the success of the top 10 performers out of 1000 who took a test than to guarantee the success of 500. If the test has any validity whatsoever, the few top performers on the test should succeed on the job. How does the proportion selected relate to success on the job? Obviously, the size of the validity coefficient is only part of it.

Taylor and Russell (1939) were the first to clarify the relationships between criterion validity and the practical problems of selection, and Figure 17-2 illustrates their analysis. The data points, showing the relationship of test scores *(X)* to criterion *(Y)*, represent a validity coefficient of about .50. The line *xx'* defines the proportion of applicants who are selected, and this proportion is called the *selection ratio*. The line *yy'* is the *base rate* and defines the proportion of unselected applicants who would be satisfactory on the job; it is the success rate on the

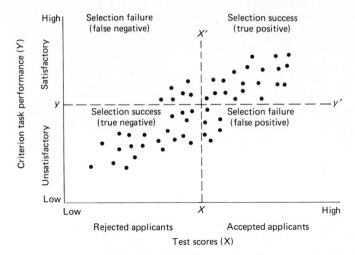

Figure 17-2 An illustration of the decision process in personnel selection. Applicants above the cutoff *xx'* for test scores are accepted, those below rejected. The line *yy'* divides satisfactory from unsatisfactory performance on the criterion task. See text for explanation.

job before the selection program began. The issue is how the validity coefficient, the selection ratio, and the base rate relate to one another.

The successes and failures of an imperfectly valid test are demonstrated in Figure 17-2. The quadrant at the upper right contains selection successes, called *true positives*. These applicants were above the *xx'* cutoff, were hired, and succeeded on the job. The quadrant of the lower right has selection failures, called *false positives*. These applicants were above the *xx'* cutoff and were hired but failed on the job. The quadrant in the lower left is the *true negative* category and represents correct decisions because these persons would have failed on the job if hired. The *false negative* category in the upper left quadrant is another case of selection failures because these applicants were rejected but would have succeeded on the job had they been hired. Thus, a valid test is a mix of successes and failures. The test is acceptable as long as the number of true positives is high. Conversely put, there should not be too many false positives. False negatives are an unpleasant fact of life to confront because they are persons who would have succeeded had they been given the chance.

Figure 17-3 shows the consequences of a small selection ratio where only a small proportion of applicants are chosen; the *xx'* line is moved to the right. The *yy'* line is kept the same in this example. All the accepted applicants (the upper right quadrant) succeed; there are no selection failures. If a psychological test with some validity was admin-

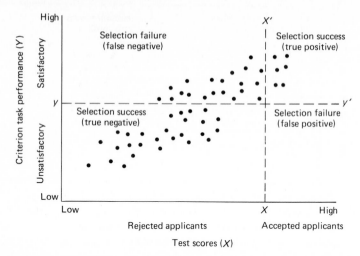

Figure 17-3 Illustration of high selection success when the cutoff *xx'* is set high for a valid test. See text for explanation.

istered to 1000 applicants, and only 10 jobs were available, it is very likely that the top 10 scorers on the test would perform well on the job.

Taylor and Russell (1939) brought these variables together and published tables which showed how they were related. To illustrate, consider a base rate of .50. When a test has a validity of zero and the selection ratio is .70, the proportion of those hired who will succeed on the job is .50, the base rate (a validity of zero is the same as unselected applicants). But with a validity coefficient of .40 and a selection ratio of .70, the proportion of successful applicants of those hired increases to .58. If, however, the selection ratio is .05 when the validity coefficient is .40, the proportion of successful applicants jumps to .82. A test of modest validity does not help much if the selection ratio is large and a high proportion of those tested are hired, but it does an impressive job when only a few applicants are hired.

Reconsider the validity coefficients of the previous section on Air Force tests in World War II. The Complex Coordination Test and the mathematics test had validity coefficients in the vicinity of .40. The selection ratios are not known, but the Air Force typically had more applicants than aircrew jobs, so selection ratios were undoubtedly small. Under these circumstances, although validity coefficients were modest, the Taylor and Russell analysis indicates that the tests would be good selection devices. Consider, on the other hand, a college that uses the ACT battery with a validity of .40. The college will not get much benefit from the ACT if it is obliged to maintain high enrollment by using a large selection ratio and accepting a large proportion of those who apply. An elite college, which accepts only a few of the many applicants that apply, will be using a small selection ratio and will find the ACT battery beneficial. Refinements of the ACT battery to improve its validity for academic performance will benefit all users, of course.

Economic and Productivity Implications of a Valid Test

The Taylor and Russell analysis is informative for personnel selection, but what is the worth of a valid test in dollars? Schmidt and his colleagues (Schmidt et al., 1979) have developed procedures for estimating the economic value of a test by comparing the dollar production of those hired by a particular selection method with the dollar production of the same number of workers hired at random. They analyzed the value of the Programmer Aptitude Test for the selection of computer programmers to work for the federal government. The test has a validity of .76. If no testing had been used and the Programmer Aptitude Test was then introduced and used for one year with a small selection ratio of .05, there would be a productivity gain of $97.2 million over the 10-year tenure of the average programmer. Even if the selection ratio was a large .80, which would be expected in a time of high demand for programmers and short supply, the gain would be $16.5 million. Schmidt et al. carried out the same analysis for computer programmers in the U.S. economy as a whole and found that a selection ratio of .05 would produce an astounding productivity gain of $1.6 billion. A selection ratio of .80 would produce a gain of $273 million. The analysis assumed that selection proceeded from the top-scoring applicant downward, and that all applicants who were offered jobs accepted them. The latter assumption is unrealistic to an unknown degree. Every applicant with a high score who rejects a job offer must be replaced with an applicant with a lower score, so the productivity of those who are accepted will be lower (Schmidt et al., 1979; Murphy, 1986).

Schmidt et al. (1984) did the same analysis for park rangers of the U.S. Park Service. They computed the productivity gain in dollars from using a test of general mental ability for one year rather than the interview that had been used. The validity of the interview was .14. The test of general mental ability had a validity of .51. The selection ratio was .10, which is typical for selecting park rangers. For 80 rangers hired in a year, which is average, the productivity gain was $1.2 million if they remained on the job five years and $2.3 million if they remained on the job 10 years.

Few people, even psychometricians, would have thought tests so valuable. In this day when analysts point ominously to the declining productivity of U.S. workers, and speculate about causes of the decline,

TABLE 17.1 Average Validities for Cognitive, Perceptual, and Motor Tests for Nine Job Families*

Job Family	Cognitive Tests	Perceptual Tests	Motor Tests
Manager	.53	.43	.26
Clerk	.54	.46	.29
Salesperson	.61	.40	.29
Protective professions worker	.42	.37	.26
Service worker	.48	.20	.27
Trades and crafts worker	.46	.43	.34
Elementary industrial worker	.37	.37	.40
Vehicle operator	.28	.31	.44
Salesclerk	.27	.22	.17

*Reanalysis of data by Ghiselli (1973).

Source: Hunter & Hunter (1984), Table 1. Reprinted by permission of the American Psychological Association.

it would pay them to ask if selection decisions can be improved. Productivity suffers when the best workers are not being placed in jobs.

What Kind of Test Has the Highest Validity Overall?

There will always be particular tests that will do the best job for predicting specialized criteria, such as a mathematics test to predict success in engineering or a motor test to predict success in mechanics school, but the viewpoint has developed in recent years that cognitive tests are the best predictors of performance in a number of job families (a job family is a category for jobs with a number of features in common). Loosely speaking, a cognitive test is a kind of intelligence test in the sense that it is a complex of items that measure mental abilities. A single cognitive test, which would have good powers of prediction for a number of job families, might include items of verbal comprehension, vocabulary, mathematical knowledge and reasoning, and problem solving. Items such as these measure the extent to which a person has developed good learning strategies and the work habits to use them, general knowledge, and problem-solving techniques. People with these mental skills are often judged intelligent.

Table 17-1 shows the validity coefficients for three kinds of tests, averaged over many studies for nine job families (Hunter & Hunter, 1984). Validities for cognitive tests, perceptual tests, and motor tests are presented. Overall, the validity of cognitive tests is the highest except for the categories of elementary industrial worker and vehicle operator, which are the least complex of the nine job families. Elementary industrial workers and vehicle operators are best predicted by motor tests. With cognitive tests ranking the highest in predictive capability, perceptual tests rank second and motor tests third. Overall, a standard intelligence test, or a cognitive test that is similar to it, has broad predictive powers for performance in a wide variety of jobs (Hunter, 1986; Hunter & Hunter, 1984; Schmidt & Hunter, 1981).

Summary

The efficiency of a system can be helped with good personnel selection. Psychological tests that measure abilities pertinent for a job are used to select the best subset from a pool of applicants.

A psychological test is a sample of a person's behavior taken under standardized conditions. The two main kinds of psychological tests are printed tests with booklet format, and motor tests for the measurement of movement skills. The rationale for psychological tests is that a job requires a particular configuration of abilities, and that some persons are more favorably weighted on the abilities than others and so perform better on the job. Tests that measure these abilities will identify the better performers.

A test is a measuring instrument and, like any measuring instrument, must be reliable. Reliability means that essentially the same measures will be obtained from one measuring occasion to the next. Beyond reliability is validity. A test with criterion validity will predict performance in another situation, such as success on a job. A test with content validity has items that sample a knowledge domain representatively. A test with construct validity, such as an intelligence test, is defined theoretically.

Tests of modest validity will select personnel who generate economic and productivity gains. Productivity gains can run to many millions of dollars when a valid psychological test is used to select personnel. Quality of personnel is one of the variables that affects productivity, and psychological testing is a way to get it.

Modern analysis has found that cognitive tests, which are similar to intelligence tests, are the best all-around predictors of performance on a wide range of jobs.

References

Anastasi, A. (1982). *Psychological testing. Fifth edition.* New York: Macmillan.

DuBois, P. H. (1970). *A history of psychological testing.* Boston: Allyn and Bacon.

Ghiselli, E. E. (1973). The validity of aptitude tests in personnel selection. *Personnel Psychology, 26,* 461–477.

Guilford, J. P. (Ed.). (1947). *Printed classification tests.* Washington, DC: U.S. Government Printing Office.

Hunter, J. E. (1986). Cognitive ability, cognitive aptitudes, job knowledge, and job performance. *Journal of Vocational Behavior, 29,* 340–362.

Hunter, J. E., & Hunter, R. F. (1984). Validity and utility of alternative predictors of job performance. *Psychological Bulletin, 96,* 72–98.

Landy, F. J. (1985). *Psychology of work behavior.* Third Edition. Homewood, IL: Dorsey Press.

Melton, A. W. (Ed.). (1947). *Apparatus tests.* Washington, DC: U.S. Government Printing Office.

Mitchell, J. V., Jr. (1985). *The ninth mental measurements yearbook.* Volume 1. Lincoln, NE: University of Nebraska Press.

Murphy, K. R. (1986). When your top choice turns you down: Effect of rejected offers on the utility of selection tests. *Psychological Bulletin, 99,* 133–138.

Schmidt, F. L., & Hunter, J. E. (1981). Employment testing: Old theories and new research findings. *American Psychologist, 36,* 1128–1137.

Schmidt, F. L., Hunter, J. E., McKenzie, R. C., & Muldrow, T. W. (1979). The impact of valid selection procedures on work-force productivity. *Journal of Applied Psychology, 64,* 609–626.

Schmidt, F. L., Mack, M. J., & Hunter, J. E. (1984). Selection utility in the occupation of U.S. park ranger for three modes of test use. *Journal of Applied Psychology, 69,* 490–497.

Taylor, H. C., & Russell, J. T. (1939). The relationship of validity coefficients to the practical effectiveness of tests in selection: Discussion and tables. *Journal of Applied Psychology, 23,* 565–578.

CHAPTER 18

Training and Training Devices

Personnel selection follows efforts to optimize design of the workplace, but there is more to do. The personnel who are selected must be trained. The best-selected personnel of the best-designed system must be trained in the skills and knowledge that will lead to efficient attainment of a system's goals.

What are the differences among training, education, and learning? Very little. "Learning" is the generic term. *Learning* refers to the plastic capability of organisms to change with experience. Learning operations increase the probability that a response will occur or that it will occur faster. The products of learning are stored in memory, which increases the likelihood of their being used on the next occasion. *Training* is the term used for learning in the context of human–machine systems. *Education* is learning in an academic setting. The learning that is training and education is closely related to transfer of training, or how behavior that is acquired in one situation applies to another. The behavior that is acquired in the training setting of a system is expected to transfer to actual system operations. Some classroom learning is knowledge for its own sake, but society expects educators to teach their students knowledge that can be transferred to society's affairs.

Transfer of Training

Definition and Measurement of Transfer

Transfer of Training Defined. Transfer of training is how behavior that is acquired in one task applies to another. *Positive transfer* is where the behavior that is learned in the first task benefits performance in the second, and *negative transfer* is where performance in the second task is impaired. *Zero transfer* is no effect. In the training of personnel for human–machine systems we must always have positive transfer, where the training teaches skills and knowledge that benefit system operations. Zero transfer or negative transfer is failure because it means that operation of the system is unaffected or hindered by the training.

Transfer of training is determined by experiment. The essential experimental design is as follows:

	Training on Task 1	Test on Task 2
Experimental group	Yes	Yes
Control group	No	Yes

After training on task 1 the experimental group is tested on task 2 and the comparison is with the control group, which is only tested on task 2. There is positive transfer if the experimental group performs better than the control group on task 2. There is negative transfer if the experimental group is worse than the control group on task 2. Zero transfer is no difference between groups on task 2. Like any experimental design, this essential design can be elaborated. There might be three experimental groups, each with a different amount of training on task 1, for example.

Measurement of Transfer. A basic index of transfer is *percent transfer of training,* and its two forms depend on the measure of performance that is used in task 2. When a lower score (e.g., errors, or time or trials to learn) represents more proficiency, the formula is

$$\text{percent transfer of training} = \frac{\left(\begin{array}{l}\text{control group}\\ \text{group score}\end{array}\right) - \left(\begin{array}{l}\text{experimental}\\ \text{group score}\end{array}\right)}{\text{control group score}} \times 100$$

To illustrate, suppose that the control group and the experimental group were both trained to a criterion of proficiency on task 2. The experimental group required 5 hours of practice to reach the criterion, and the control group required 10 hours.

$$\frac{10 - 5}{10} \times 100 = 50\%$$

The training that the experimental group had on task 1 substantially benefited its performance on task 2.

The formula is slightly different when a lower score represents less proficiency, such as the number of units produced in a period of time:

$$\text{percent transfer of training} = \frac{\left(\begin{array}{l}\text{experimental}\\ \text{group score}\end{array}\right) - \left(\begin{array}{l}\text{control}\\ \text{group score}\end{array}\right)}{\text{control group score}} \times 100$$

Suppose that factory workers in the experimental group received a special training program and were then placed on the production line. A corresponding group of workers on the production line were assigned to the control group and did not receive the training program. The production of each worker was recorded for a month. The experimental group averaged 100 units a day and the control group 80 units.

$$\frac{100 - 80}{80} \times 100 = 25\%$$

Flight Simulators and Transfer of Training

There are many ways to train—read textbooks, listen to lectures, watch training films, on-the-job training where the apprentice works alongside the master, and so on. A major way of training skills of the complexity that are required in complex systems is with *training devices*. Training devices can be relatively simple, but they can also be elaborate multimillion-dollar devices. Of particular interest are flight simulators for the training of flying skills because some of them have been evaluated experimentally for transfer of training, and research on them is a main source of knowledge about transfer of training in the practical context of a system. Flight simulators have drawn research attention because of their complexity and cost.

Flight Simulators and Flight Trainers Defined. There are two main kinds of simulators. An *engineering simulator* is a model of some of the system and the variables that influence it, and it is used for decisions about system design. A digital computer might be used to model aircraft aerodynamics in the laboratory and test pilots would fly in a cockpit mockup with two different kinds of controls for a comparison of them, for example. A *training simulator* is the second kind, and it is used for training system personnel. There is also a model of the system, usually with a digital computer, and the model drives a mockup of one or more system workplaces in which personnel train. Training simulators do not necessarily have to be realistic and complex, but often they are. A flight simulator for the training of flying skills can have a faithful cockpit layout, good aerodynamic simulation, some simulation of the external visual environment, motion effects, and an instructor's station with a capability for monitoring, recording, and reporting the trainee's performance and for structuring training exercises. Simple exercises can be structured for the beginner, or complex ones for the advanced trainee. Normal and emergency situations can be defined. Ship simulators can be equally realistic, as can simulators of the control room of nuclear power plants. Pictures of such simulators are shown in Figure 18-1 through 18-6.

A flight simulator is distinguished from a flight trainer. A *flight simulator* makes an effort to reproduce with fidelity the cockpit and aerodynamic characteristics of a particular aircraft. The simulator for the F-15 aircraft of the U.S. Air Force strives to duplicate the characteristics of the F-15 cockpit and F-15 flight characteristics as closely as possible. A *flight trainer* is faithful to a class of aircraft with similar characteristics, not to any particular aircraft. Flight trainers for light aircraft, such as pleasure aircraft, have the general characteristics of many light aircraft but of none in particular.

Although devices for pilot training were developed almost from the beginning of powered flight, Edwin A. Link, an American, is acknowledged as the inventor of the flight trainer that is the parent of the sophisticated flight simulator that we have today. Link invented the trainer in the factory of his father's piano and organ business during the period 1927–1929 (Adorian et al., 1979). No one paid much attention to Link's trainer until the Army Air Corps (as the U.S. Air Force was named at the time) was called to carry the first airmail in the

Figure 18-1 Visual Technology Research Simulator of the U.S. Navy. The purpose of the simulator is for research on the simulation of visual scenes. [*Source:* U.S. Navy brochure, "Visuals Technology Research Simulator."]

1930s. A distressing number of pilots were killed flying in bad weather, so the Army Air Corps looked for better ways to train their pilots in the use of instruments for "blind" flying. They found the Link trainer. The Link trainer attracted the interest of others in the 1930s, with some commercial airlines and military air forces around the world adopting it. During World War II hundreds of thousands of pilot trainees learned their flying skills on the Link trainer.

In the late 1930s, R. C. Dehmel, an engineer with the Bell Telephone Laboratories, became interested in the problem of flight training and designed the essentials of electronic simulation rather than simulation with the Link trainer's mechanical and pneumatic components, which had their roots in the piano and organ business. Prototypes of wholly electronic trainers were developed during World War II, and commercial mass production of them took place in the late 1940s. The analog computer was the basis of simulation in early electronic trainers, but by 1960 the digital computer entered flight simulation. Today all major flight training devices use digital computation.

Figure 18-2 Overview of a ship simulator. The simulator is used by the Maritime Institute of Technology and Graduate Studies, Linthicum Heights, MD, for the training of officers of commercial vessels in the principles of ship handling. A computerized optical system provides a 360-degree field of view on the screen surrounding the simulator. Realistic sounds are provided. Actuation of the "legs" supporting the simulator provide ship motion [*Source:* Brochure, "Maritime Institute of Technology and Graduate Studies," June 1, 1982. Reprinted by permission of the Maritime Institute of Technology and Graduate Studies.]

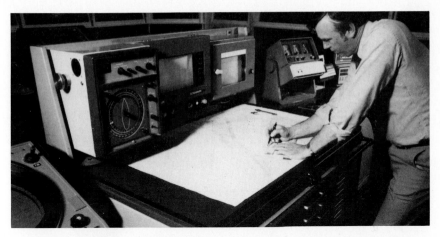

Figure 18-3 Ship simulator of Figure 18-2 from the trainee's position. The simulation is of a ship's bridge. [*Source:* Brochure, "Maritime Institute of Technology and Graduate Studies," June 1, 1982. Reprinted by permission of the Maritime Institute of Technology and Graduate Studies.]

Figure 18-4 Simulator for the control room of a nuclear power plant. The simulation is of the Maine Yankee Nuclear Atomic Power Plant. [*Source:* Link Simulation Systems Division. Reprinted by permission of The Singer Company.]

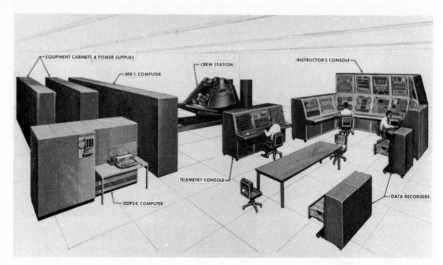

Figure 18-5 Overview of the Gemini whole-task simulator system for manned space flight. [*Source:* Felker (1964). Reprinted by permission of J. K. Felker.]

Experiments on Transfer of Training. The first properly designed and executed transfer of training experiment in the context of aviation was by Williams and Flexman (1949) at the University of Illinois at Urbana–Champaign. An account of this pioneering study in some detail will indicate how a transfer of training experiment is conducted, and how complex and costly such experiments can be even when the trainer and the aircraft are relatively simple.

The subjects of the experiment by Williams and Flexman were students enrolled in a beginning flight training course. A Link trainer was evaluated for its transfer to the light aircraft in which the students trained. Eight air exercises, as representative of skills in basic flying, were devised (e.g., effects of control movement on the aircraft, climbs and glides, etc.). Each exercise was broken down into items on which a trainee was checked pass or fail by the instructor. The criterion of learning an exercise was three consecutive errorless trials. The experimental group, with 24 subjects, practiced each exercise to criterion in the trainer, followed by practice to criterion in the aircraft. The control group, also with 24 subjects, practiced the exercises to criterion only in the aircraft. The number of trials to reach criterion was the measure of performance for an exercise. The transfer of training question was whether the experimental group would require fewer trials to learn in

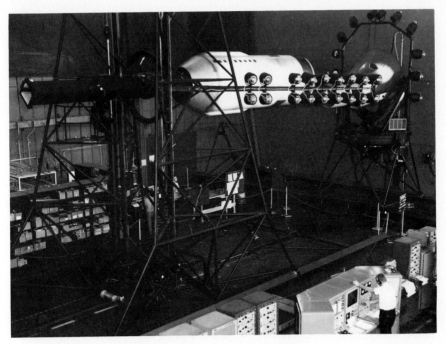

Figure 18-6 Rendezvous with a satellite was in the mission of the Gemini system. A part-task simulator for the critical docking maneuver is shown. The simulation was of the last 100 feet, where visual observation of the satellite and manual control were required. [*Source:* Felker (1964). Reprinted by permission of J. K. Felker.]

the air than the control group. The answer was "yes," although the percent transfer of training was a modest 22 percent. The measure used in the transfer of training formula was the total number of trials to reach criteria on the eight air exercises.

Williams and Flexman also computed an efficiency ratio, as they called it, but what is now called the *transfer effectiveness ratio* (Roscoe, 1971, 1972; Povenmire & Roscoe, 1971). The transfer effectiveness ratio, which is an index of how much practice on the trainer is required to achieve the saving in the air, is defined as

$$\text{transfer effectiveness ratio} = \frac{\left(\begin{array}{c}\text{control}\\\text{group score}\\\text{in the air}\end{array}\right) - \left(\begin{array}{c}\text{experimental}\\\text{group score}\\\text{in the air}\end{array}\right)}{\begin{array}{c}\text{experimental group score}\\\text{in the training device}\end{array}}$$

For the Williams and Flexman experiment the measure inserted in this formula was total number of trials to reach criteria on the eight exercises. The value of the ratio was .29. One trial in the Link trainer was worth .29 trials in the air.

Povenmire and Roscoe (1973) evaluated a relatively modern Link trainer for light aircraft, called the GAT-1 (GAT for "General Aviation Trainer"). The purpose of the experiment was to evaluate the effect of amount of training in the Link trainer on transfer to the aircraft. Different experimental groups had either 3, 7, or 11 hours of practice in the trainer. The control group trained only in the aircraft. The principal measure of performance was number of hours of practice in the aircraft to pass the final flight check in certification for a private pilot's license. The trainer that was used is shown in Figure 18-7, and the aircraft is shown in Figure 18-8.

Percent transfer for 3, 7, and 11 hours of practice in the trainer was 11, 15, and 16 percent, respectively, which is in the range of the value found by Williams and Flexman. The transfer effectiveness ratio for 3, 7, and 11 hours of practice in the trainer was 1.72, .97, and .68. The declining transfer effectiveness ratio shows that savings in the air do not increase as amount of practice on the trainer increases. Instead, there was diminishing returns. A point would be reached where additional time in the trainer would be unprofitable.

The percent transfer values found in these experiments are low, but this might be expected. The Link trainers that were evaluated were trainers, not simulators, so the match of trainer characteristics with aircraft characteristics was not high. Nor did these trainers have simulation of external visual scenes, which some simulators have. Rather, the trainers were "instrument trainers," where the trainee learns to control the aircraft by the information from cockpit instruments.

Many transfer of training experiments on flight simulators of relatively high fidelity have been conducted by now, so the training worth of more realistic, advanced devices can be appraised. What is striking about appraisals of the research on modern flight simulators is how low percent transfer values will often be. Diehl and Ryan (1977) reviewed 25 studies where the range of percent transfer values was −11 to 95, with an average value of 38 percent. Orlansky and String (1977) reviewed 35 transfer measures from 19 experiments and found a range of percent transfer values from −11 to 90, with an average of 34 percent. Semple et al. (1981) reviewed 64 transfer measures from 22 studies and found that the range of percent transfer values was from zero to 96, with an average value of 20 percent. The simulators were with and without simulation of external visual scenes.

Flight simulators have a long way to go, but that does not mean that

Figure 18-7 The Link GAT-1 trainer of the kind used in the transfer of training experiment by Povenmire and Roscoe (1973). [Reprinted by permission of Link Flight Simulation Division, The Singer Company.]

Figure 18-8 The Piper Cherokee aircraft of the kind used in the transfer of training experiment by Povenmire and Roscoe (1973). [Reprinted by permission of Piper Aircraft Corporation.]

their value is questionable. On the contrary, flight simulators are a profitable way to train. They are profitable because their operating cost is low relative to the operating cost of their parent aircraft, so even a modest amount of positive transfer can pay off. A commercial airliner, or a high-performance military aircraft, is very expensive to operate, and a flight simulator, even an elaborate one, is not. Orlansky and String (1977, p. 18) found that the median ratio of simulator cost to aircraft cost for 33 simulators and their aircraft was .12. Other analyses found the median hourly cost of operating 42 military and commercial jet aircraft to be $1066, while the median hourly cost of operating 49 simulators for aircraft of this kind was $96. The Boeing 747 commercial airliner costs $2358 per hour to operate, while its state-of-the-art simulator costs $275 per hour (Orlansky & String, 1977, p. A-4). Not much transfer of training is required to make a simulator profitable with cost differentials of this kind. A low transfer effectiveness ratio, with rather little time saved in the air relative to simulator time, can be profitable. As a caution, it should be remembered that savings in flight time cannot always be fully ascribed to the simulator in a flight training program. The quality of the training program, and the quality and experience of the pilot trainees, must be considered also. American Airlines introduce their pilots to new aircraft with flight training programs that include flight simulators (Orlansky & String, 1977, pp. 56–63). The number of hours in the air required for mastery of the new aircraft is down to 1 to 2 hours, which is impressive, but it must be remembered that commercial airline pilots are highly experienced and need to be taught only new operating characteristics of the aircraft. Moreover, the simulators of American Airlines are embedded in high-quality training programs. The U.S. space program is another example of how experienced, high-quality personnel who use simulators within the context of a well-designed training program can perform successfully. Some astronauts have gone into space with only earthbound experience and have performed commendably. A simulator never stands alone; it is always part of a larger training program which exerts its influence on flying skills also.

There are other advantages of flight simulators to consider:

1. Simulators can train responses that cannot be trained in their aircraft. Certain emergency procedures are the most notable example. How to cope with an engine fire at takeoff can be practiced only in a simulator.
2. Simulators are independent of weather, which is an important consideration for training pilots who are not qualified for flying in marginal weather.

3. Simulators often can give more intensive training in particularly critical parts of a flying task. A bomber crew may fly across the United States for a practice run on a target, receiving only one practice trial in the critical bomb-drop maneuver on a day. With only in-flight training, the crew could be undertrained in the maneuver. A simulator, concentrating only on the critical maneuver, could give a number of trials in an hour.

Evaluating a Training Device by Ratings. Testing is a routine part of system development. A new system such as an automobile or an airplane is thoroughly exercised to see that performance specifications are met. If specifications are not met, engineering adjustments are made. Further tests follow, and production begins only when all is satisfactory. Like any other system, training devices undergo hardware tests, but they do not routinely undergo transfer-of-training evaluation for their worth as training devices. Transfer-of-training experiments are almost always in behalf of research, not a test of the training value of a training device before it is delivered to a customer.

The simulator industry uses the *rating method* to estimate the training worth of a flight simulator (Adams, 1979). A pilot experienced with the parent aircraft rates the simulator for its similarity to the aircraft. The rating method is an adaptation of procedures that are used in aircraft design. In design, test pilots will rate the handling qualities and controllability of a new aircraft. The same procedures are used for rating a training simulator. Higher ratings mean higher perceived similarity of aircraft and simulator, and high similarity is assumed to mean high transfer of training. The validity of the rating method for assessing the training worth of flight simulators is unknown, and there are problems with it (Adams, 1979). Not the least of the problems is that transfer of training does not always require high similarity of simulator and aircraft. Fidelity of simulation and transfer of training need not always be closely related, as the next section will show.

Issues in the Design and Use of Flight Simulators

What are considerations in the design and use of flight simulators that relate to transfer of training? Like the parent system, the design of which must be optimized if training goals are to be efficiently met, a training simulator must be optimally designed for the attainment of training goals.

Part-Task Training. The issue for part-task training is how a complex task can efficiently be learned. Should the whole task that a simulator attempts to structure be presented again and again, or is it better for transfer of training to break the whole task down and train on the parts? Devices for part-task training are called *part-task simulators*. Some of the reasons for part-training of flying skills deal with cost and some with skill (Adams, 1960, 1961):

1. Whole-task flight simulators and aircraft can cost millions of dollars. There is always the hope of finding less expensive ways to train.
2. Parts of the flying task which are critical to mission success may receive comparatively little training within the context of whole-task training. These important parts may be difficult and of short duration, and may fail to accumulate enough training time. Intensive training on these critical parts could be given in part-task simulator deliberately designed for them.
3. Much of flight training is with experienced pilots who need only learn the new elements of the task. Part-task training devices might be devised to teach the new response requirements. Less training with expensive whole-task training devices would be needed.
4. Space flight will someday last for years, and forgetting of skills will occur. On-board training devices will be required for the retraining of forgotten skills. Part-task retraining methods are a possibility.

An experiment by Adams and Hufford (1962, Experiment I) illustrates the advantages and cautions in training with a part-task training device. Aviation was the context. A pilot must not only know the control of his aircraft but must know "procedures" as well. There are emergency procedures, and there are normal procedures such as the starting and shutting down of engines. A modern aircraft has many procedures that a pilot must know. Procedures can be taught in a properly designed whole-task simulator, and with exceptions they can be taught in the aircraft also. Should procedures be taught in these expensive whole-task devices, or can a simple, inexpensive part-task training device be developed where the controls and displays of the part trainer might be inoperative except as required for the execution of procedures? Adams and Hufford used a flight simulator as the research device. A complex maneuver was used that required the procedures of throwing 10 switches to be performed along with flight control. The experimental group learned flight control and procedures separately, as parts, and was then transferred to the whole task and compared with a control group that had the same amount of training on the whole task where flight control and procedures had to be performed together. The experimental group showed substantial positive transfer from the part training, but their performance in procedures was not quite as good as the control group on the first transfer trial. A small amount of additional practice was required on the whole task before the experimental group performed as well as the control group. Adams and Hufford (1962, Experiment II) also evaluated part training for the relearning of procedures forgotten over 10 months. Part training was effective in restoring forgotten procedures, but restoration was not complete. Some additional whole-task practice was again required. Adams and Hufford concluded that much can be learned and relearned in part trainers but that their use must be followed with whole-task practice, where the parts are integrated into a time-shared whole.

Trollip and Ortony (1977) developed a part-task simulator for teaching holding patterns to pilot trainees. We have all arrived at our destination in a commercial airliner only to have the landing delayed. When the landing is delayed, the air traffic controller assigns the pilot to a holding pattern at a specified altitude. In a no-wind condition the holding pattern is a symmetrical racetrack pattern, but with wind, which is usually the case, the pattern is distorted. A leg of the pattern is shortened when the pilot flies into the wind, and lengthened with a tail wind. Turns are affected by the wind. A pilot must understand these relationships among wind, airspeed, and holding pattern distortion so that compensatory adjustments can be made. Trollip and Ortony equipped a computer terminal with a hand control and a throttle for real-time simulation of holding patterns. Essential flight instruments could be displayed. The trainee could learn about the variables for holding patterns by pictorial presentations of flight paths for various wind and speed conditions and from instructional text on the display.

Trollip (1979) conducted a transfer of training evaluation of his device. After a trial the screen would display the actual pattern that the subject had flown overlaid on the ideal pattern for the flight conditions (kinematic knowledge of results—Chapter 5). A detailed analysis of the mistakes was also presented. Student pilots were subjects. Control subjects learned about holding patterns in a classroom setting. Experimental subjects used Trollip's training device. All subjects were then tested in a whole-task flight trainer (the Link GAT-2, which is similar to the GAT-1 trainer shown in Figure 18-7 except that it is more advanced). Two experienced pilots observed the subject in the trainer and evaluated how well a subject flew the holding patterns. Positive transfer was obtained. The experimental subjects were superior to control subjects on all counts, reaching the criterion of proficiency in fewer trials and with fewer errors. As an observation on research methodology, it is useful to notice that Trollip tested his subjects in the GAT-2 flight trainer, not an aircraft. The use of expensive aircraft as the criterion is not always desirable in the early stages of a research program. A flight trainer has many similarities to an aircraft, and a great deal can be learned by transfer to it. That Trollip obtained positive transfer to the flight trainer encourages further experiments with aircraft.

Simulation of the External Visual World. One reason that percent transfer of training is not very high is because simulation of the external visual world is absent or incomplete in flight training devices. Pilots use a vast array of visual cues for successful control of the aircraft, and simulators are deficient in them. Early trainers and simulators had static visual simulation, such as a screen with a horizon line and clouds painted on it. These crude attempts were followed by dynamic optical projection systems that were closed-loop and changed as the pilot trainee changed the orientation and position of the simulated aircraft. Today dynamic, closed-loop scenes are generated by a digital computer.

Strong advances are being made in visual simulation, and visual simulation is contributing to transfer of training (Waag, 1981), but there is little prospect soon of simulating all that the pilot needs to see (Staples, 1978). The richness of visual scenes that a pilot uses in aircraft guidance is boundless. Our processing of visual information as we move through our everyday spatial affairs is awesome, and how much more so it must be for the pilot. To deny a pilot trainee enough simulated visual information is to deny the stimuli that regulates so much of his behavior in flight. Visually impoverished simulator training requires further air training in compensation, which is a way of saying that transfer of training is low. Flight simulators with limited visual simulation have shown low positive transfer (Semple et al., 1981; Diehl & Ryan, 1977; Waag, 1981).

One deterrent to advances in visual simulation is complexity of the digital computation that is required. The need for more comprehensive, high-fidelity visual simulation imposes heavy burdens on computing equipment, programmers, and budget. Visual simulation efforts today strive to duplicate the power of the human eye, but it is unlikely that the power of the eye need be duplicated for training purposes. There is enormous redundancy in the visual scene, and we rely on only some of the cues for the regulation of behavior. A pilot should not need simulation of the wealth of detail in an airport scene for the learning of landing behavior. The runway rectangle, as an essential cue, would probably do because it is the invariant of all airports. However, simulating other kinds of scenes does not allow the luxury of reducing the simulation to essential cues because there is no way of knowing what the essential cues are. In such cases, high-fidelity simulation with as many cues as possible is desirable to guarantee that the essential ones are included. Lintern and his associates (Lintern et al., 1987) investigated several kinds of visual scenes for a dive-bombing maneuver, using the flight simulator shown in Figure 18-1. The scenes varied in their fidelity of simulation. Original learning together with transfer to other scenes in the simulator was best for the scene with the greatest detail.

The day will come when perception psychology will give us the methodology for finding the essential features of any visual scene, and when that day arrives there will be ways of prescribing simplified visual scenes for training. In Chapter 11 we saw that complex phonemes can be economically described by 13 two-valued features, and in principle the same economy should be possible for complex visual scenes. Until then, designers will have fidelity of visual simulation as high as technology and budget allow.

Cockpit Variables and Fidelity of Simulation. There is evidence that high fidelity of cockpit variables is not always required for high transfer of training. The assumption that governs the design of most whole-task flight simulators is that high fidelity produces high transfer of training. The assumption is true, but it is a half truth because high transfer of training can also be found with low-fidelity simulation.

Consider an experiment by Briggs et al. (1957). We saw in Chapter 12 that the amplitude and force requirements of a tracking control are important determiners of proficiency. Yet Briggs et al. demonstrated that training with particular amplitude and force values on a control did not affect transfer to a new set of values. Rockway (1955) and Rockway et al. (1956) found the same for the control–display ratio. Briggs and Rockway (1966) found that training on compensatory tracking did not affect transfer to pursuit tracking, and vice versa. The variables affected the momentary performance, as was emphasized in Chapter 12, but not the underlying skill that was learned and transferred. Thus, these variables should be optimized in the aircraft so that the pilot's flying performance is optimized, but they can be relaxed in a flight simulator without impairing transfer of training. It would not seem to matter for transfer of training that a tracking control has low fidelity sensitivity and "feel." However, users might complain that the control is unlike the parent system, and so might be less motivated to train. User acceptance is desirable for a training device.

Research is clear in showing that training devices for teaching procedures can be of low fidelity. Dougherty et al. (1957) found that a photographic mockup of a cockpit produced as much transfer to the aircraft as a high-fidelity simulator. An experiment on procedures training by Prophet and Boyd (1970) had the same result. Procedures that were learned in a plywood and photographic mockup of the cockpit had as much transfer to the aircraft as did a high-fidelity trainer. Cox et al. (1965) found that a 5 × 7-inch photograph was as effective as a high-fidelity simulator for teaching the 92-step procedural sequence in arming and firing a missile.

Simulator Motion and Fidelity of Simulation. A high-fidelity flight simulator can ride on top of a motion subsystem that gives quasi-realistic perception of turns, climbs and dives, turbulence, and so on. Being a mechanical device, it has limits; nevertheless, it contributes to the realism of the simulation. A motion subsystem undoubtedly contributes to user acceptance, but how necessary is it for transfer of training? There have been a number of transfer of training experiments on motion versus no motion because these motion systems are so expensive. The outcome overall for these studies is that motion probably does not contribute to transfer (Waag, 1981). Motion for the simulator may affect the performance level of the pilot such that he performs better with it than without it in the simulator, but the issue for a training device is whether its motion affects a pilot's performance in the aircraft. Apparently, it does not. The most likely reason that motion fails to affect transfer is that its cues are redundant. Instrument cues always provide guidance to the aircrew for aircraft control, and visual cues often do. Motion provides little new information.

The Instructor's Station. The instructor's station is a tool that the instructor uses for programming training exercises, monitoring the trainee's performance, and measuring it. The instructor's station has come a long way since the early days of flight simulation (Faconti, 1979; Smode, 1974).

In the early days of flight simulation the instructor's station was little more than a console for starting and stopping the simulator, with perhaps a few repeater flight instruments for monitoring the trainee's performance. As flight simulation became more advanced, after World War II, instructor stations became more elaborate, with several consoles. All the cockpit instruments were repeated. There were facilities for setting up navigation problems. Emergencies could be created and monitored. Combat exercises could be structured for military systems. Facilities for observing the trainee's performance were improved. Some performance measurement was possible. The instructor's station became complex.

The advent of the digital computer is creating a trend toward compact instructor's stations. A modern instructor's station, for the flight simulator in Figure 18-1, is shown in Figure 18-9. A computer-based console like that in Figure 18-9 makes the job easier for the instructor. The computer allows the automation of training exercises and the presentation of emergencies. Demonstrations can be automated, as can performance measurement. All variables can be recorded in the computer and played back to the trainee, reproducing all instrument indications and control movements, as well as visual and motion cues.

Figure 18-9 The instructor's station for the U.S. Navy's Visual Technology Research Simulator shown in Figure 18-1. [*Source:* U.S. Navy brochure, "Visuals Technology Research Simulator."]

Proficiency Measurement. Training simulators are built with the expectation that useful skills will be acquired for their respective parent systems, but that is not the only use that is made of them. Training simulators are often used for proficiency measurement, where the capabilities of personnel are measured. Training can occur without measurement of trainee performance, but measurement has its uses for training. Charting a trainee's progress with performance measurement allows the instructor to bring skills to an appropriate level. Training simulators are used for the relearning of forgotten skills, and measurement is used to document the severity of the forgetting and the course of relearning.

In addition to these training uses of measurement, flight simulators are used for licensing and certification. The U.S. Coast Guard uses flight simulators in certification of pilots for instrument flying, and the Federal Aviation Administration has authorized commercial airlines to use approved flight simulators for training and crew member certification when the simulator is part of an approved training program (Jones et al., 1985, pp. 20, 88–89). The process of certification implies measurement to verify that crew members can perform normal and emergency tasks at acceptable levels of proficiency. When a crew member performs well in a flight simulator the assumption is made that he will also perform well in the aircraft.

The assumption of the interchangeability of a flight simulator and its aircraft for proficiency measurement purposes deserves sympathy because measurement in the aircraft is difficult and is often not done. A pilot flying at 600 miles per hour at 40,000 feet is not accessible to much measurement, and is even less so when engaged in complex maneuvers such as those of air-to-air combat. Moreover, there are important skills that cannot be measured in the air, such as resolving the emergency of an engine failure at takeoff. In addition, an aircraft in flight is not a good environment for measurement. Unreliable measures are likely. By contrast, a flight simulator is fully accessible and in a controlled environment. Many measures can be made of many kinds of aircrew behavior, and good reliability of measurement is possible.

Measurement of a pilot's performance in the aircraft has no validity problem because the aircraft is the real thing, but when a simulator is used for proficiency measurement there is the assumption that simulator scores (test) predict air scores (criterion), which is a validity question (Chapter 17). When validity is not determined, which is usually the case, there is the unspoken assumption that a simulator which is good for training is also good for proficiency measurement. The assumption can be wrong because transfer of training and proficiency measurement can be independent. Transfer of training is concerned with raising the average level of trainee performance in the aircraft, but proficiency measurement requires validity determination, where simulator scores and air scores of individuals are correlated (e.g., Koonce, 1979).

Weitzman et al. (1979) correlated instrument flying scores on Device 2B24, a high-fidelity helicopter simulator, with corresponding scores on the UH-1H helicopter, its parent aircraft, and found the correlation to be .57. The equation that predicted air scores (Y) from simulator scores (X) was $Y = .51X + 41.17$, which means that the simulator was easier than the aircraft and gave higher scores. A score, say, of 1000 in the simulator (X) would predict a score (Y) of 551 for the aircraft. Being easier than the aircraft, this simulator would approve too many pilots if it were used uncritically for certification of proficiency in instrument flying. It is necessary to be armed with a significant validity coefficient and an equation, as Weitzman et al. (1979) used, for air scores to be predicted confidently from simulator scores.

Computer-Based Instruction

Whole-task and part-task simulators are often computer based, but computer-based instruction is a term that has come to mean something else. Simulators are for the learning of skills, but computer-based instruction is for the learning of knowledge, as a student would do in a classroom. Computer-based instruction has become a prominent approach to teaching knowledge of system functions and operations to system personnel. The strong origins of computer-based instruction were in discontent with conventional classroom practices.

Skinner (1961, 1968) saw five difficulties with classroom practices:

1. Classroom learning is so often based on the avoidance of punishment rather than positive acts such as informational feedback (knowledge of results) about the correctness of a response. The instructor often does nothing if the student makes the correct response but criticizes or punishes her for making the wrong response. Skinner saw informational feedback as a better way of learning.
2. Informational feedback is usually delayed, often for days. The student hands in a set of mathematics problems, and they are not scored and returned until a week later.
3. Informational feedback is unstructured. Instructors may say "right" or "wrong" after a student's response, but sometimes they will not bother.

4. All members of the class are expected to learn at the same pace; there is no adjustment to the slow learners and the fast learners.

5. The structure of a course is often inefficient. Some teachers are skilled in organizing complex materials so they can be learned efficiently, but some are not. Slow and fast learners alike will struggle with a poorly designed course of instruction.

Skinner contended that there is a better way. A course could be designed by instructional experts, structured in a logical, stepwise fashion. The steps could be small, easy increments of knowledge, with immediate informative feedback after each one. And could not the instructional situation be individualized, where the student could learn at his own pace? The route to these goals came to be mechanization of the learning environment.

The mechanization of programmed learning was in the *teaching machine,* as it once was called. The teaching machine was invented by E. L. Pressey of The Ohio State University (Pressey, 1926). Pressey's original interest was in the automatic administration of a multiple-choice printed test. Pressey's machine had a window for the presentation of an item and four keys with which the subject indicated her response. Pressey's insight was that his machine could teach as well as test if the subject was allowed to continue pressing keys until the correct answer was found. In the decades that followed, other mechanical devices based on Pressey's idea were built to present textbook material and for the questioning about it. *Programmed textbooks,* where the student would have text material, problems, and questions presented on one page and informative feedback on another, also appeared. All of these efforts were dwarfed in the 1960s by the digital computer as the mechanism of control and presentation. Computer-based instruction came into being.

Computer-Based Instruction Defined

A computer-based instructional system will have a digital computer in control of one or more conventional visual display units, each with a keyboard. A tutor is a private teacher who gives a student one-on-one instruction, allowing the student to proceed at her own pace. A digital computer can conduct tutorial instruction for many students simultaneously. *Computer-based instruction* (CBI) is an automatic textbook and teacher that presents lessons, tests the student's knowledge about the lessons, and gives informative feedback about test answers.

The two main kinds of lesson sequences are linear and branching. *Linear* sequencing progresses systematically from one topic to the next, presenting information on a topic and testing on it, and then moving on to the next topic. All students proceed through the same sequence, and the sequence is unaffected by the correctness of the student's response. The lesson sequence in *branching* depends on the student's response. A student who misses questions on a topic may branch backward to repeat the material. Or the student may be branched sideways to similar material on the topic. Forward branching is skipping ahead, as when a student may wish to skip a topic. Throughout all of this, graphics, animation, and sound might be used.

Tutorials are not the only use of CBI (Alessi & Trollip, 1985). CBI is used to give drills in material. Students can be drilled in arithmetic problems, spelling, or foreign-language vocabulary. CBI is used for instructional games. For example, a four-letter word game has been devised to build a student's vocabulary. Four letters are given and the student produces as many words as she can. When the student is finished the computer displays all the other words that the student failed to generate and displays a percent correct score. CBI will often be used for administering tests and keeping the records of test scores.

A Major CBI System. A large CBI system is PLATO, designed by scientists at the University of Illinois at Urbana–Champaign (PLATO stands for *Programmed Logic for Automated Teaching Operations*). Based on a large central computer, a PLATO system may have over 1000 terminals, and there are 17 PLATO systems in use around the world. PLATO is programmed to teach a variety of college courses as well as courses below the college level. Examples of other teaching that has been done with PLATO is training unemployed workers for jobs, prisoners for jobs, and farmers in better methods of farm management (Eberts & Brock, 1984). Millions of hours of instruction have been administered by PLATO since its design in the 1960s.

How Good Is CBI?

CBI seems a triumph of modern technology almost too good to be true. All of the shortcomings of the conventional classroom appear to be corrected. Ideal lessons, ideally organized by experts, are learned at a self-paced rate. But is the triumph more apparent than real? What is the evidence that CBI teaches better than the live teacher in the conventional classroom? There are four basic measures that have been used in comparing the two teaching modes: student achievement, time saved in learning, cost-effectiveness, and attitudes.

Student Achievement. For some uses of CBI it would be desirable to relate instructional method to performance on a job. The U.S. military is a big user of CBI because of the large number of technicians that they train, and it would be informative if instructional method could be related to the job. Such data are nonexistent (Orlansky & String, 1979, 1981). Instead, instructional methods are compared by using test scores as measures of academic achievement in the course.

Orlansky and String (1979, 1981) analyzed the findings of 30 studies that compared CBI with conventional academic teaching of military technicians in a variety of specialties. CBI was found superior in only about one-third of them, and the differences were not considered to be of practical significance.

Time Saved in Learning. In the same 30 studies, Orlansky and String found that CBI saved about one-third of the time required by students to complete the same courses when taught by conventional instructional methods. The saving was produced by students for whom the rate of conventional instruction was slow. A conventional classroom might set the pace so that about 90 percent of the students would complete the material in a fixed period of time. At that rate, which is geared to the slower students, a number of students would finish early, and it is these better students who profit from CBI.

Cost Effectiveness. With the power to train a large number of students simultaneously, it would seem that CBI would be more cost-effective than conventional academic training. Self-evident though this assertion might be, it could conceivably be wrong. Conventional instruction requires teachers, classrooms, books, and so on, but CBI requires one or more computers, terminals, technicians for maintenance of equipment, programming personnel, subject matter specialists, classrooms, and so on. A skilled accountant, given extensive data, is required to determine the comparative cost-effectiveness of the two methods of instruction. The answer is neither self-evident nor easy.

Orlansky and String, in their analysis of 30 studies that compared CBI with conventional instruction, found cost data too incomplete for a conclusion. They found indications of savings, and they believed that savings could be large, but they concluded that an answer must await more data.

Attitudes. Students prefer CBI to conventional instruction (Orlansky & String, 1979, 1981). Responses to attitude questionnaires are a weak criterion, however, because there can be discrepancies between attitudes and performance (Shlechter, 1986, p. 30). Students might prefer one mode of teaching but perform better in another.

Conclusion. Experiments that compare CBI with conventional classroom instruction may be asking a good question but the experiments that have been done may not answer it. To have a valid test of the question, the CBI-trained classes and the conventionally trained classes must differ only by method of training, with all else the same. The experiments that have been done have been criticized for not meeting this criterion. Critics contend that other variables are confounded with instructional method (Shlechter, 1986, pp. 13–16). Course content may not be matched for the two instructional methods, so it is not clear whether course content, instructional method, or both determine the outcome of the experiment. Similarly, the same degree of care may not have gone into the lesson materials for the two conditions, so degree of care may be confounded with instructional method. A great deal of care and attention goes into the preparation of CBI materials because a large number of explicit steps must be prepared. There is no guarantee that a live teacher gives the same attention to preparation. Also, the amount of drill, or practice time, on parts of the lessons may not be equated. It is hard to see how all these variables can be controlled for a strict comparison of CBI and conventional instructional methods.

Is the Computer Necessary?

Experimenters have been understandably occupied with the comparison of CBI and conventional instructional methods, and have been diverted from the more penetrating question of whether prominent advantages of computer-based training can be achieved without the computer. Ignoring the economics of the two methods of training, well-designed CBI has the merits of well-organized lessons, testing at the end of a topic to ensure that the student knows the topic before going on to the next one, informative feedback about the correctness of his responses, branching so that the student can study easier or different versions of a topic to help the learning of it, and self-pacing. The computer is not a requirement for any of these functions. As Montague and Wulfeck (1984) point out, the materials and the way that they are used carry the instructional functions; the computer only delivers the materials and keeps the records. There are other ways to deliver the materials and keep the records.

It was seen, above, that the principal advantage of CBI over conventional classroom training is time saved in learning (Orlansky & String, 1979, 1981). Orlansky and String also compared individualized instruction without computer support with CBI and conventional classroom training for time saved in 12 military training courses. The individualized instruction without computer support was programmed textbooks or lessons organized to require minimal instructor attention. The amount of time saved by individualized instruction without computer support was about the same as CBI. These data imply that some training programs could profitably consider individualized instruction without computer support.

The Future of CBI

The reasons for starting CBI were sound, but the research that was reviewed here cannot help but induce skepticism. The evidence in behalf of superiority of CBI is not strong, and decisive economic advantages for it remain to be demonstrated. If further research deepens the skepticism, there is a possibility that CBI will be abandoned (Montague & Wulfeck, 1984). Three factors will work against abandonment:

1. CBI can be made to work well—certainly as well as conventional classroom instruction (Orlansky & String, 1979). CBI can be a workable substitute for conventional classroom instruction.
2. CBI is widely used in schools, industry, and the military. Entrenched methods are never abandoned easily.
3. New developments for CBI may offset whatever shortcomings it has, and the goal of a superior automated instructional system may someday be realized. The strongest hope is being placed in intelligent CBI.

Intelligent CBI. Today, a well-designed CBI system can be said to have a degree of intelligence in the sense that the system does some of the things that a good instructor does—the student is advanced through items of progressive difficulty, is branched to additional material if the student's understanding requires it, and receives informative feedback. An expert instructor, however, does more. There is interaction with the student, where the student's abilities, level of knowledge, and understanding are perceived. The instructor uses the conception of the student that he has perceived to build new knowledge on the student's understanding, using appropriate material and concepts. An *intelligent CBI* is an expert system (Chapter 16) that will simulate the expert instructor and be used by a student for learning in a naturalistic way.

An intelligent CBI will have four major components (Dede, 1986):

1. *A knowledge base.* The knowledge base is the subject matter with the facts, relations, and concepts of the domain. Like knowledge of the human mind, the knowledge base must be available for reasoning, problem solving, hypothesis information, the answering of questions, and so on, as well as a resource for facts, relations, and concepts.
2. *A student model.* Based on student responses, a model is built of the user which contains her knowledge about the subject matter, information on her pertinent abilities, style of learning and solving problems, and so on.

3. *Pedagogical module.* Taking the student model that has been developed and the student's current behavior, the system generates facts, concepts, observations, ideas, hints, questions, requests, and so on, to increment the student's knowledge; it is the expert teacher in action.

4. *User interface.* Some time in the future the student will sit in front of a visual display unit and respond with a keyboard. Some day a great deal of realism will be in the instructional situation because the machine and the student will be able to have a verbal dialogue. The machine will recognize the student's voice, understand what the student says, and will reply intelligently in a synthetic voice.

Progress in intelligent CBI has been exploratory and developmental so far. As an expert system, progress is tied to advances in the field of artificial intelligence.

On-the-Job Training

Why bother with complex, expensive training programs and devices when the trainee can often learn on the job? *On-the-job training* is the assignment of a trainee to a master, where the trainee is expected to learn by watching the master while they work side by side. On-the-job training involves observational learning, but there is more to it because the master may pause and explain actions to the trainee, for example. On-the-job training is an attractive idea that is widely used. Attractive as it may be, on-the-job training does not work very well.

There are several problems with on-the-job training. The master has work assignments; he is not freed to be a teacher. The master must accomplish the job assignment while teaching at the same time. Because he is judged by his work, and because the evaluation of teaching is vague, the master is likely to concentrate on his work and teach incidentally. Incidental teaching means that the teaching is not prepared, that it does not progress from easy to difficult tasks, that informative feedback is not always given, that material is infrequently repeated when the trainee does not understand, and that teaching aids like films or slides are not used. Moreover, there is useful knowledge, like theory, that cannot be learned by watching a master who teaches incidentally. On-the-job training has little apparent cost, and classrooms, professional instructors, and training aids and devices are expensive. They get the job done, however.

Summary

The best-selected personnel of the best-designed system must be trained in the skills and knowledge that will lead to efficient attainment of a system's goal. Training devices, computer-based instruction, and on-the-job training are three main ways of training system personnel.

All training programs have the hope of transfer of training, where the skills that are acquired in the training program apply successfully to operation of the system. When this goal is achieved, there is said to be positive transfer. Much of the practically oriented research on the transfer of skills has been done with flight simulators, which are large training devices for the training of flying skills. Research has shown that the amount of transfer ordinarily obtained is cost-effective.

Training devices such as flight simulators are themselves large and expensive systems (training systems), and like any other system, they have design issues. One issue has been part-task training, where the whole task is broken down into parts and training given on them. The question that research has tried to answer is whether training on the parts will be more efficient and less costly than training on the whole task. Research has found conditions where skills can be learned in practice on part tasks. Another issue has been fidelity of simulation, or how the realism of simulation affects transfer of training. Less costly training devices is the goal of research on fidelity of simulation. The overall tenor of the evidence is that fidelity of simulation can sometimes be reduced without harming transfer of training, although the rules of attenuation are not always clear.

Training devices are for the learning of skills, but computer-based instruction is for the learning of knowledge. Computer-based instruction is an automatic textbook and teacher that presents lessons, tests the student's knowledge about the lessons, and gives informative feedback about test answers. Research comparisons of computer-based instruction with conventional classroom teaching indicates that computer-based instruction is usually as good as conventional classroom teaching but probably not appreciably better. A hope for the future lies in artificial intelligence and the development of an intelligent computer-based instructional system with an expert system that will simulate an expert instructor.

On-the-job training is the assignment of a trainee to a master, where the trainee is expected to learn by watching the master while they work side by side. On-the-job training is inexpensive, but its effectiveness relative to other modes of training is doubtful.

References

Adams, J. A. (1960). Part Trainers. In G. Finch (Ed.), *Educational and training media: A symposium.* Washington, DC: National Academy of Sciences, National Research Council, Publication 789, 129–149.

Adams, J. A. (1961). Some considerations in the design and use of dynamic flight simulators. In H. W. Sinaiko (Ed.), *Selected papers on human factors in the design and use of control systems.* New York: Dover, Pp. 88–114.

Adams, J. A. (1979). On the evaluation of training devices. *Human Factors, 21,* 711–720.

Adams, J. A., & Hufford, L. E. (1962). Contributions of a part-task trainer to the learning and relearning of a time-shared flight maneuver. *Human Factors, 4,* 159–170.

Adorian, P., Staynes, W. N., & Bolton, M. (1979). The evolution of the flight simulator. In J. M. Rolfe (Ed.), *Proceedings of Conference on 50 Years of Flight Simulation,* Session 1. London: Royal Aeronautical Society, April 23–25. Pp. 1–20.

Alessi, S. M., & Trollip, S. R. (1985). *Computer-based instruction.* Englewood Cliffs, NJ: Prentice-Hall.

Briggs, G. E., & Rockway, M. R. (1966). Learning and performance as a function of the percentage of pursuit component in a tracking display. *Journal of Experimental Psychology, 71,* 165–169.

Briggs, G. E., Fitts, P. M., & Bahrick, H. P. (1957). Effects of force and amplitude cues on learning and performance in a complex tracking task. *Journal of Experimental Psychology, 54,* 262–268.

Cox, J. A., Wood, R. O., Jr., Boren, L. M., & Thorne, H. W. (1965). Functional and appearance fidelity of training devices for fixed-procedures tasks. Alexandria, VA: Human Resources Research Office, George Washington University, Technical Report 65-4, June.

Dede, C. A. (1986). A review and synthesis of recent research in intelligent computer-assisted instruction. *International Journal of Man–Machine Studies, 24,* 329–353.

Diehl, A. E., & Ryan, L. E. (1977). Current simulator substitution practices in flight training. Orlando, FL: U.S. Navy, Training Analysis and Evaluation Group, February.

Dougherty, D. J., Houston, R. C., & Nicklas, D. R. (1957). Transfer of training in flight procedures from selected ground training devices to the aircraft. Port Washington, NY: U.S. Naval Training Device Center, Human Engineering Technical Report NAVTRADEVCEN 71-16-16, September.

Eberts, R., & Brock, J. F. (1984). Computer applications to instruction. In F. A. Muckler (Ed.), *Human factors review: 1984.* Santa Monica, CA: Human Factors Society. Pp. 239–284.

Faconti, V. (1979). Evolution of flight simulator instructional capabilities: The first fifty years. In J. M. Rolfe (Ed.), *Proceedings of Conference on 50 Years of Flight Simulation,* Session 2. London: Royal Aeronautical Society, April 23–25. Pp. 52–61.

Felker, J. K. (1964). Project Gemini spacecraft training. Symposium paper presented at the annual meeting of the American Psychological Association, Los Angeles, September 6.

Jones, E. R., Hennessy, R. T., & Deutsch, S. (Eds.) (1985). *Human factors aspects of simulation*. Washington, DC: National Academy Press.

Koonce, J. M. (1979). Predictive validity of flight simulators as a function of simulator motion. *Human Factors, 21,* 215–223.

Lintern, G., Thomley-Yates, K. E., Nelson, B. E., & Roscoe, S. N. (1987). Content, variety, and augmentation of simulated visual scenes for teaching air-to-ground attack. *Human Factors, 29,* 45–59.

Montague, W. E., & Wulfeck, W. H., II. (1984). Computer-based instruction: Will it improve instructional quality? *Training Technology Journal, 1,* 4–19.

Orlansky, J., & String, J. (1977). Cost-effectiveness of flight simulators for military training: 1. Use and effectiveness of flight simulators. Arlington, VA: Institute for Defense Analyses, IDA Paper P-1275, August.

Orlansky, J., & String, J. (1979). Cost-effectiveness of computer-based instruction in military training. Arlington, VA: Institute for Defense Analysis, IDA Paper P-1375, April.

Orlansky, J., & String, J. (1981). Computer-based instruction for military training.. *Defense Management Journal, 17,* 46–54.

Povenmire, H. K., & Roscoe, S. N. (1971). An evaluation of ground-based flight trainers in routine primary flight training. *Human Factors, 13,* 109–116.

Povenmire, H. K., & Roscoe, S. N . (1973). Incremental transfer effectiveness of a ground-based general aviation trainer. *Human Factors, 15,* 534–542.

Pressey, S. L. (1926). A simple device which gives tests and scores—and teaches. *School and Society, 23,* 373–376.

Prophet, W. W., & Boyd, H. A. (1970). Device-task fidelity and transfer of training: Aircraft cockpit procedures training. Alexandria, VA: Human Resources Research Organization, Technical Report 70-10, July.

Rockway, M. R. (1955). The effect of variations in control-display during training on transfer to "high" ratio. Wright-Patterson Air Force Base, OH: Wright Air Development Center, Air Research and Development Command, WADC Technical Report 55-366, October.

Rockway, M. R., Eckstrand, G. A., & Morgan, R. L. (1956). The effect of variations in control-display ratio during training on transfer to a low ratio. Wright-Patterson Air Force Base, OH: Wright Air Development Center, Air Research and Development Command, WADC Technical Report 56-10, October.

Roscoe, S. N. (1971). Incremental transfer effectiveness. *Human Factors, 13,* 561–567.

Roscoe, S. N. (1972). A little more on incremental transfer effectiveness. *Human Factors, 14,* 363–364.

Semple, C. A., Hennessy, R. T., Sanders, M. S., Cross, B. K., Beith, B. J., & McCauley, M. E. (1981). Aircrew training devices: Fidelity features. Brooks Air Force Base, TX: Air Force Human Resources Laboratory, Air Force Systems Command, Technical Report AFHRL-TR-80-36, January.

Shlechter, T. M. (1986). An examination of the research evidence for computer-based instruction in military training. Fort Knox, KY: U.S. Army, Research

Institute for the Behavioral and Social Sciences, ARI Field Unit, Technical Report 722, August.

Skinner, B. F. (1961). *Cumulative record.* New York: Appleton-Century-Crofts.

Skinner, B. F. (1968). *The technology of teaching.* New York: Appleton-Century-Crofts.

Smode, A. F. (1974). Recent developments in instructor station design and utilization for flight simulators. *Human Factors, 16,* 1–18.

Staples, K. J. (1978). Current problems of flight simulators for research. *Aeronautical Journal,* January, 12–32.

Trollip, S. R. (1979). The evaluation of a complex computer-based flight procedures trainer. *Human Factors, 21,* 47–54.

Trollip, S., & Ortony, A. (1977). Real-time simulation in computer-assisted instruction. *Instructional Science, 6,* 135–149.

Waag, W. L. (1981). Training effectiveness of visual and motion simulation. Brooks Air Force Base, TX: Air Force Human Resources Laboratory, Air Force Systems Command, Technical Report 79-72, January.

Weitzman, D. O., Fineberg, M. L., Gade, P. A., & Compton, G. L. (1979). Proficiency maintenance and assessment in an instrument flight simulator. *Human Factors, 21,* 701–710.

Williams, A. C., & Flexman, R. E. (1949). Evaluation of the School Link as an aid in primary flight instruction. Urbana, IL: Institute of Aviation, University of Illinois, Aeronautics Bulletin 5.

Author Index

Subject Index